# The Metallography of Early Ferrous Edge Tools and Edged Weapons

R. F. Tylecote and
B. J. J. Gilmour

BAR British Series 155
1986

# B.A.R.

5, Centremead, Osney Mead, Oxford OX2 0DQ, England.





B.A.R. 155, 1986: 'The Metallography of Early Ferrous Edge Tools and Edged Weapons'





ISBN 9780860544012 paperback
ISBN 9781407317694 e-book
DOI https://doi.org/10.30861/9780860544012
A catalogue record for this book is available from the British Library
This book is available at www.barpublishing.com

CONTENTS

## Part I - List of Figures

FIGURE PAGE

Part II - List of Figures

## TABLES



## Results

# The Metallography of Early Ferrous Edge Tools and Weapons

by R.F. Tylecote and Brian Gilmour

## 1. General Introduction

Some of the objects examined in this work arise from a continuous programme by one of us over about 20 years. This work could have gone on a lot longer but it was felt that in view of the increasing numbers of objects now arising from British excavations and the increasing number of people working in this field, it would be useful to bring all the results together, so that others could benefit from the results and compare them with their own. Also, compared with other European countries, Britain appeared to be very backward in the examination of such artifacts especially when our work is compared with the work of Kolchin on the material from Novgorod, Piaskowski on the Polish material and Pleiner on Czech and other European material. In actual fact some of the results presented here have been published before in the archaeological reports dealing with their excavation.

While the principal author has been concentrating on the domestic tools, the second author has been investigating weapons. His results will eventually be incorporated in a thesis to be presented to the Institute of Archaeology in the University of London. It was thought that the inclusion here of some of these results would balance the work and allow comparison between the civil and military aspects of the smith's work.

In the two sections on results, the objects examined have been divided into domestic and agricultural tools such as knives and axes, and military weapons such as swords, daggers etc. Naturally there was a certain overlap in technique and it became obvious that the single edged sword known as seax or scramasax was based upon the design and production of the domestic knife.

We have not examined enough axes to know whether there was a difference in technique between the construction of military and civil axes. But the pattern-welded sword appears to have been designed in its earliest phase as an ornamental or prestige weapon and its military usefulness seems to have taken second place to its appearance. If it was not for the fact that we know that such swords were used for fighting (Beowulf etc) we would have supposed that its purpose was like the ceremonial sword of today.

Most of the artifacts examined were found in Britain but it is uncertain if they were also manufactured in Britain, as, from the Roman period onwards, an extensive international trade was being carried on in some domestic tools like knives, and in prestige weapons.[2] We have included a few artifacts from countries outside Europe, mainly because they illustrate techniques that are different but which may have been used in Europe.

Early edge tools were made in three ways; by surface carburising,[1] by welding steel to iron or by work-hardening wrought iron.[3] The most common method, after the Roman period, was the welding of steel to iron. The wide use of of the process of welding steel to iron rather than the use of solid steel was mainly for two reasons. The first is that hardened carbon steels - and carbon steels were the only steels used before the latter half of the nineteenth century - are brittle in their fully hardened state. Judging from the knives examined here, after the Roman period the fully hardened state was very much sought after and therefore to get a long lasting implement that did not break when flexed, it was necessary to weld an edge of hardened steel to an unhardenable iron or soft steel back.

The other reason for welding the two was that from the earliest Iron Age, steel was more costly than iron. This is still the case, remembering that so-called "mild" steel is little more than iron. It is difficult to arrive at a comparative costing in Britain for the Medieval period because so much steel was imported, whereas much of the iron used was of local origin. This is largely due to the fact that, to get steel by the solid state diffusion of carbon into iron, one must use fairly pure iron to get the maximum diffusion rates which are not very great. A good deal of the early iron produced in Britain contained phosphorus which slows down the diffusion of carbon but on the other hand gives a relatively hard iron when cold hammered. This, however, is not hard enough for the best knives for which carbon steel is needed and which must be made either by carburizing high quality local iron which was at a premium, by carburizing imported iron, or by using imported steel. About 1300 A.D., steel was selling in Britain at about £3 per ton, while wrought iron was only about £.60 per ton, thus giving a relative price of 5 times.

The incentive to use steel economically gave rise to much ingenuity on the part of blacksmiths and cutlers. Clearly there were a great number of solutions to the economic and technological problems and it is possible that individual smiths had their own personal way of dealing with the problem.

In this way we might be able to detect the work of different individuals of different schools. On the other hand there were fashions in these things, and of course improvements in technique which invoked changes in design. We see for example, the use of pattern welding in Anglo-Saxon sword smithing - a process which could be used to give a hallmark and therefore some indication of quality to justify the high cost. Pattern-welding, as used in the Anglo-Saxon period, did not give a hard edge and for the best blades had to be combined with the straight-forward welding-on of a hardenable steel edge. (See below, Part 2). But the pattern-welded part looked attractive and no doubt was so much prized that its use was carried forward to later Medieval times.

It was not until the nineteenth century that steel-making processes had improved so much, and steel had become relatively so much cheaper, that it was possible to make a cheap and satisfactory knife out of homogeneous carbon steel. Even so, the quality was not much better and for some purposes, such as scythes, the old fashioned method of sandwiching a piece of steel between two pieces of iron was still used.

Therefore, for most of the last 2000 years, knives and other edge tools have been made by combining the minimum of steel with the maximum of iron consistent with a good cutting edge. This can be done basically in four ways as shown in Fig. 1.

1. By having a layer of steel covered with two plates of iron in such a way that the steel projects at the thinned

cutting edge. Sharpening such a tool will always give a steel edge to the blade.

2. Welding-on a steel strip to the edge of a piece of iron, sometimes made by piling. The steel edge will wear away with sharpening until eventually there is no steel left and the tool becomes useless and is discarded.

3. Using a piece of piled material which consists of alternate layers of iron and steel. Given sufficient heat after welding, these layers may be homogenised so that the tool consists of a piece of homogeneous steel.

4. By having an iron core around which a piece of steel has been wrapped i.e. the reverse of Type 1. This seems comparatively rare perhaps because it uses more steel than 1, in most cases and, more importantly, the lasting qualities of the tool are not much better than in the case of 2.

While the simple Type 1 sandwich is easy to categorise, there are in fact a wide range of structures giving much the same results going from the 'English' sandwich to the 'Danish' sandwich in which a layer of steel is welded to one side of a piece of iron. This one-sided sandwich may be very limited in its use of steel so that the steel forms a thin edge more like the Type 2 structure. This is used for wood-cutting chisels right up to the 19th century. In other cases the steel layer may be applied to the whole width and length of the blade. It is not always possible to differentiate the type when only limited section length is allowed and so, when we are in doubt as to the type it is returned as Type 1/2. The one-sided sandwich is referred to as Type 1(a), and more complicated multidecker sandwiches are referred to as Type 1(b). It is not always easy to differentiate this type from a Type 3 blade in which the various layers have not been fully diffused.

If we include work-hardened ferrite with or without phosphorus and call it Type 0 we have 5 basis types shown in Fig. 1 in which Type 1 in particular has a number of subdivisions. Type 2 can be subdivided into scarf welded in which case the iron-steel weld is at an angle, or butt-welded in which case it is at right angles to the section.

A knife for example, has a number of different profiles that to a certain extent define its use but not necessarily so. A typology dealing with shape has been devised by Mr. P. Ottaway based on his examination of the mainly medieval knives from York.[4,5] These are lettered alphabetically A,B,C,D etc. and it is hoped to use this typology for shape in future work where reference is made to this parameter. By using alphabetical letters for form, and numbers for metallurgical structure confusion is reduced. But a stroke should be used to separate <u>form</u> from structure i.e. A1/2a.

## 2. Preparation of Specimens

Pieces were cut from the knife blades by means of an 0.48 mm thick abrasive cut-off wheel as they were usually too hard to cut with a hacksaw. In most cases pieces from 6 to 12 mm thick were removed by cutting right through the blade, but in some cases where the specimens were well preserved and had some aesthetic value, 'V' shaped pieces were removed by making two cuts from the blade edge meeting somewhere in the middle of the blade.

From two to six sections were mounted together vertically in a cold or thermosetting resin. The maximum temperature obtained in this mounting process is about 150°C and has been found not to change the structure of carbon steels in the short time normally used (about 10 minutes). The mounts were then polished using the standard metallographic techniques of dry or wet grinding followed by diamond polishing. The polished sections were normally etched in 2-3% nitric acid in alcohol (Nital) although other standard etching treatments were used occasionally to clear up specific points.

## 3. Examination

The mounts were first examined by a lower-power (X10) binocular microscope, and outline sketches made; the structural details were then filled in using a normal metallurgical microscope with magnifications of 50 - 950 times.

In the sections shown, the structural phases have been shown in brief using the following system of symbols:

| | |
|---|---|
| As | Light-etching ferrite bands on a weld junction usually due to the high arsenic content of the iron or steel. |
| Cem | Cementite; when present with F it is usually in the grain boundaries. |
| F | Ferrite. |
| F(+P) | Ferrite with some pearlite. |
| F + P | Ferrite and pearlite (about equal amounts). |
| gbc. | Grain boundary carbide. |
| M | Martensite. |
| M + F | Martensite and ferrite due to quenching at too low a temperature for the carbon content of the steel. |
| M + T | Martensite and troostite. |
| N | Nitride or carbide needles. |
| P | Pearlite. |
| P(+F) | Pearlite with some ferrite. |
| SC | Spheroidised carbides. |

T Troostite.

TM Tempered martensite.

B Bainite.

Most of these terms are in normal use amongst metallographers and are international (see appendix for definitions). The term 'troostite' has gone out of use in some countries and in some areas of the U.K. Here it is retained for the very common nodular dark-etching phase which is, when seen under a very high magnification, a form of pearlite which forms under continuous cooling conditions (CCT) at rates between those giving lamellar pearlite and those giving martensite.

Until recently the quite common occurrence of 'needles' in ferrite was put down to the presence of iron nitride ($Fe_2N$). More recently, however, with the increasing use of the electron microscope it has been found that many that were thought to be nitride were, in fact, carbide, so the letter N is here used to indicate both types of needle.

Very often the junction between steel and iron is well marked by a light-etching band of ferrite. We now know[7] that this is due to the surface enrichment of iron or steel in elements such as arsenic when the metal is being heated in the smith's hearth for welding. When a piece of steel is welded to iron and both have low arsenic content (less than 0.015%), a diffuse joint is obtained unless the diffusion of carbon across it is hindered by a lot of slag entrapped in the weld or the weld is carried out especially fast. Where the 'white bands' are extremely prominent this fact has been indicated by the symbol 'As' and is an indication of high arsenic, most likely in the iron and therefore indicative of a certain type of iron ore as a source for the metal. So far we do not have much data on the As content of British, or indeed other, iron ores; but no doubt in the course of time this situation will be rectified. Meanwhile these white lines help to measure the degree of piling that has gone into the production of the iron or steel.

Arsenic is a much more common element in early iron and it is not unusual to have as much as 2% As giving appreciable solid solution hardening, as with phosphorus. (vide infra).

In their work on the corrosion resistance of wrought iron, Chilton and Evans[6] found evidence for two types of segregation (1) microsegregation (Q zones) which was due to the solid state reduction of the ores (differences between iron and its impurities) and (2) macrosegregation (V zones) which was caused by enrichment on the welding planes. Both processes might be responsible for the type of enrichment of Ni and Cu in iron but for the moment we will consider process (2). Here in the days before the introduction of the electron scanning microscope (1955), Chilton was able to obtain fine micro-analysis by A. Berger in Belgium which gave results showing that modern wrought iron had been enriched from 0.19% Ni to 2.7% along the welding planes during the piling process. This is the order of increase one gets with arsenic[7], and there is no doubt about the process giving rise to it. The width of the enriched zone was of the order of 0.1 mm (= 100 $\mu$m) and these zones could be clearly seen in the microstructure as parallel lines showing the regularity obtained in piling. 'White' lines of similar regularity can be seen in the ferrite that forms the back of a

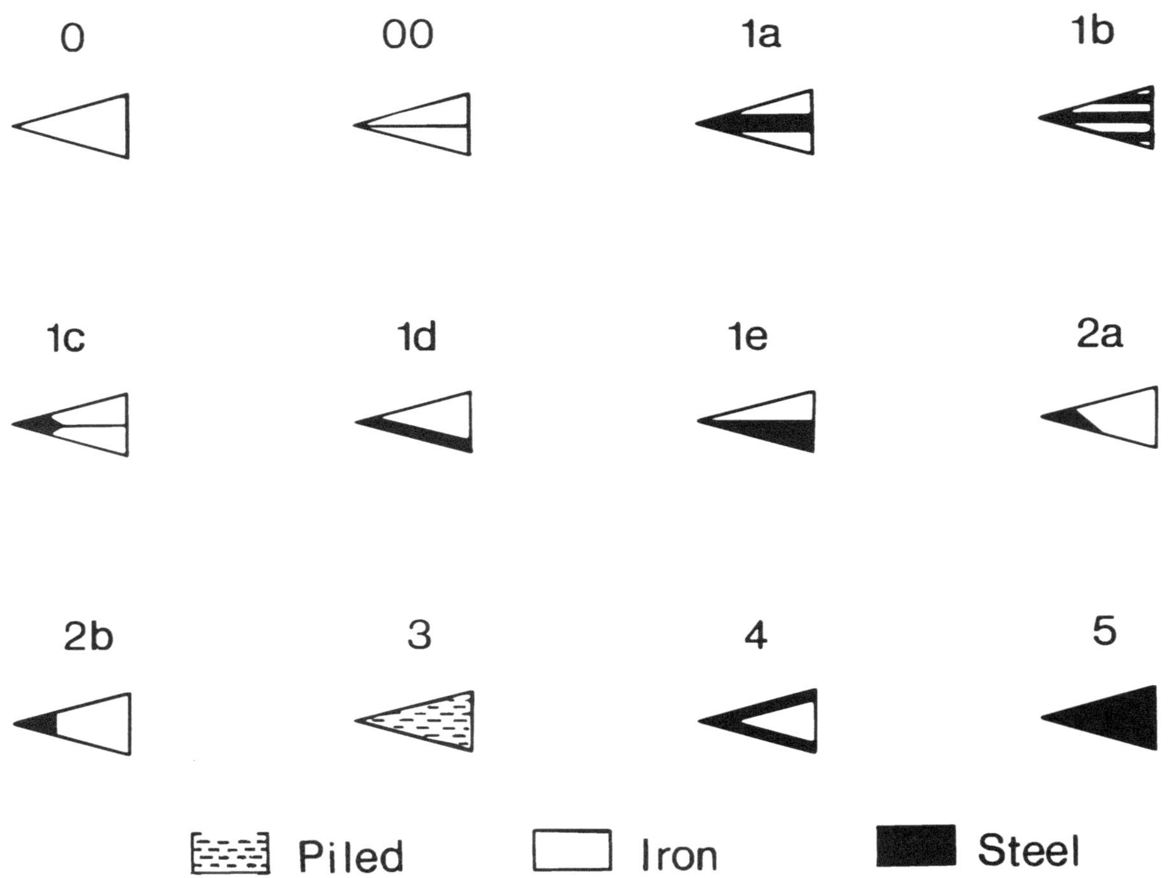


Fig. 1. Methods of welding iron to steel to make blades.

knife from the medieval site of Goltho in Lincs. (Fig. 21). An electron probe scan showed Ni enrichment here too (Fig. 22). This amounted to 1-2% max. over a width of about 100 mm.

## 3.1 Hardness and its Measurement

Hardness measurements were made to check the phases present and to evaluate the quality of the heat-treated blades.

The normal way today of measuring hardness is to indent the specimen with a ball-shaped steel or pyramidal diamond indenter under a controlled load. This leaves an impression the size of which is a measure of the hardness. In the case of a homogeneous material the load applied has very little affect on the hardness reading obtained which is itself a product of the load applied divided by the area of the indentation ($kg/mm^2$). But as the load applied is reduced below about 100 g, i.e. over the microhardness range, the readings tend to get higher and the results obtained with different measuring instruments cannot be accurately compared. On the whole, measurements obtained with the Vickers pyramidal diamond over the range of loads from 30 kg to 0.5 kg are comparable (Fig. 2), and indeed may be roughly comparable with those made by the steel ball indenter, the Brinell system, up to a hardness of about 300 HB.

The best use of the microhardness range (5 to 100 g) is for determining the phases present in the microstructure, e.g. for differentiating ferrite from cementite without etching. Readings in this range which are normally made by pyramidal diamond indenters, or special-shaped diamond indenters (Knoop Process), are not accurately comparable with the higher load measurements. Low load microhardness measurements may be as much as 50% higher than high load measurements on the same material.

In this work we have used the British Standard (BS 860) for expressing the results of hardness measurements. This means that readings obtained with a pyramidal diamond indenter are expressed as HV with the load following in kg. (i.e. readings obtained with a 1 kg load are returned as HV1). Measurements using the Brinell system are given as HB. In Part II the measurements, unless otherwise stated, are microhardnesses using 100 g loads (HV 0.1).

The figures in brackets following the phase symbols are the hardness readings and indicate the carbon content and efficiency of heat-treatment of the blade.

## 4. Some Aspects of the Physical Metallurgy of Iron and Iron-carbon Alloys.

In this paper the structures found relate to the iron-phosphorus, iron-arsenic, and iron-carbon alloys. The first two are somewhat unusual in modern ferrous metallurgy although there is now an increasing interest in re-phosphorised mild steel. But there is a fair amount of material available in the literature. As for iron-arsenic alloys these are not used today, nor were they recognised until comparatively recently as a group that needed to be reckoned with in archaeometallurgy.

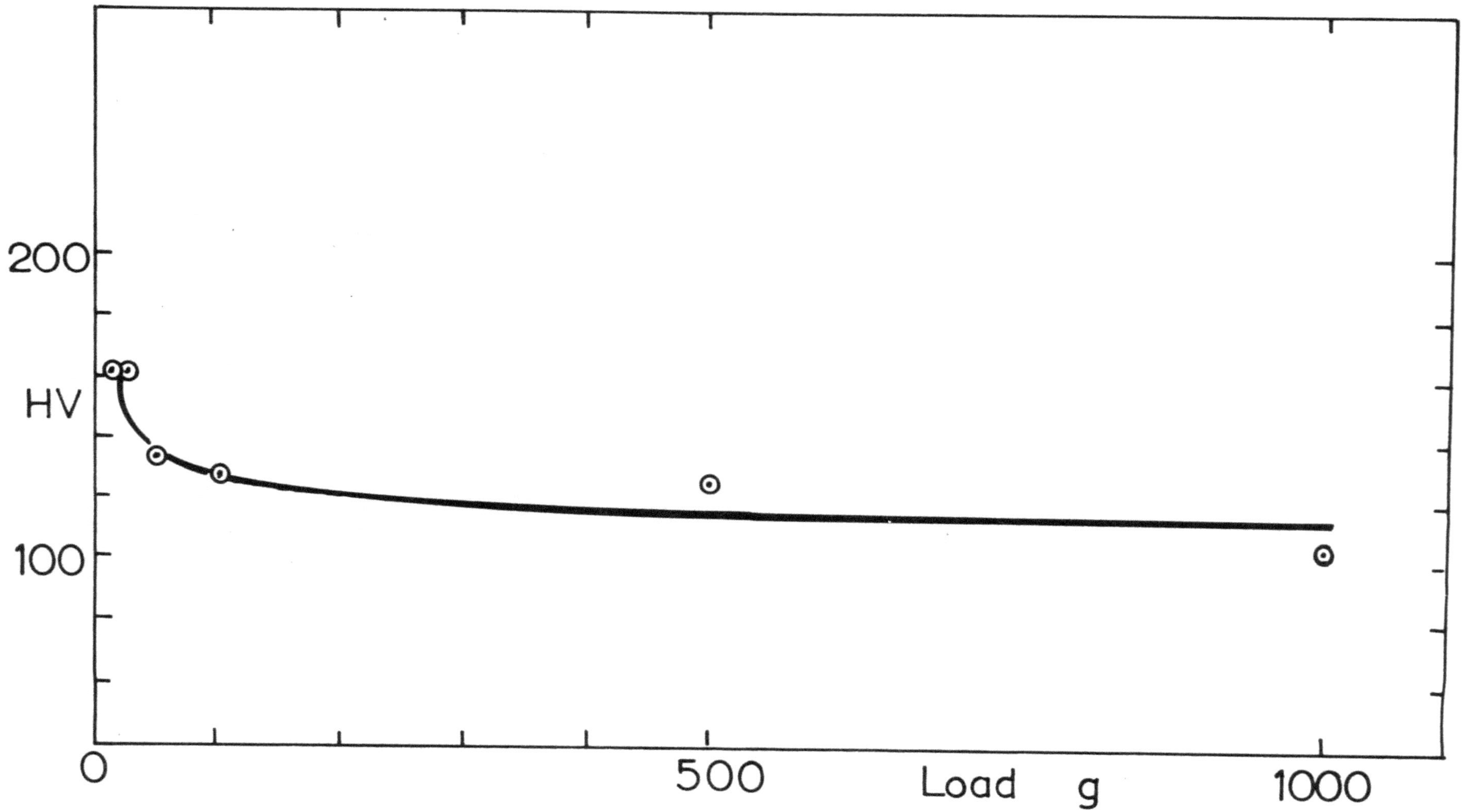


Fig. 2. Effect of load on hardness HV.

Recently several examples of very high hardness, exceeding 300 HV have been obtained in wrought iron free of carbon, but possibly high in phosphorus. The cold working of mild steel, like most other metals, will increase the hardness. One source[8] shows an increase in hardness from 120 to 190 HV as the deformation, expressed as a reduction in thickness, increases to 50% (Fig. 3).

But the addition of phosphorus to iron has a very great solid solution hardening effect as seen in Fig. 4, where the increase is seen to be much greater than any other solid solution element.[9] The % P figures are taken from Hopkins and Tipler,[10] whose UTS values have been converted to hardness by dividing by 0.23, and their figures for high % P extrapolated from 0.31% P.

The iron-phosphorus system is one that has a closed gamma loop. This means that on cooling alloys with up to 0.3% P, there is a phase change from ferrite through austenite and back again to ferrite. But alloys with more than 0.5% P stay in the ferritic state throughout their cooling. This means that in practice, at room temperature, one can often have, in the range 0.3 - 0.5% P, both ferrite and austenite. We might note in passing that the diffusion rate of iron in ferrite is two orders of magnitude greater than in austenite.

The reason for phosphorus having the greatest influence on strength is perhaps, because it has the smallest atomic radius (1.09Å) (Table 1). If this is the reason it also explains the reduction in the diffusion rate of the interstitial element, carbon, when both are present together.

It is also known that arsenic hardens ferrite, as hardnesses as high as 380 HV have been obtained in retained ferrite 'white lines' believed to be high in As. These lines are common in piled iron high in As. But according to the atomic radii given in Table 1, one would not expect arsenic to have such a great solid solution hardening effect as phosphorus.

Further work has been done on the hardening effect of large amounts of phosphorus above those investigated by Hopkins and Tipler[10] and the results of these are incorporated in Tables 2 and 3 and which compare the relative effects of cold work on iron - 0.1%C and iron-phosphorus alloys.

It would appear that the combined effect of work-hardening and phosphorus could account for ferrite hardness in excess of 300 HV.

### 4.1 The Carburisation of Iron

A high-phosphorus iron can be work-hardened to give a cutting edge with a hardness of the order of 250 HV which is as good as any bronze tool. To get a blade harder than this the iron has to be carburised to convert it to steel which can then be further hardened by quenching and tempering.

Phosphorus slows down the diffusion of carbon into iron (Table 4), so in order to get the most efficiently carburised iron, a good-quality ore has to be selected; such an ore was, on the whole, rare in these islands. For this reason, by the medieval period, steel was largely imported from Sweden, Russia, and Spain. No doubt a small amount was a by-product of the bloomery process which tended to produce rather heterogeneous metal from which the high-carbon parts could be removed.

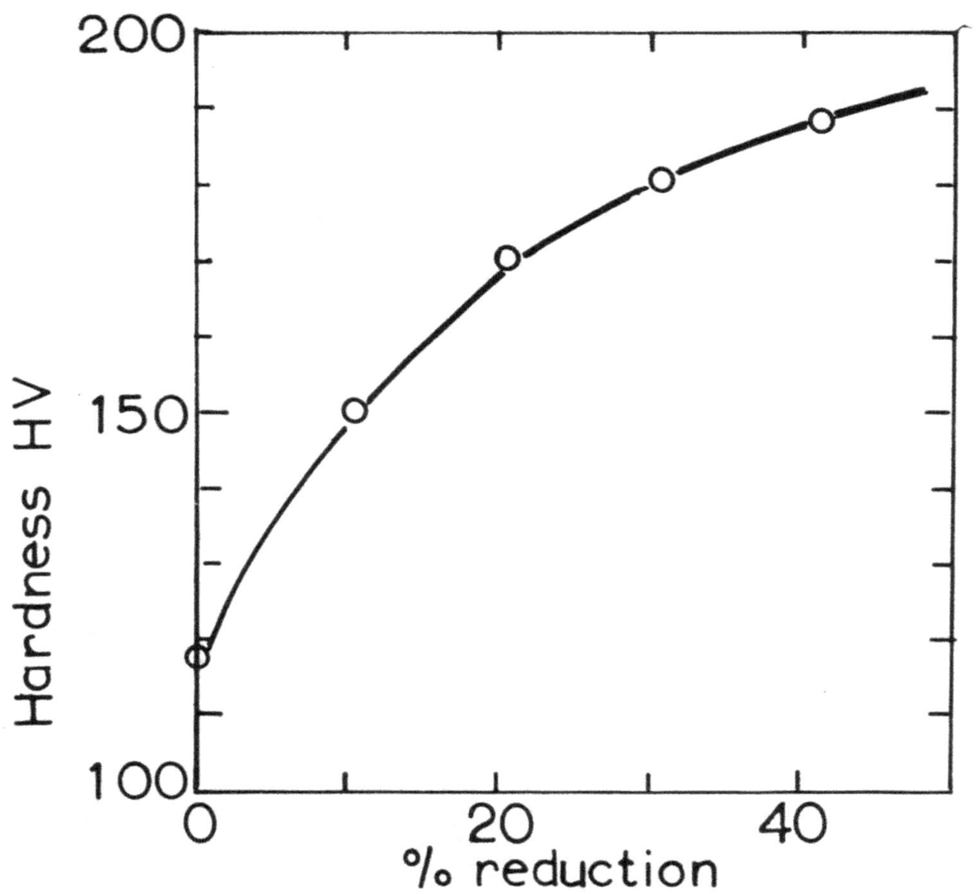


Fig. 3. Effect of working on the hardness of low-carbon iron. (after Williams et al[18]).

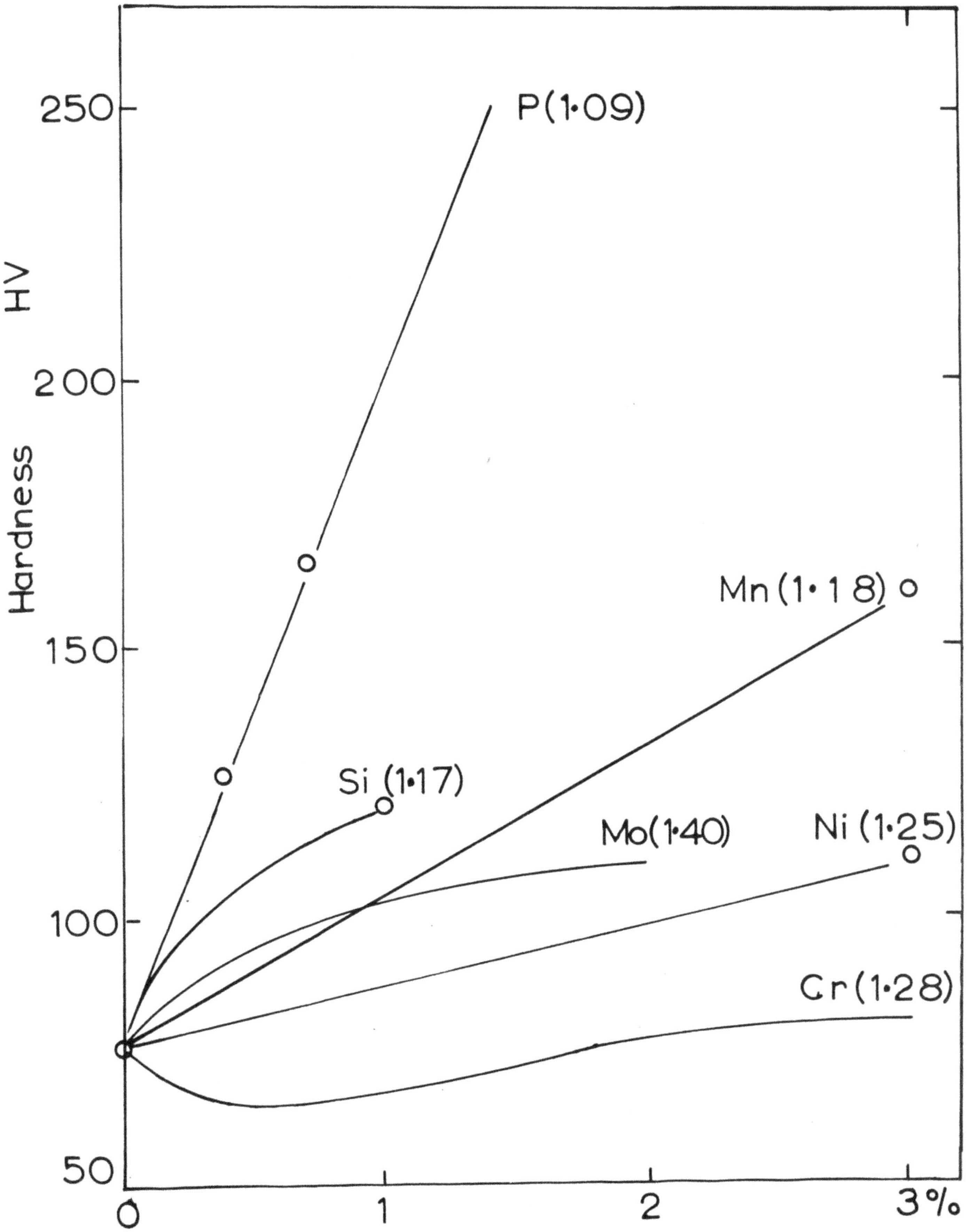


Fig. 4. Hardening of iron by solid solution elements. (After Rollason et al[9]).

TABLE 1

Atomic Radii of the elements (Å) and their position in the Ferrite lattice

| | | | | | |
|---|---|---|---|---|---|
| | | Cu | 1.28 | | |
| Ti | 1.47 | Fe | 1.27 | B | 0.97 |
| Al | 1.43 | Co | 1.26 | C | 0.77 |
| W | 1.41 | As | 1.25 | N | 0.71 |
| Mo | 1.40 | Ni | 1.25 | O | 0.60 |
| U | 1.36 | Mn | 1.18 | H | 0.46 |
| Cr | 1.28 | Si | 1.17 | | |
| | | Be | 1.13 | | |
| | | P | 1.09 | | |
| | | Substitutional | | Interstitial | |

TABLE 2

Effect of Carbon and Phosphorus on the Hardness of Annealed Iron

| % | HARDNESS - HV | |
|---|---|---|
| | Carbon ǂ | Phosphorus + |
| 0 | 75 | 75 |
| 0.05 | - | 78 |
| 0.15 | 100 | 91 |
| 0.31 | - | 113 |
| 0.23 | 110 | 145 |
| 0.60 | 180 | 216 |
| 0.90 | 200 | 250 |
| 1.20 | 260 | 287 |
| 1.50 | - | 270 |

ǂ After E.C. Rollason[9]

\+ First four values after Hopkins and Tipler[10] on pure iron and phosphorus.

Remainder based on experimental work by J. G. McDonnell of the University of Aston using less pure materials.

Table 3

Effect of cold-work on the hardness of Iron and Phosphorus-containing Iron

| COLD WORK | HARDNESS - HV | | |
|---|---|---|---|
| % reduction in thickness | Iron + 0.1%C (Williams[8]) | Iron + 0.5%P (McDonnell[+]) | Iron + 1%P |
| 0 | 120 | 216 | 250 |
| 10 | 150 | 300 | 320 |
| 20 | 170 | 320 | 340 |
| 30 | 180 | - | - |
| 40 | 188 | - | - |
| 50 | 192 | - | - |

\+ Based on experimental work by J.G. McDonnell of the University of Aston.

TABLE 4. EFFECT OF PHOSPHORUS ON THE DIFFUSION OF CARBON

| Starting Material | %C Absorbed in a Given Time |
|---|---|
| Fe | 0.94 |
| Fe+0.6% P | 0.60 |
| Fe+1.2% P | 0.52 |

(Stead[11])

Later on, in the 18th century, the bloomery and the finery could be adjusted to give steel directly, so giving 'natural' steel. It is of course possible that this process was used much earlier.

Other processes exist such as the Brescian process which mixes cast iron and wrought iron, the latter, a solid, absorbing carbon from the former in its liquid state. A very skilled smith could make his own steel from imported good-quality iron by carburising it in his hearth, but this would be slow.

The diffusion of carbon in iron even at temperatures as high as 900°C is very slow (Table 5; Fig. 5). To a certain extent the process is sensitive to the medium used, and some accelerating effect is probably obtained from the alkali carbonates of the wood ash that form as charcoal is consumed. The metal to be carburised is immersed in fine charcoal a slight distance away from the tuyere to avoid rapid oxidation of the charcoal, but with enough consumption of the charcoal in front of the tuyere to maintain a temperature of 900-1100°C. Under these circumstances, it will take five hours to raise the carbon content at a position 0.10 in (2 mm) below the surface to 0.45%C (Fig. 5). To do this consistently requires a very skilled smith; that is why carburising is best done first and a piece of satisfactory steel obtained before a tool is made. Then the good piece of steel can be built into or onto the tool that is being made. Because the diffusion of carbon into steel is so slow it was necessary to beat the iron thin and then, when the carbon had been absorbed, to fold it over and weld it into a piece thick enough to make the artefact required. Evidence of this is shown in the 'piled' structure of both iron and steel. Subsequent heating after folding can even out the variations in carbon content, thus eliminating evidence of piling except for linear slag inclusions and arsenic enrichment lines (see below). This gives a homogeneous piece of steel.

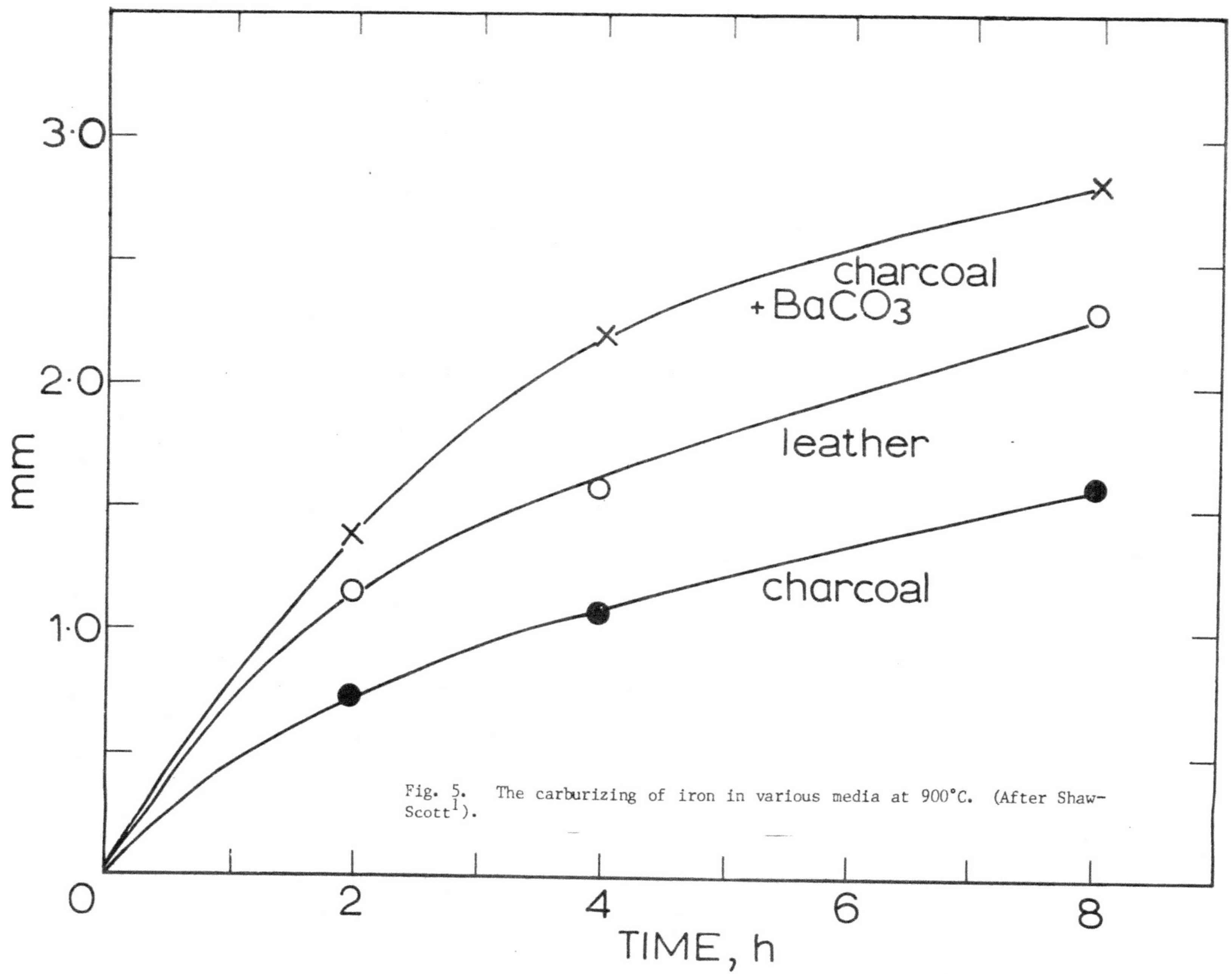


Fig. 5. The carburizing of iron in various media at 900°C. (After Shaw-Scott[1]).

TABLE 5. EFFECT OF TEMPERATURE ON THE THICKNESS OF THE CARBURISED LAYER AFTER 2 hrs.

(minimum carbon = 0.3%)

| Temperature, °C | Thickness, mm |
|---|---|
| 950 | 1.14 |
| 1000 | 1.64 |
| 1050 | 1.90 |
| 1100 | 2.66 |

(Bramley and Lawton[12])

4.2 Heat-treament

To get the best out of a steel edge, it has to be heat-treated. With carbon steel this involves heating it up to a temperature between 700 and 850°C which is quite a good red heat and plunging (quenching) it into a suitable liquid which cools it rapidly. Nowadays this can be oil, water or brine, but in medieval times we read of blood, urine and other liquids being used. One would not think that these had any advantage over the first three. The final hardness after quenching would depend on the carbon content of the steel and the rate of cooling; for the same carbon content one would expect the hardness to be in order of the first three liquids mentioned, and oil giving the lowest reading. But the temperatures are critical and no hardening would result on any steel if the temperature of heating was below 700°C. Above this a steel containing 0.85% carbon would be fully hardened at 750°C while a 0.4% carbon steel would need 850°C to be fully hardened. The hardness of the former might be too high for efficient use i.e. it might be too brittle and crack or break off in use. In this case it would be tempered by heating to a temperature between 200 and 600°C for a suitable time, probably a few minutes to half an hour depending on the temperature and the type of use for which the knife was intended.

It is clear from the hardness figures that the majority of the knives were given double treatment of this type resulting in the structure known as tempered martensite (TM). Treatments of this sort have resulted in hardnesses ranging from 313 to 900 HV, and in conjunction with a low carbon back would give a very satisfactory blade.

When a medium carbon steel is quenched at too low a temperature, a reduced hardness is obtained as not all the steel is converted into hard martensite but some is left in the ferritic state. In early times when the carbon content was not accurately known this was all too easy and some of the blades show evidence of this (M+F).

Slow or slack quenching, i.e. into liquids with lower thermal capacity than water, such as oil, usually result in mixtures of hard martensite and the less hard phase, troostite (M+T). In many cases slow quenching is done intentionally as it is one way of achieving in one operation the same hardness as the combined operations of water quenching and tempering. In many cases it results from inefficient water quenching and was not intentional.

The other process which was probably used in early times is gradient quenching in which part of the blade is protected by an envelope of clay which is relatively thick in areas where the slowest cooling rates are desired, i.e. maximum toughness, and thin or non-existent when the fastest cooling rates are needed near the edge. This is one way of obtaining a tough back and a hard edge from the same piece of steel. But it does not seem to have been used in many, if any, of the tools examined.

If the whole blade is heated to a suitable hardening temperature but only the edge is quenched then when the edge is removed from the quenching bath a form of auto-tempering is obtained due to the heat flow from the unquenched part of the blade tempering the martensite of the edge; this is another way in which tempered martensite (TM) can be obtained.

When a blade of Type 1 construction is made for example with a 0.8% C core and outer layers of 0.2% C, the structure after smith welding and slow cooling would be expected to be pearlite in the core and ferrite + pearlite in the outer layers. The correct heat-treatment for the core would be to heat it to 750°C for 15 to 30 mins and quench. Under ideal conditions this should give a martensitic core and ferrite + martensite in the outer layers. But the structures seen in the outer layers are more complicated than these and have been described in the foregoing as F + P. To some extent they resemble the structures seen in the low carbon core of modern case carburised and hardened steels.

## 5. Results of Examination of Edge Tools

### 5.1 Cutlery

#### (a) Knives

Very few objects of the pre-Roman period in Britain have been metallographically examined. This is partly because there are not many and partly because those that we have are not in a conditon to repay examination. (Table A1).

The site at Huckhoe,[15] in Northumberland has yielded an early form of broad-bladed knife (Fig. 6) from a level now dated to the 7th - 6th century B.C. The blade has been made from piled bloomery iron with a variable but low average carbon content (about 0.1 - 0.2%). The hardness of most of the section is 110 - 143 HV1 which suggests that the metal is low in phosphorus. One surface has been carburised to give a carbon content of about 0.8% and a hardness in the range 230 - 257 HV. The structure of the carburised layer is mainly coarse lamellar pearlite, but the ferrite+pearlite areas in the centre show a Widmanstätten distribution suggesting that the blade has been coolled fairly rapidly in air after carburising one surface at 900 - 1200°C, but that the cooling rate has been much slower below 750°C.

It seems that one surface was protected during carburising by a clay envelope so that at least half the section would remain soft and ductile to support the harder and more brittle side which acted as the cutting edge.

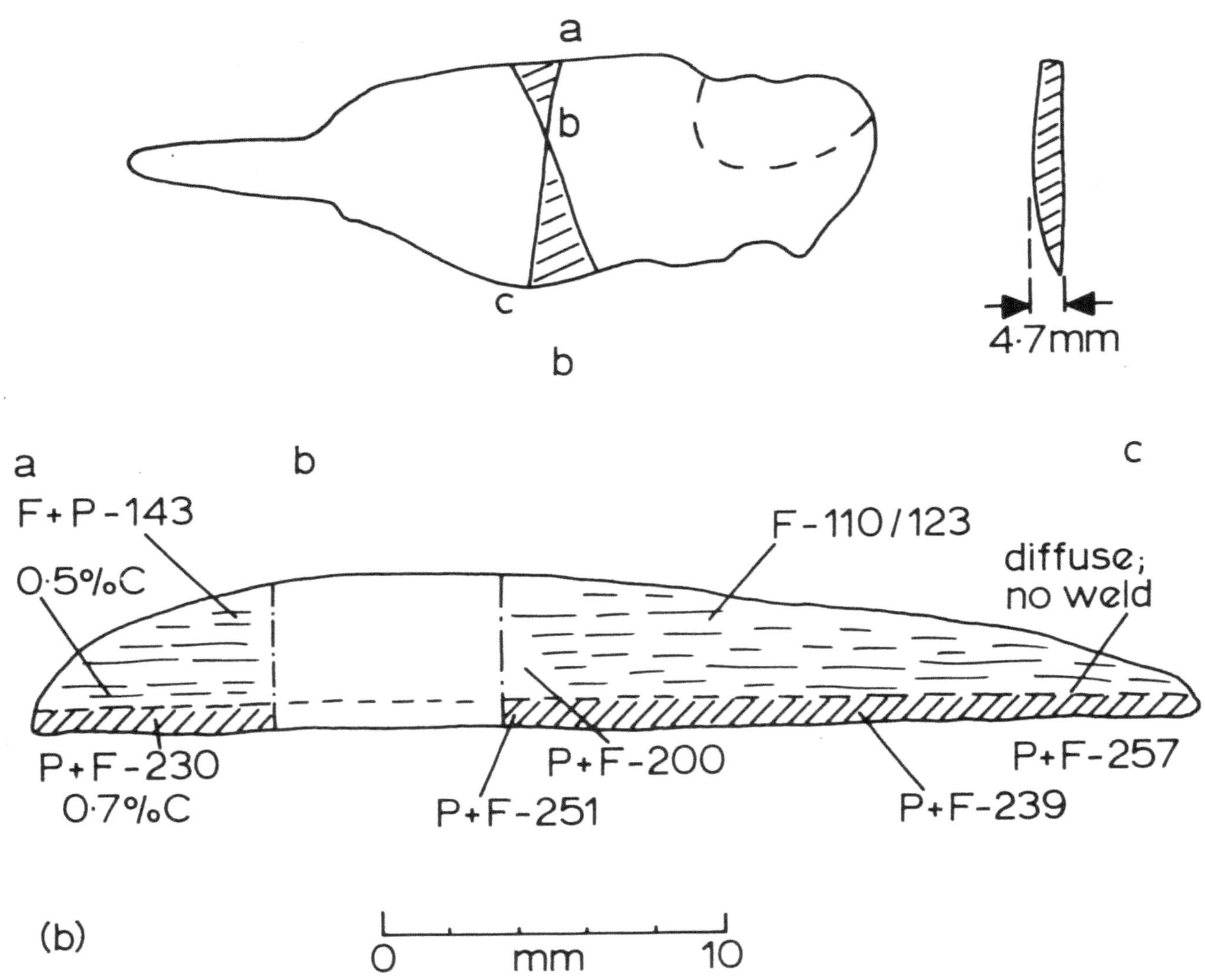


Fig. 6. Knife blade from Huckhoe, Northumberland and metallographic section.

TABLE A. Knives and Cleavers.

| Object | Site | Site/Lab Ref. No. | Date | Structure Edge | Hardness HV Back/Core | Type |
|---|---|---|---|---|---|---|
| **Pre-Roman A1.** | | | | | | |
| Knife | Winklebury, Hants. | 2615/2618 | 3rd-1stC BC | FMT/500 | F+T/193 | 3 |
| Knife | Huckhoe, N/d | Fig. 11,4 | 7th-6th | P/257 | F/110 | 1a |
| **Romano-British A2.** | | | | | | |
| Knife | Winterton, RV | HW/GU/17 | R-B | TM/360 | F/131 | 1 |
| " | Ware, Herts | 1849 | R-B | TM/135 | F+P/189 | Carb. |
| Cleaver | " " | 1747 | R-B | - | F/123 | 0 |
| Knife | " " | 325 | R-B | - | F(P)/162 | 0 |
| " | " " | 630 | R-B | - | F+P/136 | 0 |
| " | Wanborough | 707253 | R-B | TM/720 | F+P/148 | 3 |
| " | (Wilts) | 771180 | R-B | TM/369 | F/168 | 4 |
| Cleaver | " " | 780856 | R-B | - | F+P/158 | 0 |
| " | " " | 790723 | R-B | - | F+P/251 | 0 |
| Socketed knife | " " | 790296 | R-B | - | F/185 | 0 |
| Knife | " " | 780163 | R-B | - | P+F/240 | 3 |
| Knife | " " | 790725 | R-B | - | F+P/175 | 0 |
| Ring-handles | " " | 790726 | R-B | - | F/133 | 0 |
| Knife | | | | | | |
| Knife | " " | 790268 | R-B | - | P(F)/240 | 3 |
| Knife | " " | 684039 | R-B | - | M+P/269 | 3 |
| Knife | Gestingthorpe, Essex. | 70 | R-B | - | F/164 | 0 |
| Knife | " " | 116 | R-B | - | F/122 | 0 |
| Knife | " " | 91 | R-B | - | F/159 | 0 |

TABLE A2. Knives and Cleavers. (continued)

| Object | Site | Site/Lab Ref. No. | Date | Structure Edge | Hardness HV Back/Core | Type |
|---|---|---|---|---|---|---|
| Knife | Gestingthorpe | 16 | R-B | - | F/153 | 0 |
| Knife | " " | 26 | R-B | - | F/205 | 0 |
| Knife | " " | 73 | R-B | - | F/160 | 0 |
| Knife | " " | 88 | R-B | - | P+F/168 | 0 |
| Knife | " " | 25 | R-B | P+F/153 | F/- | 4 |
| Knife | " " | 105 | R-B | P+F/256 | - | 4 |
| Knife | " " | 93 | R-B | P+F/223 | - | 3 |
| Knife | Chelmsford, Essex. | 777303 | R-B | F/305 | F/220 | 0 |
| Knife? | Poundbury, Dorset | 820567 | Late R. | - | F+P/135 | 0 |
| Knife | " | 820497 | 3-4thC. | - | F/201 | 0 |
| " (RH) | " | 820264a | " | - | S/131-229 | 0 |
| " (RH) | " | 820264b | " | - | F/134 | 0 |
| " | " | 820364 | R-B | P+F/205 | F/150 | 2 |
| " | " | 820210 | R? | TM/401 | TM/401 | 3 |
| " | " | 820550 | R? | F+P/156 | F/125 | 1a |
| " | " | 820002 | Late R | F/205 | F/137 | 0 |
| " | " | 820000 | R-B | F+P/185 | F/148 | 1a |
| " | Brancaster, Norfolk | 773302 | R-B | F+P/230 | F/178 | 1a |
| Cleaver | " | 773715 | R-B | - | F/270 | 0 |
| Knife | Catsgore, Som. | 466 | R-B | P+F/235 | F+P/153 | 3 |
| " | " | 510 | R-B | P/250 | S/162 | 1a |
| " | " | 300 | R-B | F(P)/200 | F/153 | 1a |
| " | " | 635 | R-B | - | F/107 | 0 |
| " | " | 878 | R-B | - | F/159 | 0 |
| Cleaver? | " | 155 | R-B | - | F/214 | 0 |
| Knife | " | 153 | R-B | - | F/225 | 0 |
| " | " | 375 | R? | - | TM/250 | 3 |
| " | Winchester | CACP358 | R | F+M/269 | F+P/143 | 2 |

It is possible that the original intention was to harden the carburised surface by reheating the whole knife to a temperature of 750°C and quenching in water, but somehow this was omitted and the knife left in a carburised but unhardened condition.

Another knife comes from a 3rd - 1st century B.C. level of the hill fort at Winklebury, Hants, from which a currency bar has also come.[16] This was a narrow-bladed single-edged knife (Fig. 7) and a section was cut near the tang end of the blade.

This had fairly uniform carbon content and appeared to have been made by piling. The overall hardness varied from 345 to 418 HV5 and there was an arsenic enrichment band down the middle near the sharp end. High-power examination showed a ferrite-martensite-troostite structure which clearly indicated heat-treatment. Micro-hardness readings showed that the hardness of the martensitic regions was 500 HV (0.1), the trootsite 380 HV (0.1), and the residual ferrite matrix about 193 HV (0.1).

It would appear that this was a medium carbon steel which had been heated to between the upper and lower critical temperatures, i.e. between 700°C and 820°C, and slowly quenched. The carbon content was fairly uniform for an early steel but there were variations in the relative amounts of the three phases. This clearly shows a much improved technique over the Huckhoe knife and resulted in a very serviceable implement.

Romano-British Cutlery

Two knives came from the Winterton, Lincs, Roman Villa.[17] These were numbered 1 and 17. The first was found to have been completely rusted through the blade but the second had still some residual iron near the tang (Fig. 8). A section was taken through the blade at the point shown in Fig. 9. (Table A2).

This section is clearly an example of Type 1. Due to corrosion the steel central layer is no longer central; it has indeed been made a little off-centre at the back but there is every reason to believe that the cutting edge was protected by two pieces of wrought iron in its original state. The iron is in the ferritic state with some slag and had a hardness of 131 HV1. The central steel layer consists of tempered martensite (TM) but this has been rather over tempered or has been of rather low carbon content so that the average hardness is as low as 240 HV giving a maximum of 360. This is a good deal lower than the average for this type of blade.

On one side of the join between the steel and the iron there is an arsenic enrichment band suggesting that the adjacent piece of iron had a fairly high As content ( $>$ 0.015%). The other side did not show this effect suggesting that the iron here had a slightly different composition.

The Romano-British site at Ware, Herts was excavated in the 70s by Clive Partridge.[18] It was a settlement of some importance in which smithied objects as long as 2m were produced in a number of hearths. The vicinity of the hearths yielded hammer scales showing that tools had been made in them. The following items of cutlery were submitted for metallographic examination.

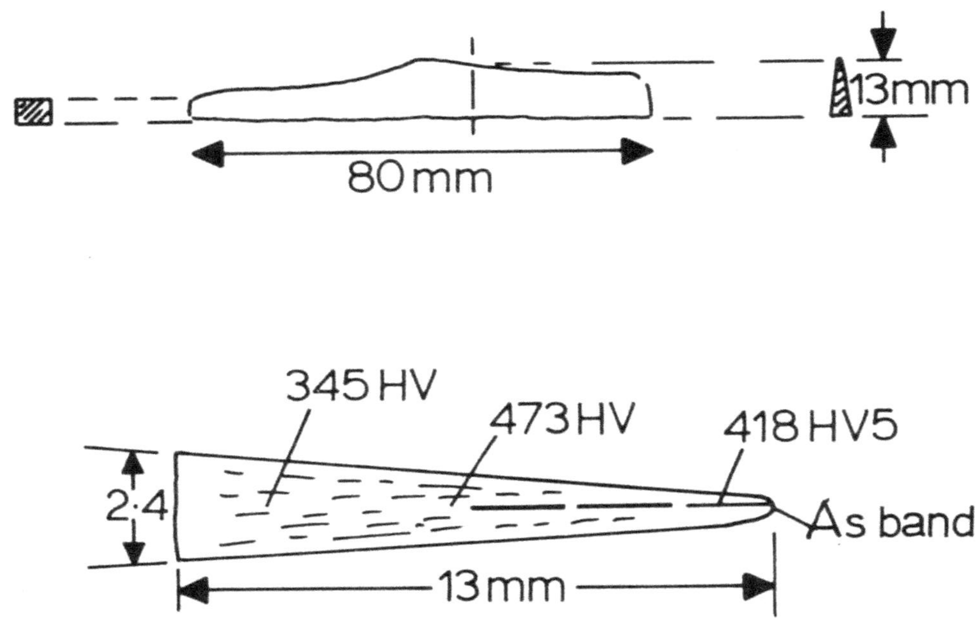


(a)

Fig. 7. Knife blade from Winklebury, Hants and metallographic section.

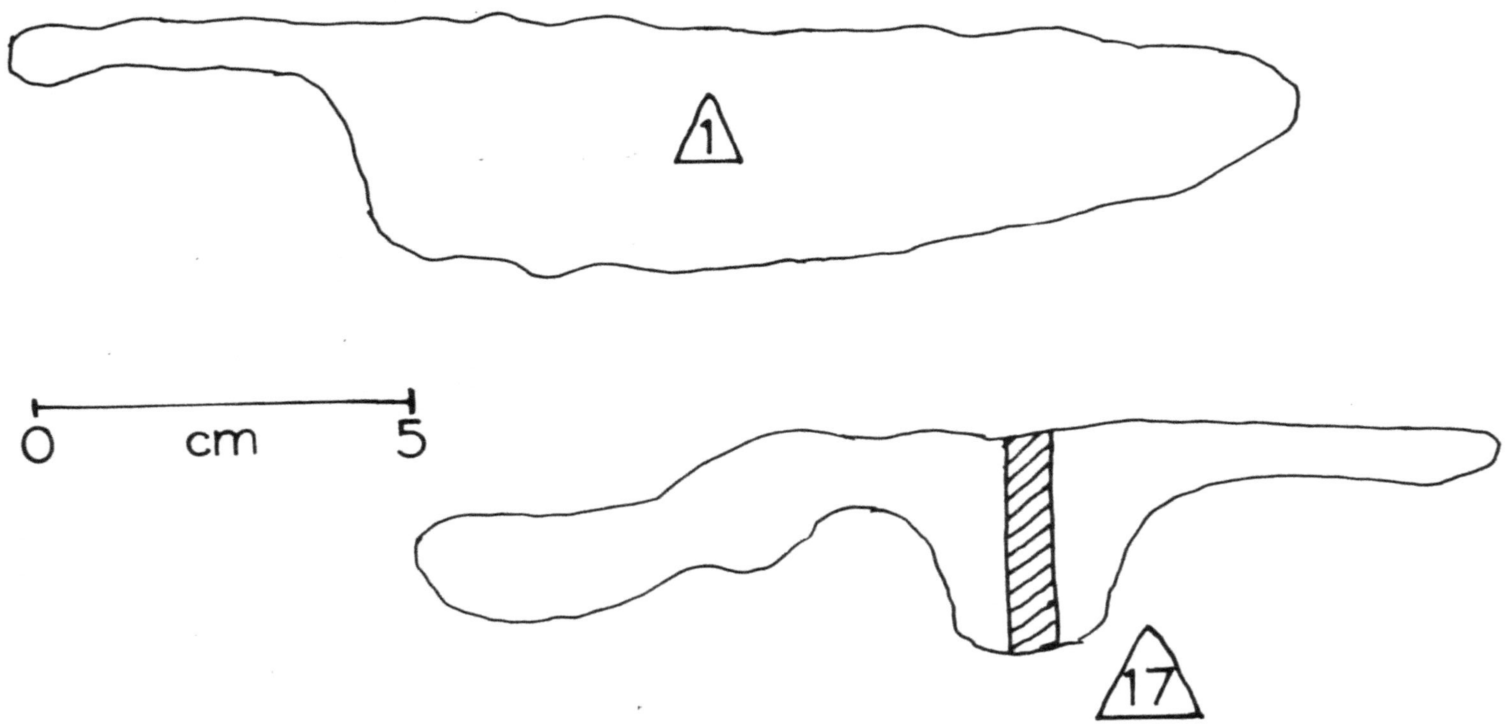


Fig. 8. Outlines of two knife blades from Winterton, Lincs. Roman Villa.

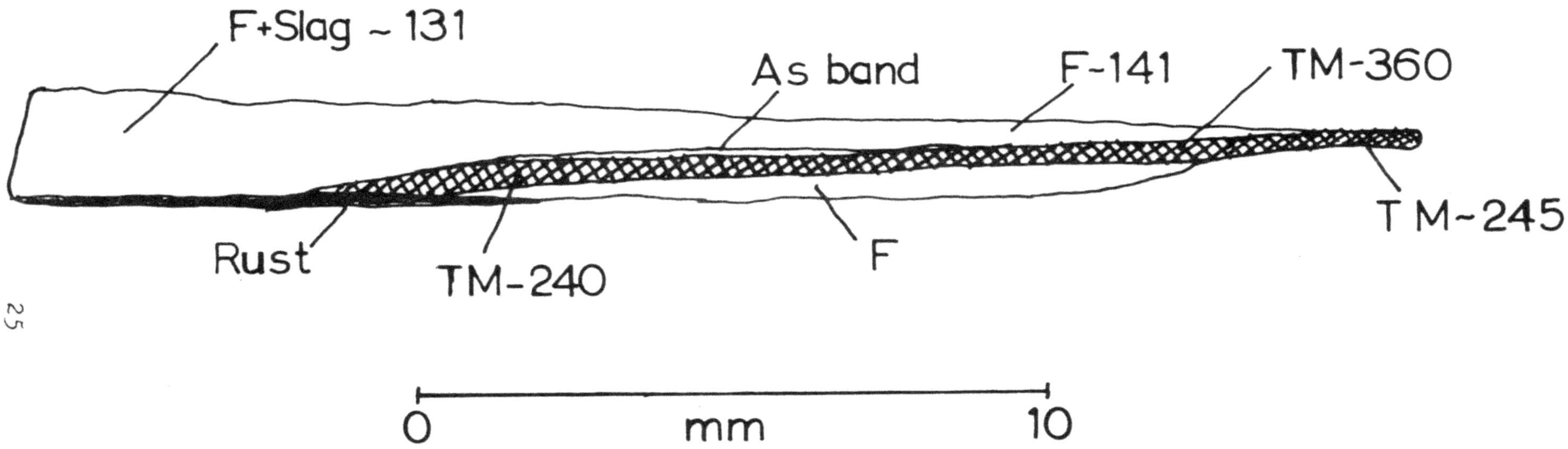


Fig. 9. Section of knife 17 from Winterton, Lincs. Roman Villa.

1849 - Knife Blade

A high carbon cutting edge has been formed by carburising a piece of wrought iron with a hardness of 189 HV1 and a ferrite and pearlite structure. The junction is diffuse and there is no sign of welding. The cutting edge has a hardness of 315 HV1 and consists of irresolvable pearlite or tempered martensite with fine ferrite grain boundaries.

1747 - Chopper or Cleaver

This thick blade was mostly mineralised but had a residual lamination of wrought iron along one surface. The hardness of this was 123 HV, but the fact that it remained whiled the rest of the edge was corroded away suggests that it consists of a different composition, i.e. contains some phosphorus, while the rest did not. High-phosphorus iron is more resistant to rusting than pure non-phosphorus-containing iron. The metal was clean and free from slag.

325 - Knife Blade

Contains longitudinal slag streaks, i.e. running towards the edge. Mostly wrought iron with a little spheroidised pearlite indicating a carbon content of 0.1%. The hardness is 162 HV1.

630 - Knife Blade

Contains fine elongated slag streaks. A medium carbon steel (0.3%C) with a ferrite and pearlite structure. The pearlite is lamellar to spheroidised suggesting that the blade had a fairly long time at 700°C while it was being forged and that it was finally slow cooled. The hardness is 136 HV1 suggesting very low phosphorus.

WANBOROUGH

A Romano-British site, now in the greatly expanding town of Swindon in the Borough of Thamesdown on the borders of Wiltshire, has produced a large amount of interesting iron work. (Fig. 10) The cutlery examined from this site is listed in Table 6. The results of the metallographic examination were as follows:

790723 - Cleaver or chopper

Uniform and homogeneous ferrite + pearlite with about 0.1 to 0.15% C. The carbon content is at its maximum near the edge where the hardness is 251 HV1. But such a high hardness is not typical of a simple F+P structure and implies considerable phosphorus.

780856 - Knife blade (cleaver)

Fine slag stringers down the centre. This has been piled up and consists of 3-4 major layers varying from pure ferrite to 0.3% C steel. The structure is that of ferrite and pearlite and the hardness is 158 HV.

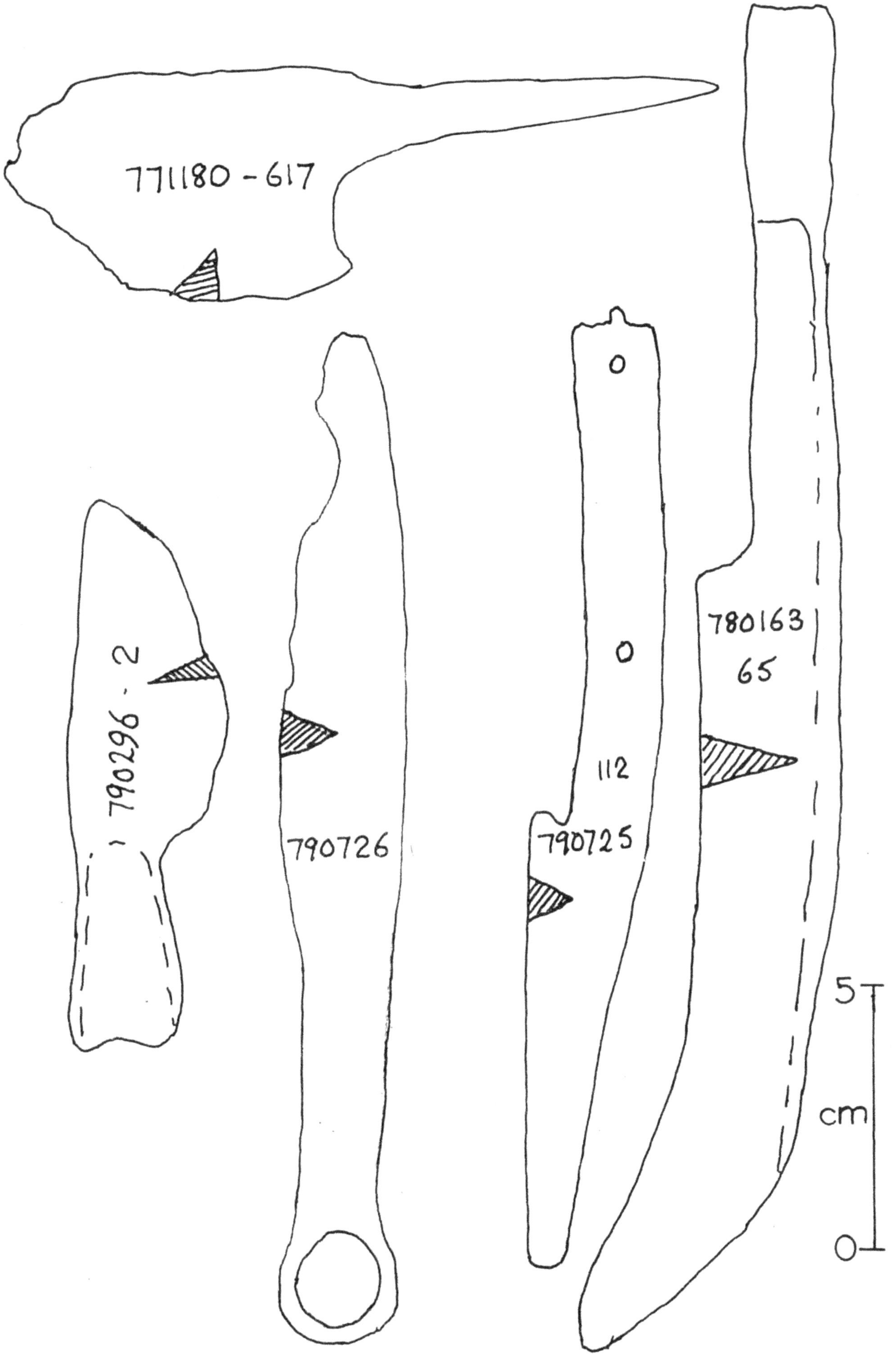


Fig. 10. Outlines of Roman cutlery from Wanborough, Wilts.

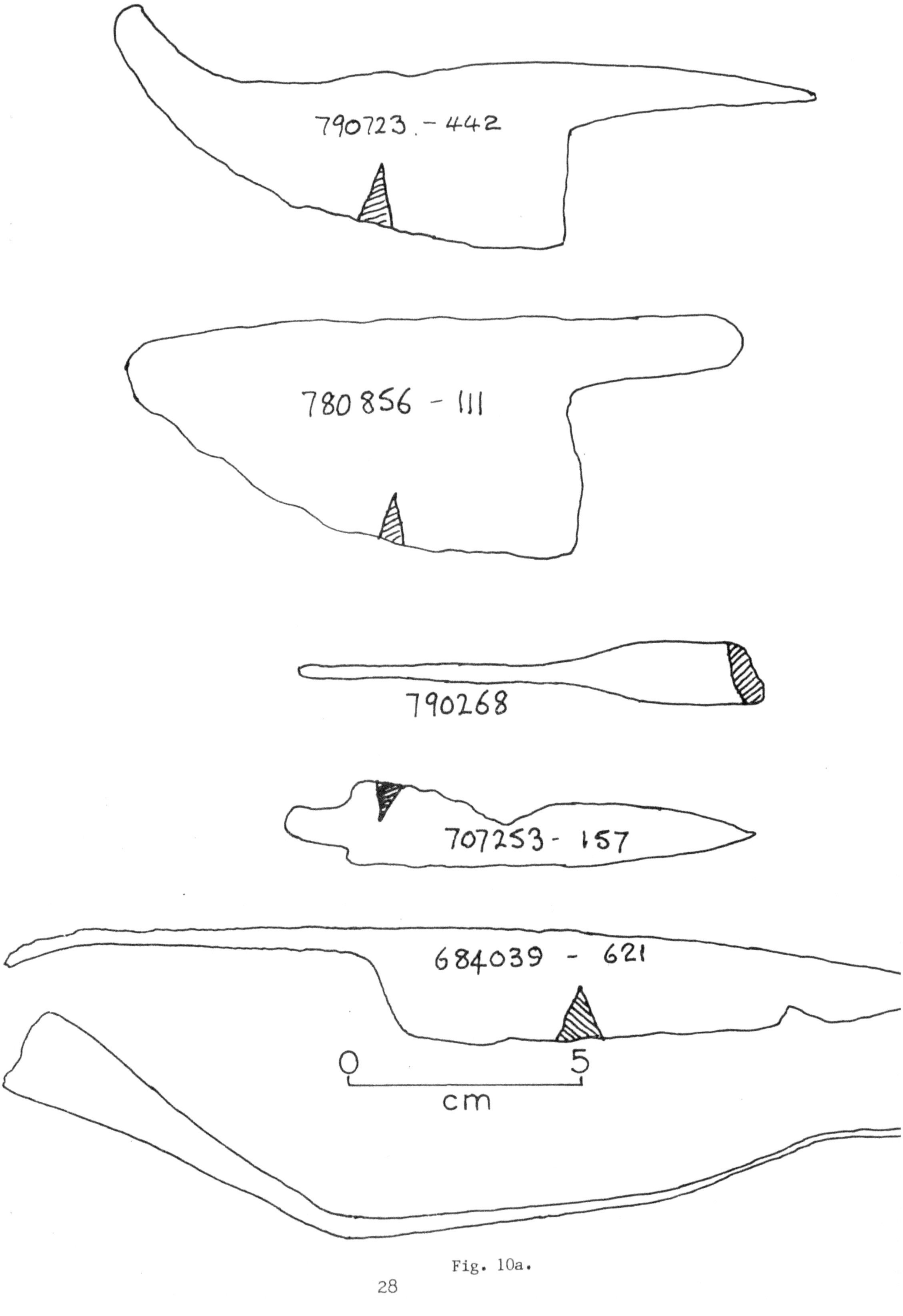


Fig. 10a.

TABLE 6.

Cutlery from the Romano-British Site at Wanborough, Wilts.

| Object | Site Ref. No. | Date, AD | Structure/Hardness HV Edge | Back |
|---|---|---|---|---|
| Chopper | 790723 | 325- +400 | F+P/251 | |
| Cleaver | 780856 | 325- +400 | F+P/158 | |
| Knife | 707253 | 388-402 | TM/ 720 | F/148 |
| Knife | 780163 | + 345 | F+P/240 | |
| Knife | 684039 | 3rd-4th C | P+M/269 | |
| Knife | 790725 | 230-325 | F+P/175 | |
| Knife | 790296 | 3rd-4th C | F+P/185 | |
| Knife | 790628 | Late 4th C | P+F/240 | |
| Knife | 790726 | 325- +400 | F/133 | |
| Knife | 771180 | Late 4th C | TM/369 | |

707253 - Knife blade

Very clean metal. This would seem to have been intended to be a homogeneous steel. But diffusion has not been complete and the carbon is very variable. The edge is a very nice example of tempered martensite with a hardness of 720 HV. The inside varies from ferrite to ferrite + pearlite and has a hardness of 148 HV. A well-made, well heat-treated implement (Fig. 11).

780163 - Knife

Some slag delineates the weld lines. It is more or less homogeneous carbon steel with about 0.6% C in the form of ferrite and pearlite. The hardness is 240 HV1 which agrees with the structure. The back of the knife carries a heavy reinforcing rib.

684039 - Knife

Very clean with a group of slag stringers near the back. Light etching shows only carbide in an unetched matrix, but the hardness of 269 HV suggests that the matrix is not ferrite. Deeper etching darkens the background and reveals a two-phase structure consisting of finely divided light and dark areas probably martensite and pearlite (bainitic). Thus, this would appear to be a slack-quenched blade. (Fig. 11).

790725 -Knife Blade

Contains very fine slag stringers. The carbon content varies from 0 to 0.15% showing a ferrite and ferrite + pearlite structure. Where the carbon is low the pearlite tends to break down into stringers of spheroidised carbides. The hardness is 175 HV and the mean carbon content is about 0.1%.

790726 - Knife

Very coarse ferrite and slag with a hardness of 133 HV which suggests a low phosphorus content.

790296 - Knife

This was a well-piled structure with multiple banding. The overall carbon content is in the range 0.1 - 0.15%. There is more slag near the cutting edge and the ferrite grains show some signs of distortion. The hardness is 185 HV1 which indicates some phosphorus.

790268 - Knife

Homogeneous high carbon steel. The structure is pearlite and ferrite and the carbon content is about 0.6%. The pearlite is fine but resolvable at X 400. The hardness is 240 HV which is what one would expect of this carbon content with low phosphorus.

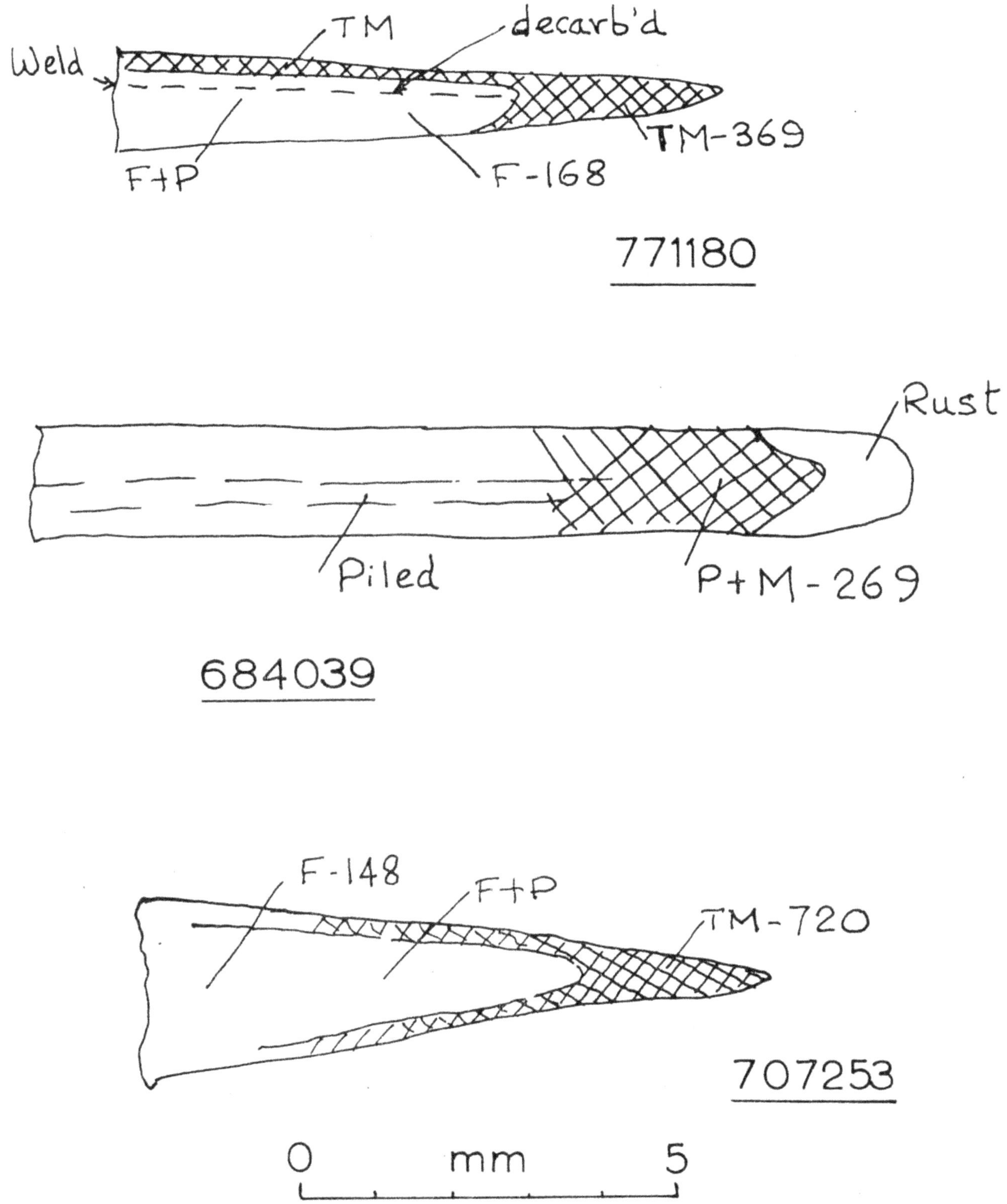


Fig. 11. Metallographic sections of Wanborough cutlery.

771180 - Knife

Coarsely piled but mostly ferrite of medium grain size. The carbon content in this region does not exceed 0.15% and the hardness is 168 HV suggesting some phosphorus. But one surface has had welded to it a piece of high carbon steel which extends round the cutting edge (see Fig. 11). This consists of tempered martensite with a hardness of 369 HV. The weld line is now in the F+P area suggesting that the high carbon steel was slightly decarburised during heating for welding. The F+P structure at the centre is Widmanstätten which probably arose during heating for welding. If so, the high carbon side has been reheated to a hardening temperature of 800°C which has left the widmanstätten centre more or less unaffected. This would have been easy with an oxyacetylene flame but is difficult to do in a smith's hearth and one wonders whether it was intentional.

Two Romano-British sites in Essex, Gestingthorpe[20] and Chelmsford, have yielded knives which on the whole are of poor quality but fairly typical of Roman production. The results are as follows:

Gestingthorpe

70 - Knife

The edge consists entirely of large equiaxed grains of ferrite with a hardness of 164 HV 0.5. This suggests an appreciable phosphorus content.

116 - Knife

With flanged socket. Well rusted along one side. Seems to be mostly ferrite with a little spheroidised pearlite and slag. The hardness is 122 HV1.

91 - Knife

Consists entirely of fine grained ferrite and slag. The hardness is 159 HV1.

16 - Knife

Another piece of ferrite with a hardness of 153 HV1.

26 - Knife

Yet another ferritic blade. The hardness in this case is 205 HV1.

106 - Knife

The edge has a ferrite and pearlite structure with about 0.2%C and a hardness of 240 HV1. The composition and structure are very uniform and the pearlite is spheroidal rather than lamellar. The high hardness and the low carbon suggest an appreciable phosphorus content.

73 - Knife Blade

Fragment. A section through the blade consists of very coarse ferrite and some intergranular corrosion. Contains no weld lines and very little slag. The hardness is 160 HV1.

88 - Knife

The blade has been formed by folding over leaving a slaggy area in the fold. The structure is Widmanstätten ferrite and pearlite with a carbon content in the range 0.1 to 0.4%. The hardness of the 0.4% area is 168 HV1.

25 - Knife

This blade has been made by folding a thin sheet of low carbon steel with a carbon content of 0.15% over pure ferrite. The border is slaggy and the carbon has diffused. The pearlite is spheroidal to near divorced and the hardness is 153 HV1.

105 - Knife

A steel cored blade of Type 1. The core consists of ferrite and pearlite with a carbon content of about 0.5% - 0.8% and a hardness of 256 HV1. On one side there is pure coarse ferrite and on the other ferrite and pearlite. The pearlite is fine and non-lamellar indicating a fairly fast cooling rate through the 700-600°C range. What a pity this blade was not heat-treated.

93 - Knife Blade

The structure is a very uniform pearlite with ferrite grain boundaries. There are some thick slag stringers, some running the whole width of the blade together with white lines indicating arsenic. In the thick areas of the blade the lamellar structure of the pearlite is resolvable, but in the thin section it is finer. At the edge the carbon content is about 0.6% and the hardness is 223 HV1.

CHELMSFORD

777303 - Knife

A piece about 10mm long was removed from the cutting edge of the blade about half way along. The first 4mm from the edge were fully rusted. The rest was ferrite and slag with a hardness of 220 HV1 which suggests a high phosphorus content.

Repolishing revealed a piece of residual metal in the rusted portion. This had a micro-hardness of 305 (25g) and 295 (100g), which shows some signs of hardening. The structure, however, was still ferritic with heavily elongated grains denoting work-hardening. This type of edge hardening is still practiced on edge tools in Africa and has been seen being carried out in the market at Kabushiya in the Sudan.

This type of hardening has been used on the shears found at Silchester (see below). Work-hardened ferrite is much more corrodable than the unworked material and so would be expected to be preferentially rusted.

Another site which yielded Romano-British and later ironwork is Poundbury in Dorset excavated by C.J.S. Green.[21] Cutlery from this site gave the following results:-

820567

Object, possibly a knife blade from Roman cemetery. Almost all steel with a carbon content in the range 0.1-0.2%. The structure is of the Widmanstätten type with rather spheroidal pearlite. It is fairly homogeneous with very little slag. The hardness is 135 HV1 which suggests low phosphorus.

820497

A knife from the same cemetery. Made of very coarse-grained ferrite with little slag. A hardness of 201 HV suggests appreciable phosphorus.

820264 a and b

Two possible ring-handled knives from 3rd-4th cent. Late Roman levels; (a) contains a good deal of carbon but it is distributed in a very uneven way not suggestive of an edge tool. There are areas of sorbite or tempered martensite with a hardness of 229 HV, and lower carbon areas of coarse-grained pearlite with ferritic grain-boundaries having a hardness of 131 HV. The pearlite is spheroidised rather than lamellar showing that it had been held for some time between 600° and 700°C. (b) is all ferrite with very little slag. The hardness is 134 HV.

Fe 583

A pointed object from the Late Roman settlement. This is almost entirely ferrite but of very variable grain size. The slag stringers and the grain structure show evidence of piling. The hardness is 140 HV suggesting low to moderate phosphorus. There is a little pearlite in places but the carbon content does not amount to more than 0.1% anywhere.

820364

A knife from the Roman cemetery No. 3. Very heterogeneous but clearly steeled. The join is diffuse. The back is mainly ferrite with a hardness of 150 HV. The cutting edge is ferrite+pearlite with a Widmanstätten structure indicating that it has been at a high temperature, cooled rapidly, and held at 600-700°C which spheroidised the pearlite. The hardness obtained by such treatment is 205 HV. Why did the smith fail to quench it?

820210

Unstratified but in a Roman or post-Roman context. This is a sharp homogeneous knife which seems to be entirely tempered martensite with a hardness of 401 HV.

820550

Large knife unstratified but in a Roman or post-Roman context. An interesting and unusual structure. A strip of iron has been carburised at one edge to give ferrite-and-pearlite with about 0.3% carbon and a hardness of 156 HV1. This has been welded along most of the width of the blade to piled wrought iron in which the vary coarse ferrite has a hardness of 125 HV. The pearlite is spheroidised and no heat-treatment has been given.

820002 A Piece of a Blade - (Late Roman)

Contains a medial weld with a lot of slag; otherwise the metal is relatively clean. Etching showed that it was entirely ferritic of moderate grain size. The hardness varied from 205 near the sharper end to 137 HV1 inside the blade. This artifact appears to have been made from a very few strips of ferrite of varying phosphorus content.

820000. Fe 1095 - A Knife - (Early Roman)

This was cut from the blade of a knife (Fig. 12). It had an adherent and continuous film of rust; but within it there was evidence of complex welding and piling. Basically there was a medial weld line on one side of which the metal was ferrite while on the other it was ferrite and pearlite. The metal was slaggy with two-phase slag inclusions. The carbon content of the higher carbon side was about 0.5% and the pearlite was resolvable at X400. The hardness of the high carbon side was 185 HV1, while that of the ferrite was only a little less at 148 HV1. There was no sign of martensite. Clearly then, this blade has been made by the welding on of a piece of carbon steel to a blank of medium phosphorus iron and no attempt has been made to heat-treat the higher carbon piece.

The Roman site at Brancaster, Norfolk has also yielded some cutlery:-

773302

A long tapered object; it could be a blank for a knife. Two distinct layers. One was ferrite with a hardness of 178 HV. The other was steel, ferrite and pearlite corresponding to a carbon content of about 0.4%. This layer had a weld down the middle showing that it had been 'doubled' before welding to the iron. The hardness was 230 HV.

773715

A cleaver with a broad blade. Heavily laminated but mostly ferrite with variable grain size. It is very slaggy and there is no grain boundary carbide but the hardness is 270 HV which indicates a very high phosphorus content.

A number of knives were examined from Catsgore Roman Villa,[23] Somerset:-

Cat. F.466 (12)

This knife has a variable carbon content of about 0.6% at the back decreasing to 0.3% at the centre and tip. There is a well defined slag stringer starting at the tip and finishing about half way up the blade suggestive of welding. All the pearlite-ferrite structure is Widmanstätten. The hardness at the centre is 158 HV and at the tip 235 HV.

Clearly, the carbon distribution is unintentional but is some way the maker has managed to get a higher hardness at the tip than at the centre - possibly by cold work or a high phosphorus content in the iron. Not a very efficient tool.

F.510 (16) Cat. 1063

A small piece from the blade of a knife. This has a completely and densely rusted cutting edge with a hardness of 575 HV1 (in rust). The rest consists of two steels joined down the centre. One side is 0.7-0.8% pearlitic carbon steel (250 HV1), and the other 0.2-0.3%C which consists of spheroidised carbides in ferrite (162 HV1).

F.300 Cat. 402

Fragment of a blade with a weld down the centre. On one side the structure is ferrite with a hardness of 153 HV1; and other side contains about 0.2%C in the form of fine pearlite giving a hardness of 200 HV. The edge on both sides is almost wholly ferrite and would not give a good cutting edge. The iron has a moderate phosphorus content.

F.408 Cat. 635

A blade consisting wholly of ferrite and slag. The section cuts the slag stringers end-on but they are round showing that the blade has only been worked along its length and has not been worked towards the cutting edge. The hardness of 107 HV1 indicates a very low phosphorus level.

F.489 Cat. 878

A piece from another blade. The iron was difficult to etch and is therefore probably phosphoric. The hardness is 159 HV1. There seem to be some signs of carbide in the grain boundaries.

F.155 (6) Cat. 1092

A small piece from a large blade (probably a cleaver) that was difficult to etch and therefore consists of high phosphorus ferrite. The hardness is 214 HV1.

F.153 (10) Cat. 349

A piece from a blade consisting entirely of coarse-grained ferrite with a hardness of 225 HV1 which suggests a high phosphorus content.

F.375 Cat. 593

Iron knife with lead-wrapped handle. A very thin blade consisting entirely of tempered martensite and slag with a hardness of 520 HV1.

MEDIEVAL KNIVES (Table A3)

Most of our knowledge of knives of this period come from Martin Biddle's excavations at Winchester.[24] But these do not start before the 9th century and we have comparatively few of the centuries immediately following the end of the Roman occupation.

Some of the levels at Poundbury[21] date from destruction levels in a post-Roman settlement, and other are Saxon or later. We also have five knives from the Saxon site at West Stow in Suffolk[25] and one dated to the 8-9th century site at Ramsbury, Wilts.[26] The results of this small group are as follows:

Poundbury; Post-Roman Settlement

Fe.125

This consists predominantly of tempered martensite with a small piece of low carbon slaggy ferrite welded to the back; this has a hardness of 160 HV. The tempered martensite is very homogeneous and has a hardness of 520 HV1 but contains one large lens of slag. A good knife.

Fe.126

An interesting example of pattern welding; after etching in Nital it was found to consist of thick alternate streaks of a dark phase with thin streaks of a light phase. The thin streaks are chevron-shaped (Fig. 12), showing that after being put in as laminations they received forging pressure in such a way that they were bent at the centre line of the blade. The light phase is a pure ferrite and it has a hardness of between 276 and 330 HV. It could be a ferrite with high levels of As and P. No martensitic structure was developed. The dark phase which had a hardness of 185-245 HV is probably tempered martensite or pearlite. A good knife.

TABLE A3. Medieval knives

Post-Roman to the 15th century

| Object | Site | Site Ref. No. | Date | Structure Edge | Hardness HV Back/Core | Type |
|---|---|---|---|---|---|---|
| Knife | Poundbury, Dorset | 603 | Saxon | TM/615 | F/138-193 | 2 |
| " | " " | 809 | " | F+cem/214 | - | 3 |
| " | " " | 786 | Saxon? | F+cem/210 | F/105 | 0 |
| " | " " | 125 | " | TM/520 | F/160 | 2 |
| " | " " | 508 | Post-Roman | TM/330 | F+P/172 | 2 |
| " | " " | 605 | Saxon? | TM/553 | F/186 | 2 |
| " | " " | 126 | Saxon? | F+P/185-330 | F+P/185-330 | 1a |
| " | West Stow, Suff. | 716216 | Saxon | F/- | P+F/180 | - |
| " | " " | 716210 | " | P+F/- | F/- | 2 |
| " | " " | 716248 | " | F/- | F+P/- | 0 |
| " | " " | 716232 | " | - | F/165 | 0 |
| " | " " | 716300 | " | M/300 | F/- | 3 |
| " | Ramsbury | 14 | Late Sax. | M/880 | F/- | 3 |
| Knife | Winchester | CY276 | 9-10th Cent. | M+T/533 | F+P/219 | 1 |
| " | " | BS271 | 10th Cent. | TM/636 | F/113 | 2 |
| " | " | BSSC6223 | " " | TM/439 | F/127 | 2 |
| " | " | CY194 | 11th cent. | M/633 | P+F/232 | 1 |
| " | " | BS128 | " " | F/113 | | 0 |
| " | " | ACN127 | " " | M+F/313 | - | 1 |
| " | " | BS6250 | " " | F+SC/102 | F/95 | 2 |
| " | " | CG1884 | " " | TM/290 | - | 1 |
| " | " | BS4085 | 11-12th cent. | TM/571 | - | 1 |
| " | " | BS6458 | " " | TM/426 | F/117 | 2 |
| " | " | BS4200 | " " | SC/229 | F/165 | 1a |
| " | " | BS4263 | " " | TM/551 | F/116 | 2 |
| " | " | BSSC112 | 12th cent. | TM/536 | F/116 | 2 |
| " | " | BSSC329 | " " | M+F/536 | F/262 | 1 |
| " | " | LBS41 | 11-12th cent. | F+P/143 | - | 3 |
| " | " | LBS25 | " " | F+P/99 | - | 3 |
| " | " | BS1354 | 12-13th cent. | M/426 | P+F/313 | 2 |
| " | " | BSSC471 | " " | - | F+P/237 | 2 |

TABLE A3. Medieval knives (continued)

| Object | Site | Site Ref. No. | Date | Structure Edge | Hardness HV Back/Core | Type |
|---|---|---|---|---|---|---|
| Knife | Winchester | BS4128 | 12-13th cent. | F+P/197 | P/210 | 2 |
| " | " | BS6490 | " " | TM/251 | F/162 | 1 |
| " | " | BSSC279 | 13th cent. | M/386 | F/198 | 2 |
| " | " | BSSC609 | " " | - | F/140 | 2 |
| " | " | BSSC461 | " " | - | F/186 | 2 |
| " | " | CG1176 | " " | P/313 | - | 3 |
| " | " | BSSC613 | " " | TM/557 | F+SC/210 | 2 |
| " | " | BS2807 | " " | TM/290 | F/131 | 2 |
| " | " | BSSC81 | " " | F+SC/203 | F/160 | 2 |
| " | " | BS2802 | " " | M+F/467 | F/113 | 2 |
| " | " | BS2668 | " " | F+Cem/169 | - | 3 |
| " | " | CG2554 | 13-14th c. | TM/575 | F+P/201-225 | 2 |
| " | " | BSSC597 | 14th cent. | TM+T/229 | - | 3 |
| " | " | BS2061 | " " | TM/440 | F/137 | 2 |
| " | " | BS2580 | " " | TM/239 | F/119 | 2 |
| " | " | BS542 | " " | SC/239 | F/165 | 2 |
| " | " | BS1698 | " " | M/401 | F/117 | 2 |
| " | " | BS659 | " " | F+P/210 | F+P/186 | 1 |
| " | " | BS2242 | " " | SC/205 | F/136 | 1 |
| " | " | WP666 | 14-15th cent. | M/379-214 | | 3 |
| " | " | LBS134 | 13-16th cent. | TM/390 | F/182 | 1a |
| " | Chingley | 9E1 | 13th cnt. | TM/533 | TM/532 | 3 |
| " | Goltho, Lincs. | G13 | 12-13th cent. | TM/407 | F+P/ - | 3 |
| " | " | G22 | 12-14th cent. | SC/175 | F/141 | 2 |
| " | " | A120 | 12-15th cent. | TM/557 | F/151 | 1 |
| " | " | G19 | 14-15th cent. | TM/341 | F/ - | 4 |
| " | Barton Blount, Derbyshire. | House 24 E206 | 15th cent. | M/274 | F/124 | 2 |
| " | " | Croft G | 12-15th cent. | TM/701 | F/148 | 2 |
| " | Winchester | WP3398 | 15-16th cent. | M/571 | F+P/186 | 4 |
| " | " | WP2972 | " " | M+F/482 | F+P/263 | 1 |
| " | " | BS701 | " " | F+P/202 | F/ - | 2 |

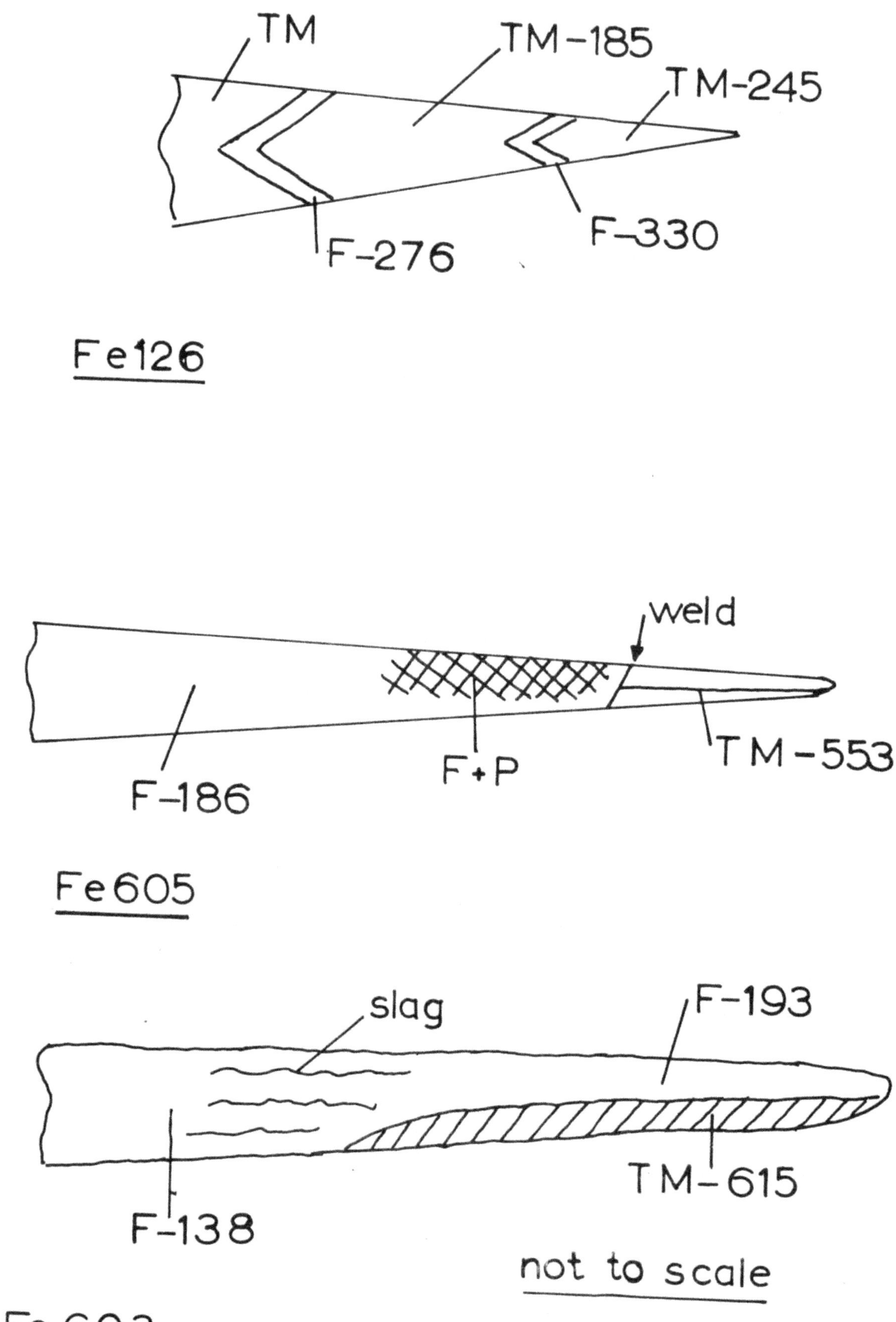


Fig. 12. Sections of knives from Poundbury, Dorset.

Fe.508; E.9

A well-made, steeled, knife, from the post-Roman settlement. The back consists of ferrite-and-pearlite with a carbon content of between 0.2 and 0.4% and a hardenss of 172 HV1. The edge is tempered martensite (sorbite) with a hardenss of 330 HV1. The join is diffuse with no sign of arsenic enrichment.

Fe.605; E.86

Knife. This is a very sophisticated piece of work. The back of the knife is predominately coarse-grained ferrite with a hardness of 186 HV1 (Fig. 12). The cutting edge is a carbon steel, folded and welded to the back near the cutting end; the weld is indicated by white decarburised bands which do not seem to have resulted from arsenic enrichment. The mid-section is ferrite-and-pearlite with about 0.4%C; the pearlite is not resolvable at X600. The hardness of the cutting edge of tempered martensite with some excess grain boundary carbide is 553 HV1. It seems that the heat treatment has been done by gradient heating with the edge of the knife heated to 800°C or so and then quenched into water while the rest of the knife was protected and therefore allowed to cool more slowly. This knife would make a very efficient tool. The tempering was probably done by the heat-flow back from the main body of the knife after the edge had been cooled to below 100°C.

Fe.809; E.189

Knife. A fairly homogeneous but piled structure consisting of ferrite with partially spheroidised cementite. The carbon content varies from about 0.2 - 0.3% at the cutting edge to 0 - 0.2% at the back. The maximum hardness is 214 HV1. Unless supported by some hammer hardening at the cutting edge, this would not be a very efficient knife.

Unstratified, possibly derived from the Post-Roman settlement.

Fe.736; E, U/S

Knife. Unstratified. Ferrite with very fine and evenly distributed carbide with a carbon content varying from about 0.1% at the back to about 0.6% at the cutting edge. The corresponding hardnesses are 105 and 210 HV1. It would seem that the carbon content had been achieved by long-term carburising at about 750°C or, possibly, high temperature carburising at over 900°C followed by spheroidising during working at 700 - 750°C.

Fe.603; E, U/S

A knife of post-Roman type. A piece of comparatively homogeneous steel has been welded along part of the width of the blade and this has been effectively heat-treated to give a tempered martensite with a hardness of 615 HV1 (Fig. 12). The weld shows the white lines of arsenic enrichment. The rest of the blade is very slaggy ferrite with a hardness of 138 HV.

West Stow, Suffolk; Saxon. (The blade outlines and positions of Sections are given in Fig. 13).

716210

Mainly equiaxed ferrite with about 0.1% C in the form of pearlite. On the edge, however, a piece of steel of about 0.6% C has been welded on. Here the carbon is in the form of pearlite showing that no heat treatment has been attempted. The geometry is such that the steel edge only extends to half the thickness of the blade; to expose it as a cutting edge would need the removal of the other half by suitable grinding.

716232

This consists entirely of coarse ferrite. If there was a welded-on edge this has been rusted away. Hardness: 165 HV1.

716248

This consists mostly of fine ferrite but on one side, near the middle, a piece of 0.6% C steel has been welded-on as shown by well-marked slag inclusions. This has missed the edge of the blade and would not improve its cutting characteristics.

716216

This is rather an odd piece. The centre of the blade consists of 80% pearlite with a hardness of 180 HV5 while the sharper edge is all ferrite. But this is so badly rusted that it may have been originally edged with steel. The back has a piece of pure ferrite welded to it - again shown by well-marked slag inclusions.

716300

After polishing this consisted of a minute fragment. But its resistance to etching suggested either pure ferrite or untempered martensite. Further etching showed that it consisted of two external layers of martensite, somewhat decarburised on the external surfaces, with a layer of ferrite sandwiched between. The martensite had a hardness of 300 HV1 and was untempered. This suggests that it was of low carbon content. The structure in the central area was confused; it had probably been quenched between the upper and lower critical temperatures (700 and 900°C) and there was probably a small amount of martensite present amongst the ferrite.

Ramsbury - Late Saxon

14. A Tanged Knife; Period 3b, Layer 55

A section was cut through the centre of the blade of what was clearly a very well heat-treated knife. This showed that it was an example of Type 2, with a steel edge welded to a mainly ferritic back. The cutting edge

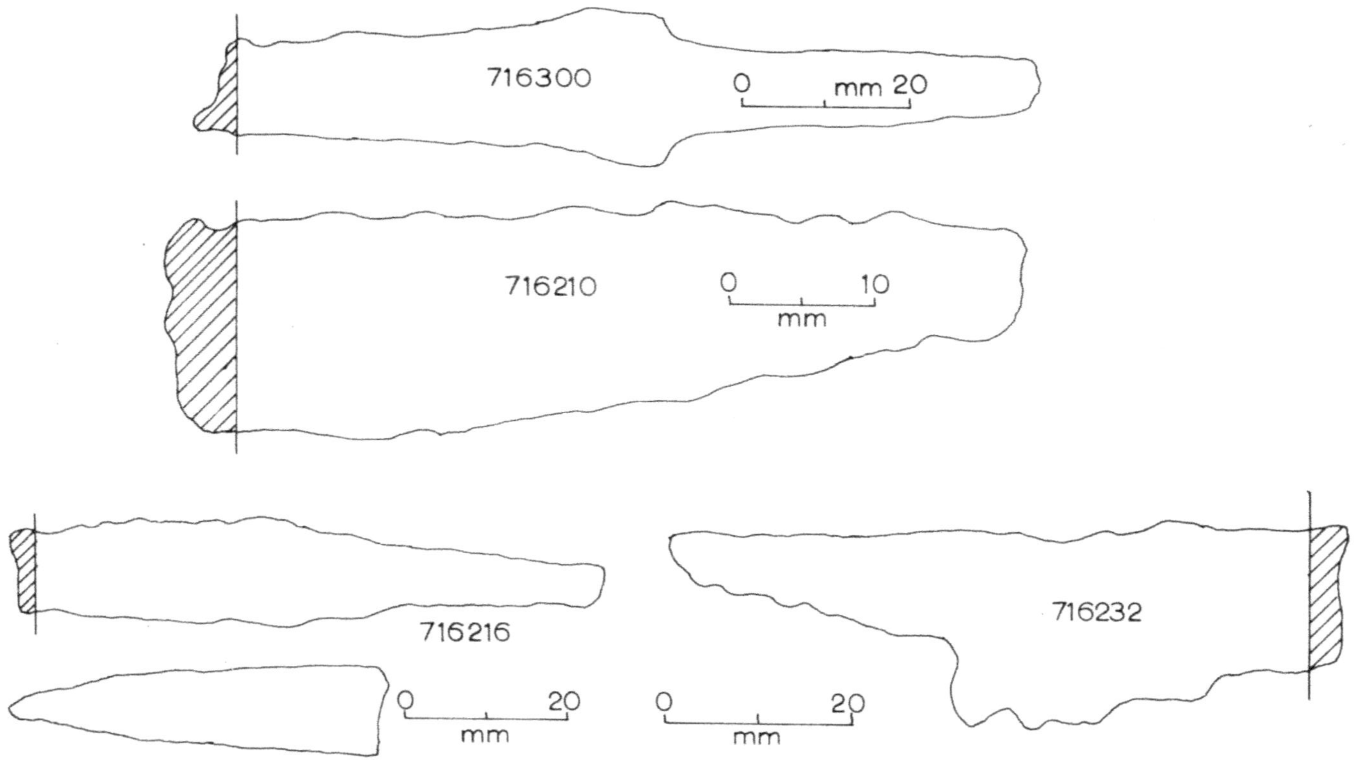


Fig. 13. Outlines of four blades from West Stow, Suffolk showing positions of sections.

was tempered martensite with a hardenss of 880 HV1, joined with a wide ferritic zone through which some carbon had diffused to give some martensite with a hardness of 270 HV. One side of the back was ferrite with a hardness of 200 HV1 while the other was ferrite + pearlite with a hardness of 178 HV1.

We can already see a great improvement in quality here over the Romano-British product. Out of 13 knives examined, 7 had been carburised and heat-treated. The main exceptions were in the group from West Stow which had been well corroded.

Winchester

We have examined 43 knives from this town covering the period of the 9th - 14th centuries. The outlines and the points of sectioning are not given here but will be found in the appropriate section of the Winchester Studies, 7. A summary of the results is given in Table A3.

In this table the knives have been arranged chronologically. All four of our main groups are represented; those with pattern-welded backs turn up in all periods between the 12th and 14th centuries but of course are not numerous as they represent a very special and highly priced artifact. It is clear that, with the exception of the 13th century specimen CG1176, the homogeneous specimens are unstratified or late.

Examples of the structures are shown in the photomicrographs Figs. 14 to 20. Fig. 14 shows a steel-cored blade of Type 1 (WP 1516). Both the steel core and the iron plates on either side were made by piling and there is not a lot of difference in overall carbon content between the core and the rest although the former contains more material of higher carbon content. Fig. 15 shows a nice example of pattern-welding in a 13th - 14th century blade BS 1031. The edge has been made of a piece of steel bent double and corrosion has shown up the not very good weld made in the doubling process. Arsenic enrichment lines are also very clear here and indicate that the steel itself contains comparatively high As; the weld between the pattern-welded back and the carbon steel edge is also delineated in this way. The back is made of alternate bands of ferrite and ferrite+pearlite. The straight butt joint between the back and the edge is very well done and it would seem that there has been quite a lot of grinding of the sides of the blade to clean off the upset from the butt joint.

Fig. 16 shows two of the steel-edged blades. The bottom one, No. BS 271 would have a very much longer life than the top, BS 4263. In both cases the weld has been made by over-lapping the steel and the iron and welding from one side, thus giving an angle to the steel-iron joint. Diffusion is minimal and there are no white lines. (Type 2(a)).

Fig. 17 shows a typical steel-cored blade; again both elements are made from piled material. But in this case, we get the tell-tale white lines of As enrichment. The steel core itself by virtue of the slag shows that it has been made of 4 to 6 thicknesses of steel and is well diffused with low As. On the other hand the iron and low carbon steel on either side show indications of white As lines showing that it was high in arsenic.

Fig. 18 shows the only iron-cored knife from Winchester and one of the few known from Britain. Here a piece of high carbon steel has been wrapped

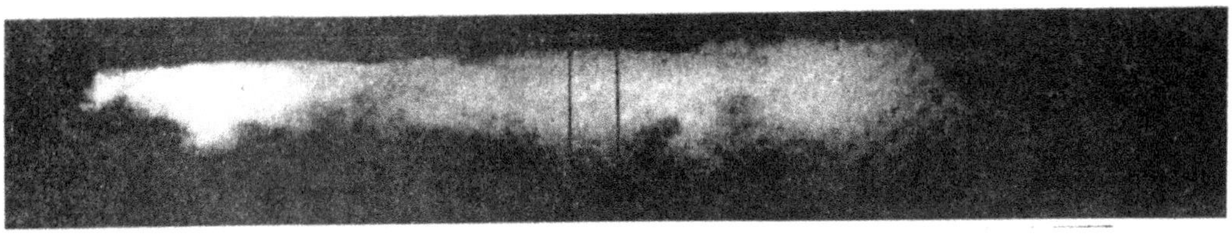

(a) X-ray of steel-cored blade of the fifteenth century from Winchester, × 2

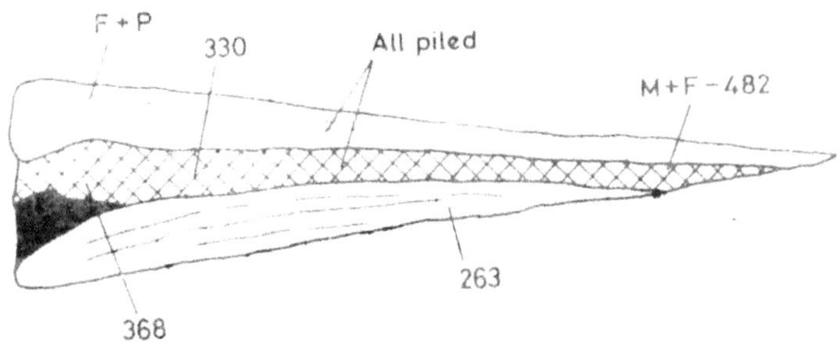


(b) Section through edge of blade at position indicated above showing metallic structure, × 12. F + P ferrite and pearlite (about equal amounts), M + F – martensite and ferrite due to quenching at too low a temperature for the carbon content of the steel; numbers indicate hardness in units HV.

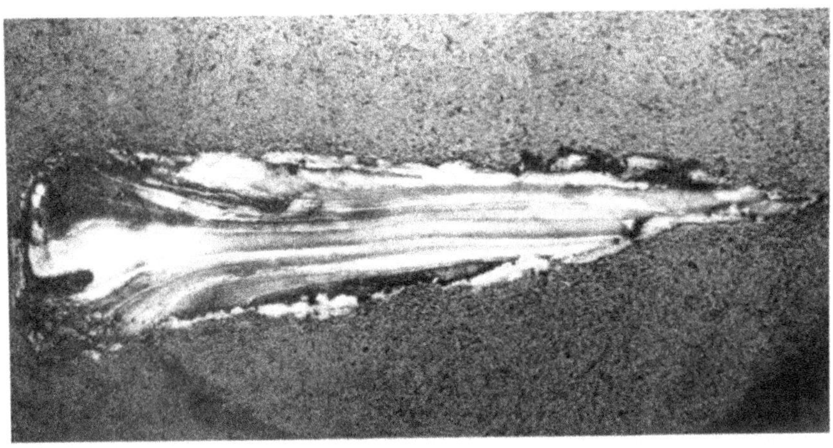

(c) Photomicrograph of section through blade showing variations in carbon content,

2mm

Fig. 14. X-radiograph and photomicrograph of a 15th cent. knife blade from Winchester (WP 1516).

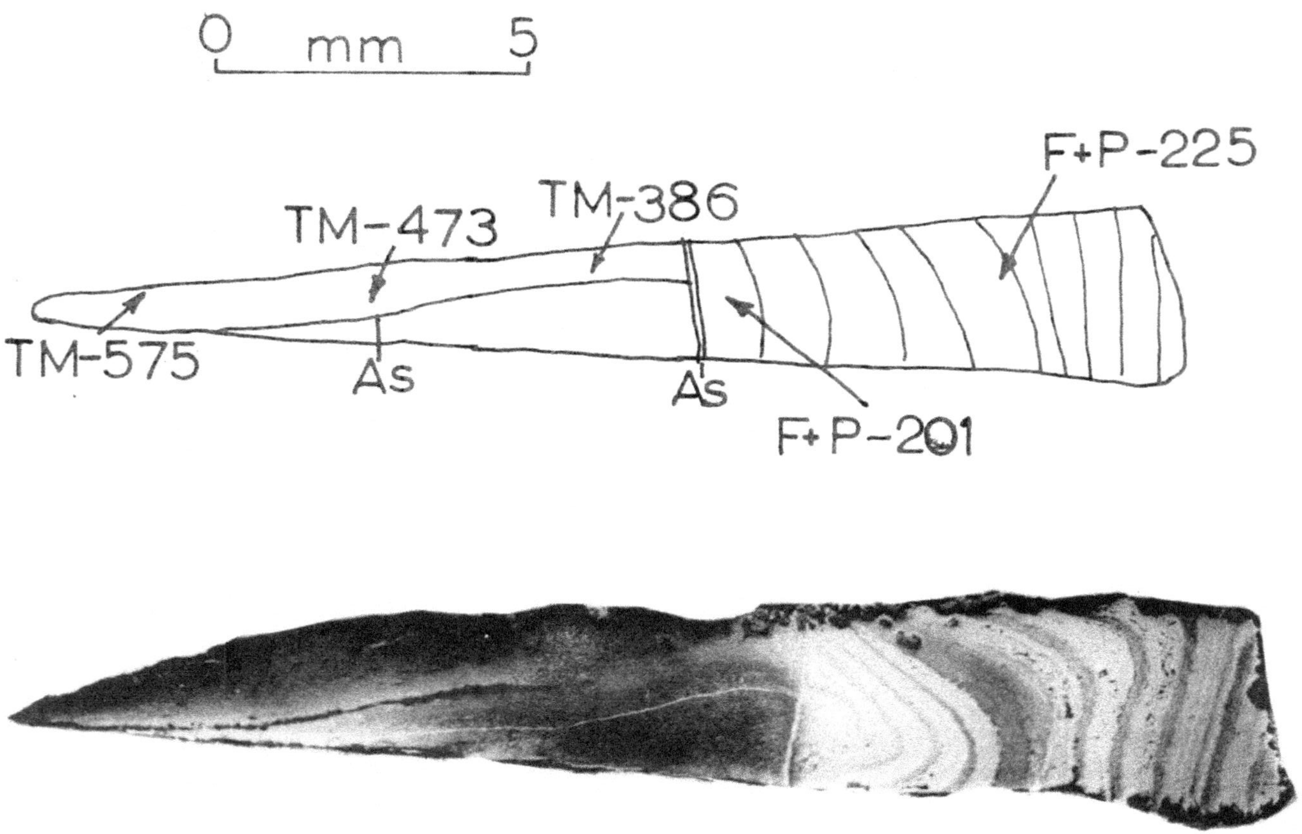


Fig. 15. Section through a pattern-welded knife blade from 13-14th. cent. Winchester (BS1031).

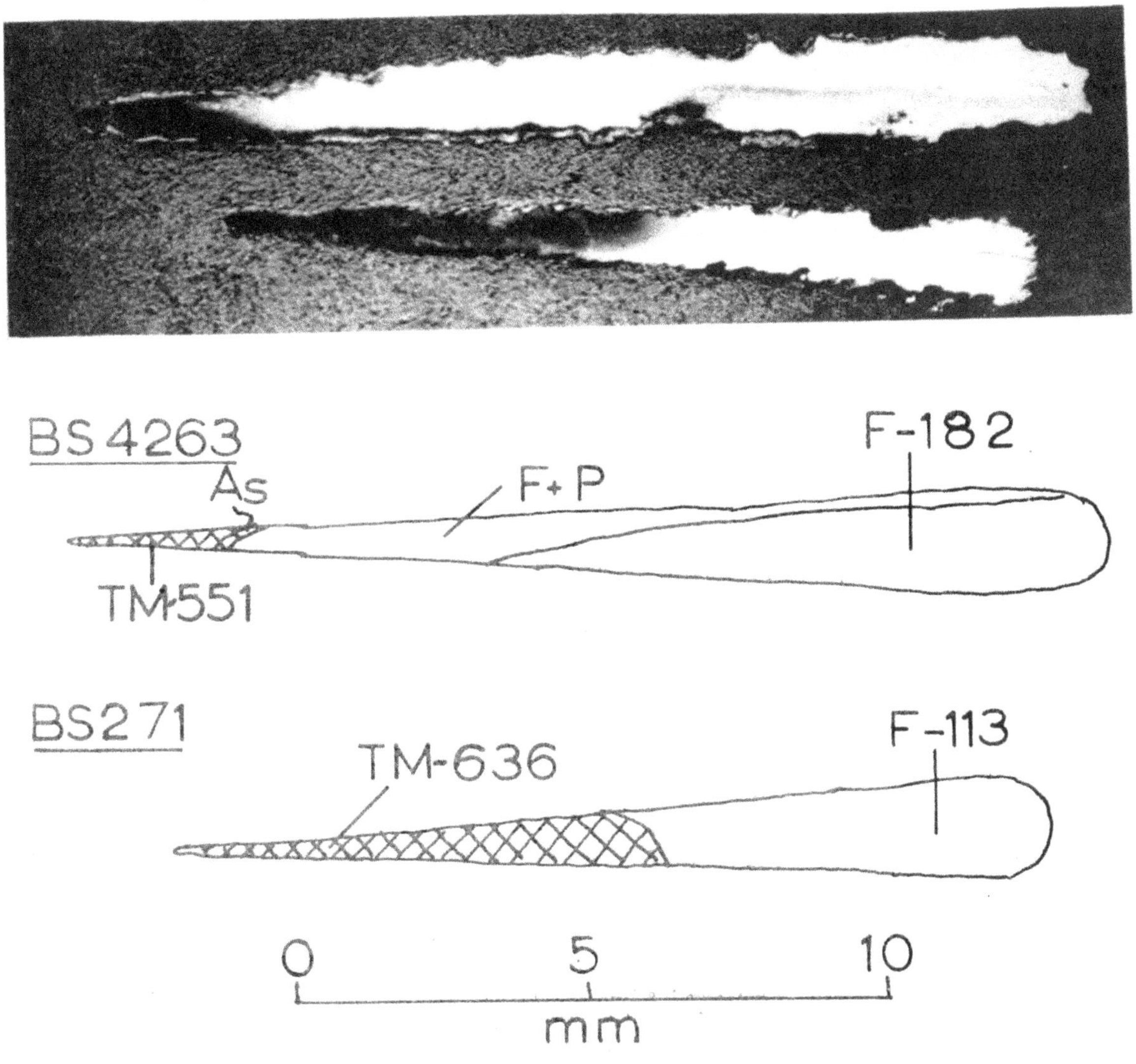


Fig. 16. Section through two scarf-welded steeled medieval knife blades from Winchester. BS 4263 is 11-12th century and BS 271 10th cent.

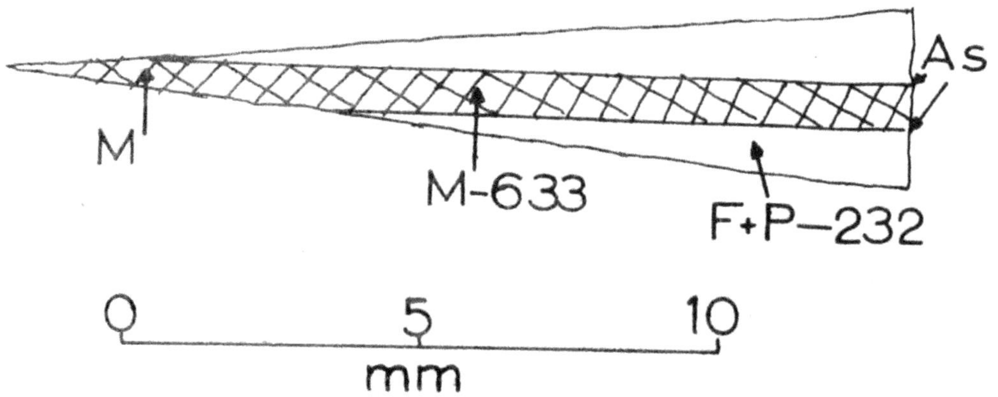


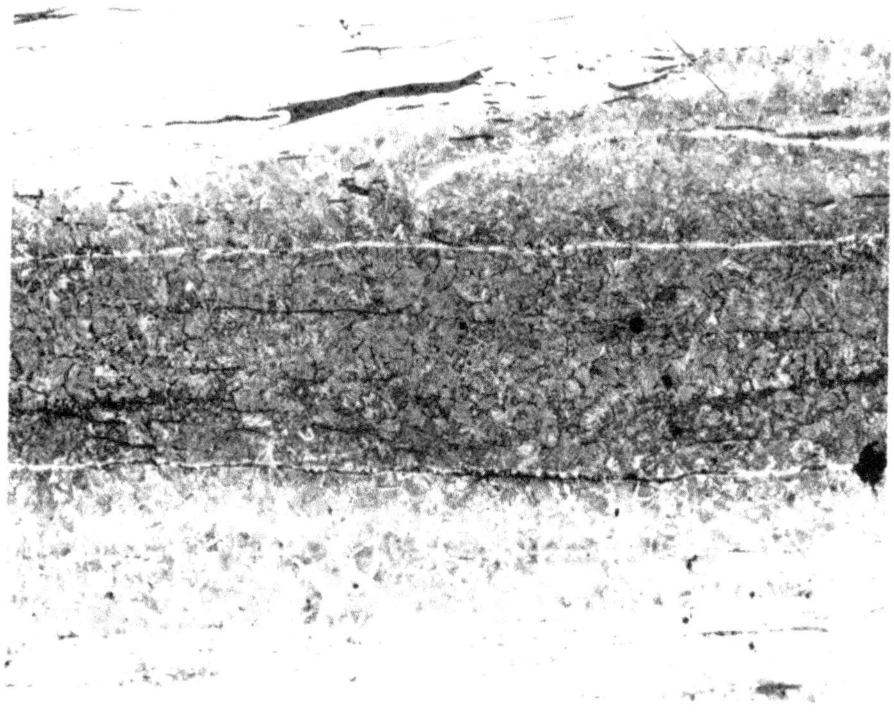

Fig. 17. Section through knife blade from Winchester dated to the 11th century. Top; section showing structure and hardness; bottom, photomicrograph showing high carbon martensitic core with arsenic enrichment lines. X50 (CY 194).

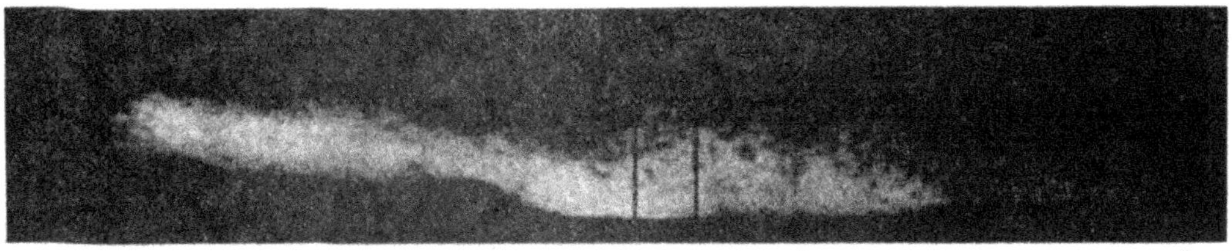

(a) X-ray of sixteenth century iron-cored steel blade from Winchester, × 2

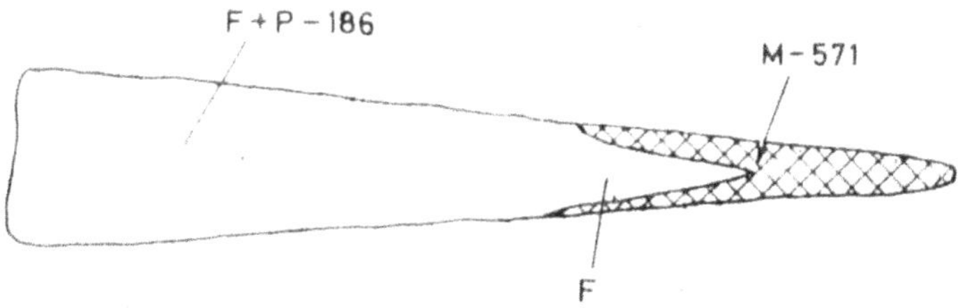


(b) Section cut from same blade at position indicated, × 10. F = ferrite, F + P = ferrite and pearlite (about equal amounts), M = martensite; numbers indicate hardness in units HV.

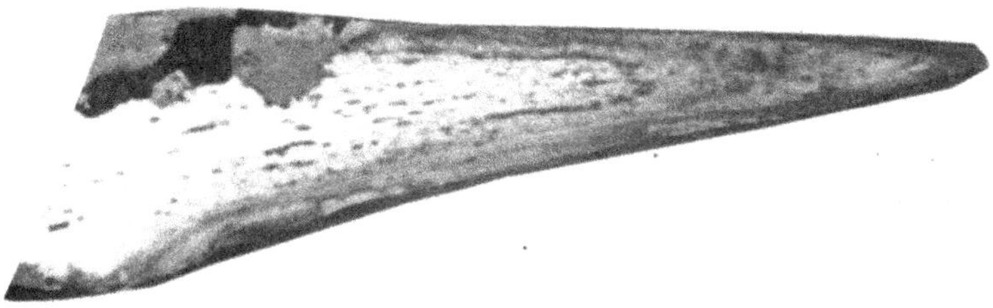

(c) Photomicrograph of cutting edge of the blade showing iron core (light) and steel edge (rusty and darker), × 30.

Fig. 18. X-radiograph and photomicrograph of a 15-16th century iron cored steeled blade from Winchester. (WP 3398).

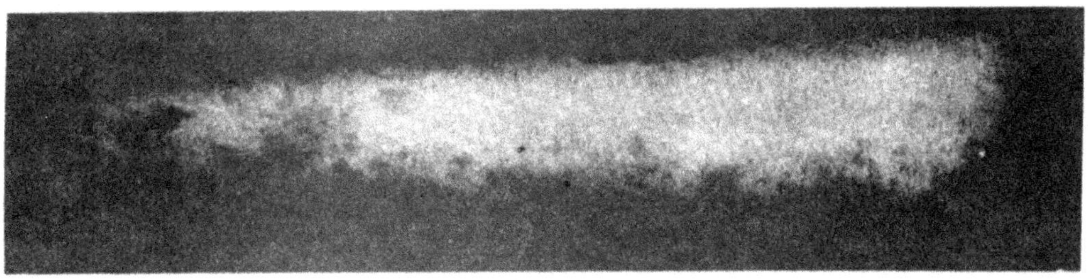

(a) X-ray of eighteenth century homogeneous steel blade from Winchester

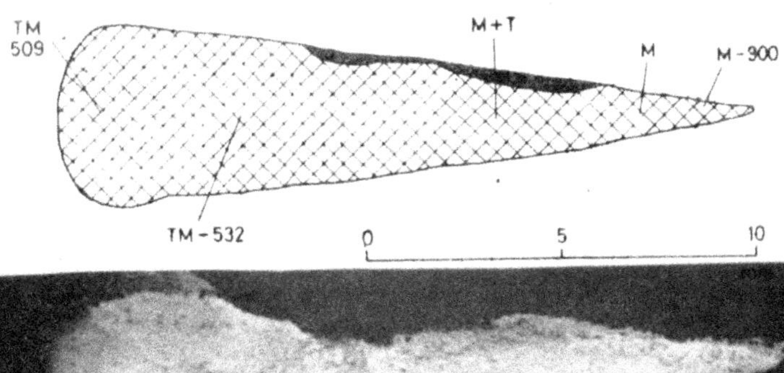


(b) Section of above blade. M = martensite, M + T = martensite and troostite, TM = tempered martensite; numbers indicate hardness in units HV.

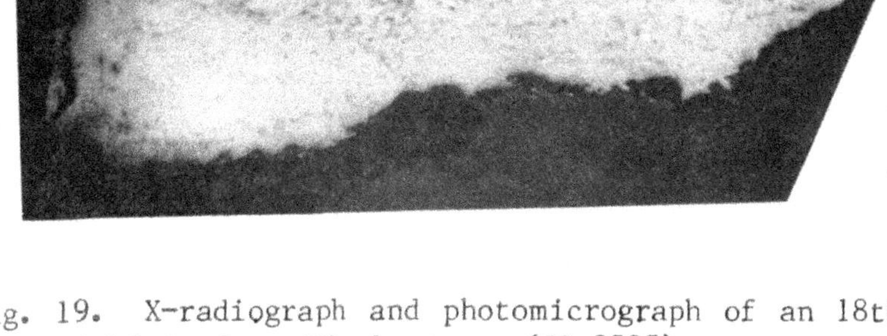

(c) Photomicrograph of blade edge showing hardened region (dark), × 6.

Fig. 19. X-radiograph and photomicrograph of an 18th century homogeneous steeled blade from Winchester. (CG 2585).

around a mainly ferritic core. Fig. 19 shows a homogeneous carbon steel knife from an 18th century level. This knife shows no obvious evidence of piling but one suspects that this was indeed the process used and that the evidence has been obscured by much heating and forging. It has been hardened throughout to give martensite with a hardness of about 500-900 HV, the maximum of which is higher than that of the modern stainless steel table knife. (550 HV).

The blades found to be of iron have most likely been originally steel-edged but the edge has been worn or corroded away. Considering the general high level of the majority of the blades examined it seems very unlikely that an iron blade would be offered for sale in this period.

An unstratified blade, BS 3494 (Fig. 20) which was shown to be pattern-welded by radiography was in many ways like Fig. 15 but the edge and the back were reversed so that the back was hardened carbon steel (742 HV) while the edge consisted of pattern-welded laminations of martensite and ferrite (mean 334 HV). This was certainly not intentional and it would seem that in the process of forging the smith forgot which was meant to be the edge and which the back, and thinned the back rather than the cutting edge.

Other Medieval and Post-Medieval Knives (Tables A3 x A4)

A 13th century whittle-tang knife from the forge site at Chingley, Kent,[27] was found to have been made from a homogeneous carbon steel with the cutting edge hardened to give martensite with a hardness of 533 HV.

The deserted medieval village of Goltho produced four knives dated from the 12th to the 15th centuries. No. A120, although incomplete, was found to be an excellent example of a Type 1 knife with a very well heat-treated steel blade (TM-557 HV) and a back of piled ferrite with a hardness of 151 (Fig. 21). Further examination with the SEM of the piled back showed that it consisted of at least 20 laminations with marked nickel enrichment (Fig. 22).

The Tudor site of Holyoak in Leicestershire produced 6 knives worthy of examination. Most of these were of Type 2 with hardnesses in the range 330-640 HV, all tempered martensite.

Later, 16th-17th century levels, at Chingley produced four knives of mixed type. The best was of Type 2 with an edge hardness of 857 HV; while a Type 1 (? pruning) knife only had a hardness of 239 and another homogeneous Type 3 knife had a hardness of 391 HV. The fact that one was only a mood or blank suggests that knives were actually being made at Chingley in the 16th-17th centuries.

Summary - Knives

Tables A1-4 summarise the results of the metallography of the knives and cleavers from the pre-Roman period up to the 18th century. The majority are made by the welding of steel to iron in one way or another. Many of the Roman blades are no more than ferrite with some phosphorus. Others have been made by welding-on a piece of steel to an iron back but rarely has this been hardened effectively.

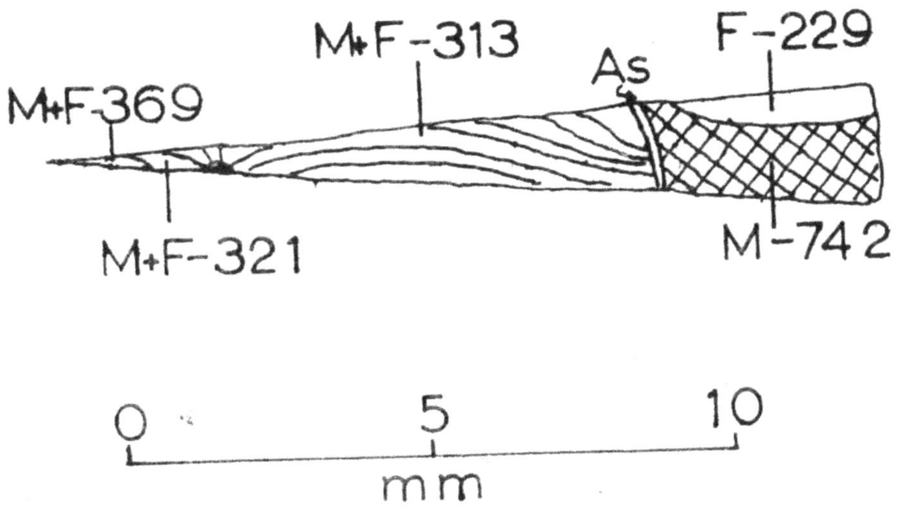


Fig. 20. An unstratified but clearly medieval knife from Winchester in which the pattern-welded section has become the cutting edge rather than the back as was intended (BS 3494).

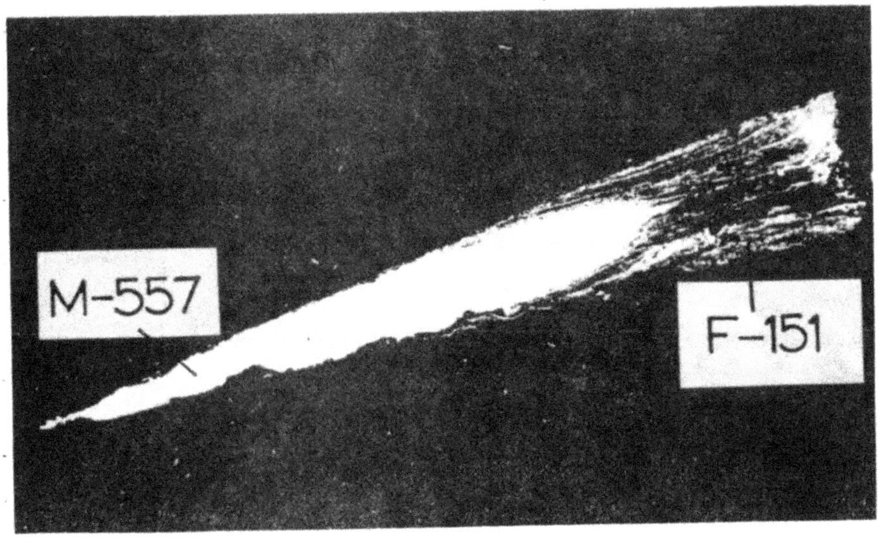


Fig. 21. 15th century steeled blade from Goltho, Lincs.; the back is made of 10-20 laminations.

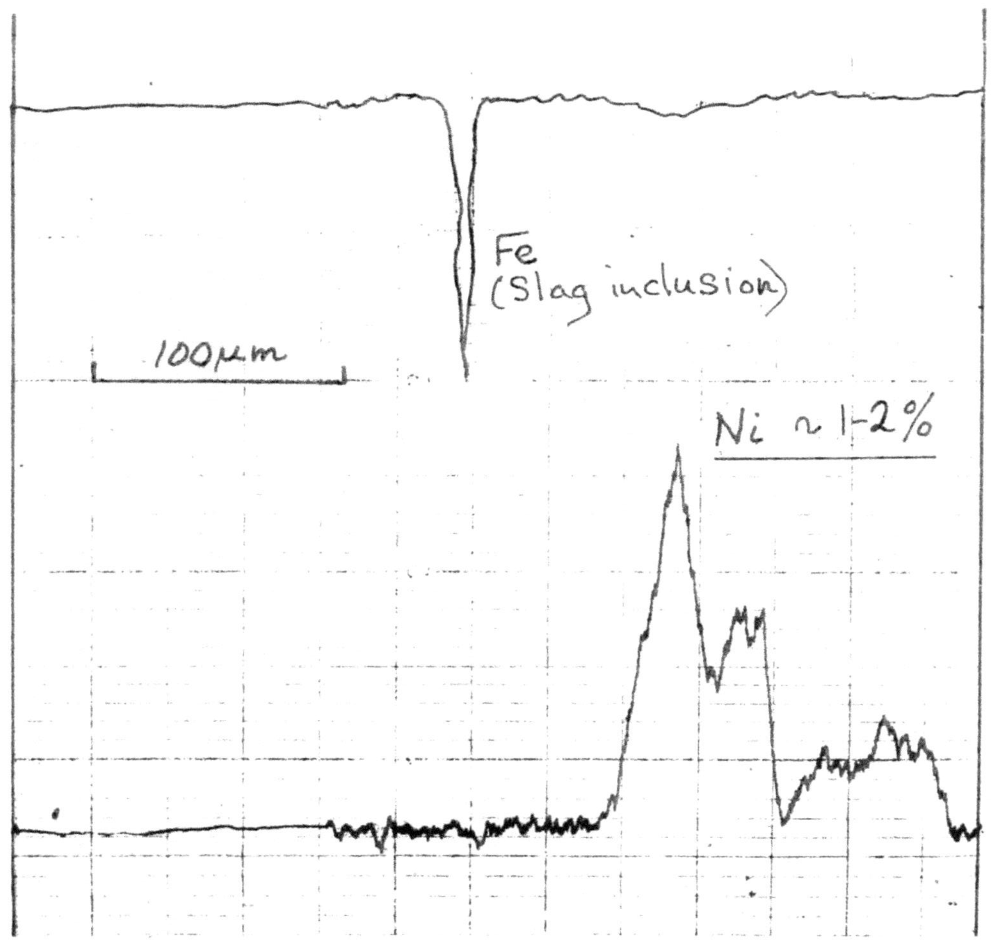


Fig. 22. SEM trace across laminations of above blade showing nickel enrichment at the interface of two laminations.

TABLE A4. Post-Medieval knives

| Object | Site | Site Ref. No. | Date | Structure Edge | Hardness HV Back/Core | Type |
|---|---|---|---|---|---|---|
| Knife | Holyoak, Leics. | 5 | 1550 | TM/640 | F/148 | 2 |
| " | " " | 2 | " | TM/550 | F/130 | 2 |
| " | " " | 3 | " | TM/430 | F/115 | 2 |
| " | " " | 4 | " | TM/420 | F/127 | 4 |
| " | " " | 6 | " | TM/290 | F/148 | 2 |
| " | " " | 1 | " | TM/330 | F/129 | 2 |
| " | Chingley, Kent | Fig. 29,1 | 16-17thC. | TM/857 | F+P/192 | 2 |
| " | " " | Fig. 29,2 | " | M+F/391 | F/218 | 3 |
| Pruning knife | " " | Fig. 29,14 | " | S/239 | F/92 | 1 |
| Knife (?mood) | " " | Fig. 29,5 | " | F/124 | | 0 |
| Knife | Winchester | CG2585 | 18th cent. | M/900 | TM/509 | 3 |

The post-Roman knives are much better. Only four out of 62 are ferritic and the carbon steels have usually been heat-treated effectively. While there is no preference for one type at any one time, Type 1 becomes less common in the later periods. Of the blades examined in the post-Roman period, up to the end of the 15th century, the numbers in each type are as follows:

| Type | No. | % |
|---|---|---|
| 0 | 4 | 7 |
| 1 | 15 | 24 |
| 2 | 32 | 52 |
| 3 | 10 | 16 |
| 4 | 1 | 1 |
| | 62 | 100 |

There is a tendency for the sandwich, Type 1, to be more common at Winchester in the earlier periods, 9th-12th centuries. But there is no tendency for the welded construction to give way to the homogeneous, Type 3, as we move towards the later periods. Type 2 is common throughout the medieval period, no doubt because of its cheapness.

The change of technique would not occur until the invention of crucible cast steel in the 1750s and this, because of its high cost, was not to become popular in the cutlery industry until the 19th century.

(b) Shears and Scissors

While these serve much the same purpose, the latter were not introduced until the medieval period. The Romano-British shears from Silchester[31] appeared to have no welded edge but the metal was clearly of high phosphorus content and work-hardened to give 290 HV, a hardness comparable with unhardened carbon steel. The structure was entirely ferritic but evidence of severe work hardening is shown by the large numbers of Neumann lamellae.

The other shears examined come from the Salisbury drainage ditches[32] which were in use from the 14th century to about 1850 when they were cleaned out. (Fig. 23). Those from Crane and Fisherton Bridges have steel welded to one side (the cutting side) of the blade. Fisherton is more typical - Type 2 - with only a small piece of martensite welded to the blade. But this has a hardness of 516 HV which is good (Fig. 24). The Crane blade has a piece of steel welded to the whole side (Fig. 24). The back is ferrite while the edge is again martensitic with a hardness of 570 HV. The shears from Castle Street are homogeneous steel (Type 3) and may well be later than the other two from Salisbury. But in this case the hardness is only 400. (See Table B).

When we come to the scissors we expect to find that Type 2 is the more usual type (see Figs. 23 and 24). The Salisbury example is fairly typical with a martensitic edge with a hardness of 660 HV. Two of the 16th-17th century Chingley ones are of the same Type (2) with hardnesses of 221 and

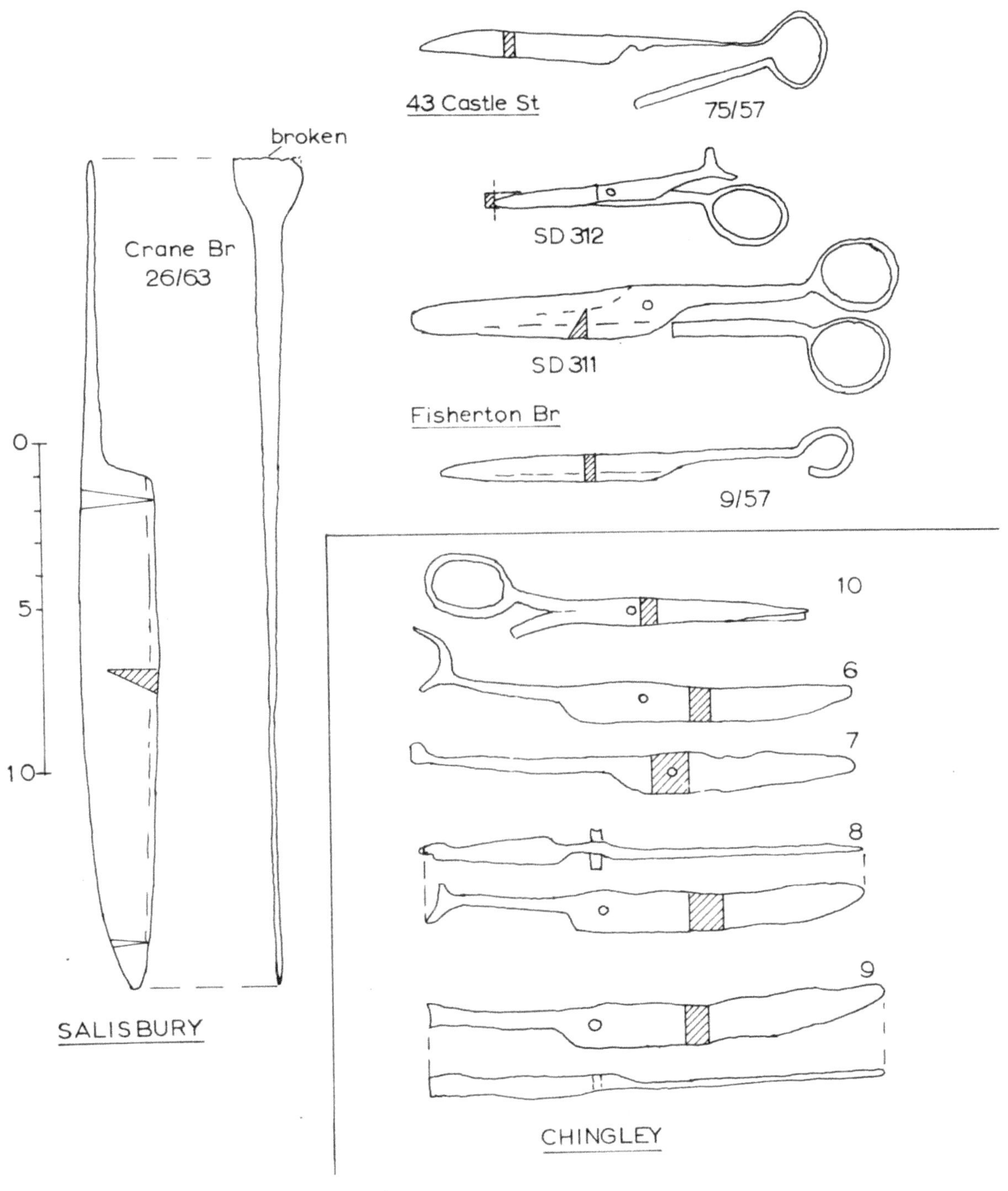


Fig. 23. Outline of shears and scissors from Salisbury, Wilts. and Chingley, Kent.

TABLE B. Scissors and Shears.

| Site | Site/Lab.Ref.No. | Date | Structure/Hardness HV Edge | Structure/Hardness HV Back | Type |
|---|---|---|---|---|---|
| **Shears** | | | | | |
| Silchester, Herts. | | R-B | F/282-290 | F/ | 0 |
| Salisbury, Fisherton Br. | 9/57 | 14thC-1850 | M/516 | F+M/214 | 2 |
| Salisbury, Crane Br. | 26/63 | " " | TM/570 | F/119 | 1a |
| Salisbury, Castle St. | 75/57 | " " | TM/400 | | 3 |
| **Scissors** | | | | | |
| Salisbury | SD/311 | " " | F+C/220-224 | | 3 |
| " | SD/312 | " " | TM/660 | F/245 | 2 |
| Chingley, Kent | Fig. 29/9 | 16-17thC. | TM/460 | F/139 | 1 |
| " " | Fig. 29/6 | " | SC/221 | F/182 | 2 |

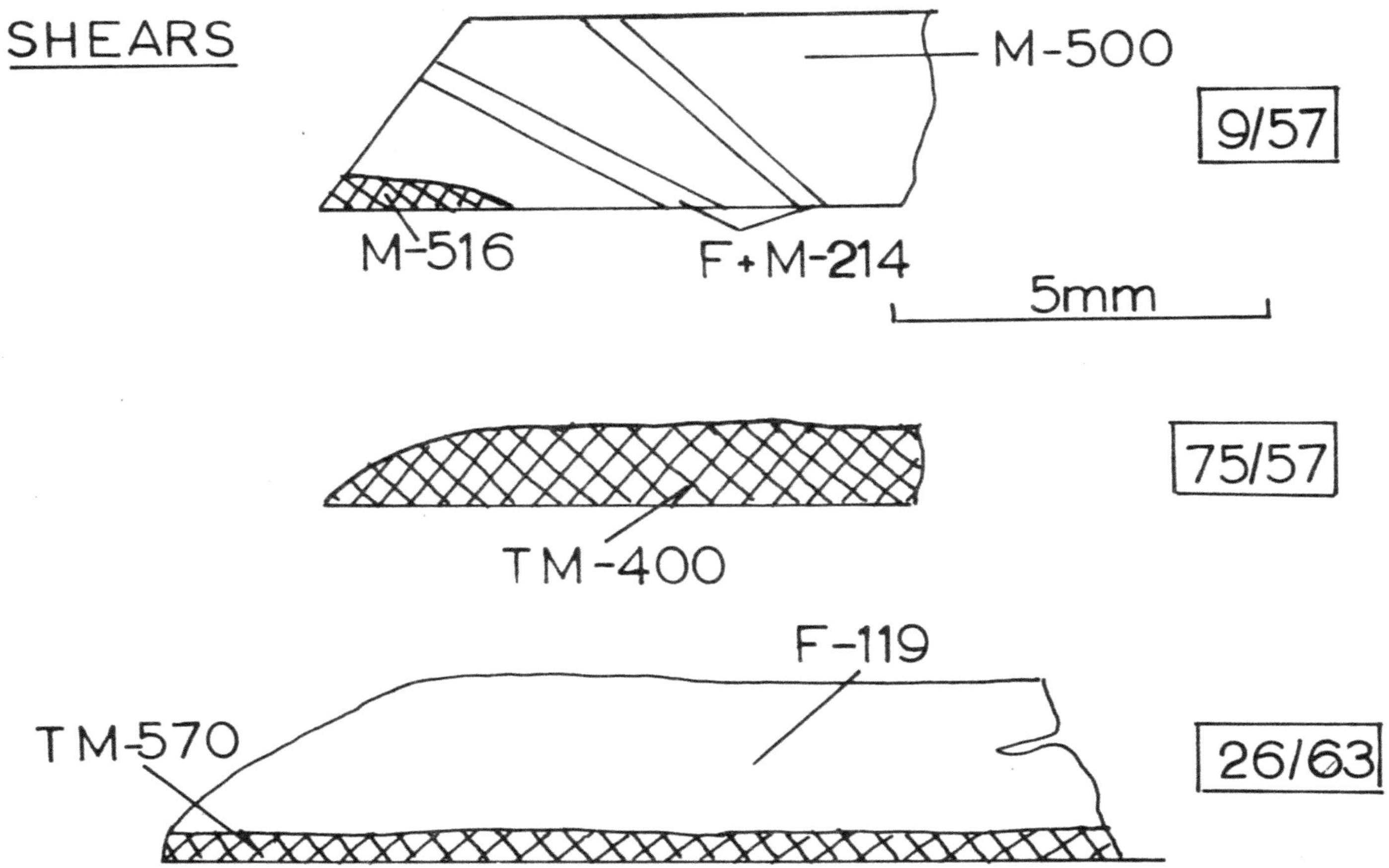


Fig. 24. Sections of shears and scissors from Salisbury and Chingley.

| | | | |
|---|---|---|---|
| A | Shears, | Salisbury | 9/57 |
| B | Shears, | " | 75/57 |
| C | Shears, | " | 26/63 |

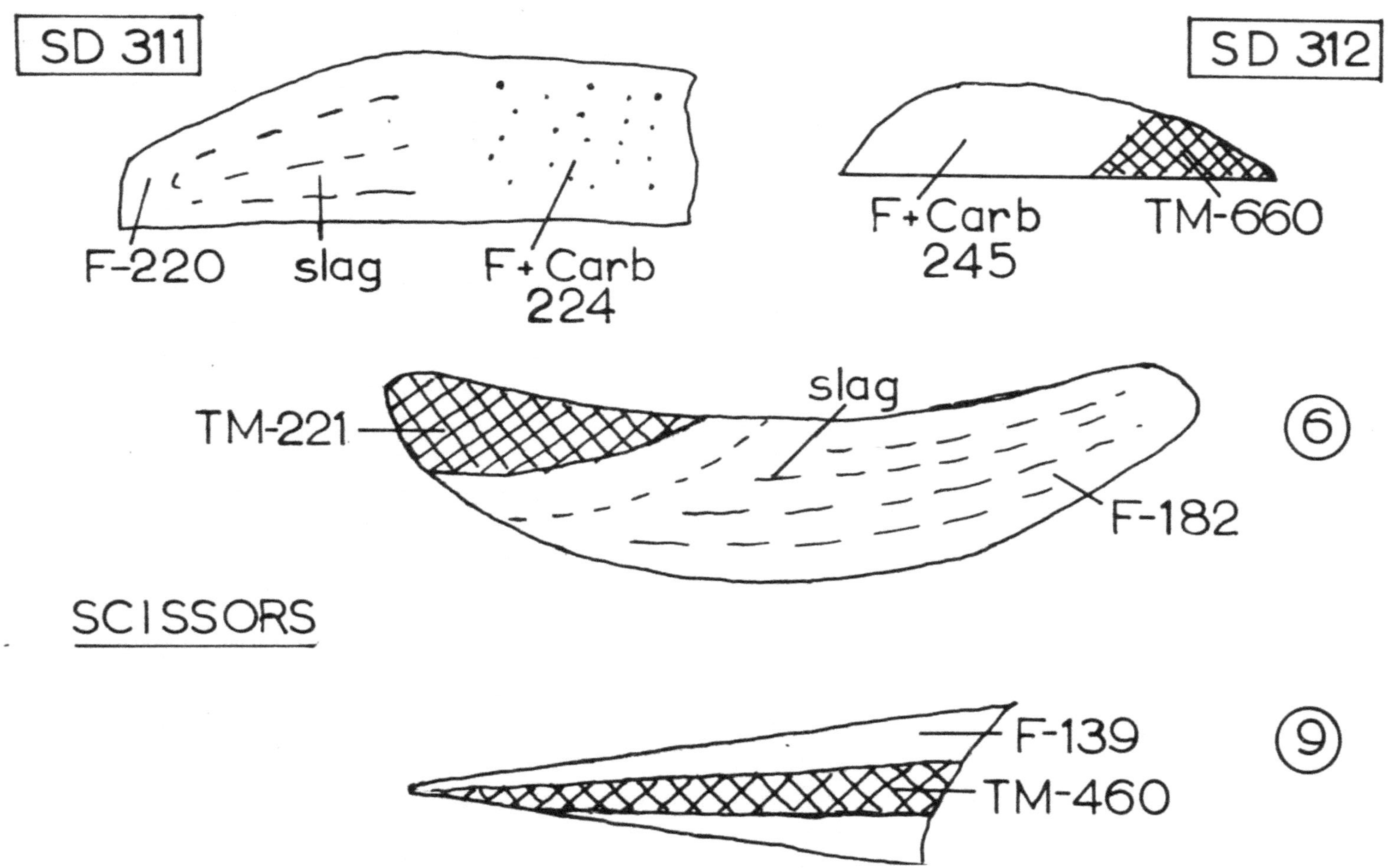


Fig. 24a. D Scissors, Salisbury SD 311
E " " SD 312
F " Chingley 6
G " " 9

857 respectively. But one of the Chingley blades is a typical Type 1 and would seem to be a re-used knife blade. The core hardness was 460 HV. Finally, we have a Salisbury blade which is unhardened steel with a hardness of 220 HV and consisting mainly of ferrite and granular carbide. This has been well forged on the cutting edge (Table B).

### 5.2 Agricultural and Wood Cutting Tools

#### (c) Axe heads

The only pre-Roman axes metallurgically examined from the British Isles are those reported by Scott from Lough Mourne,[33] Ireland, and that from Rahoy, Scotland, by Desch for Childe.[34] None have been examined from England although more than 20 socketed axes are known. One has been X-rayed and found to have its loop welded on.[35]

We have now examined 5 shaft hole axes from the Roman period (Table C). Two came from Catsgore, Somerset.[23] One of these (F 176) consists of piled strips of low phosphorus iron of various carbon contents with a central area reaching 0.4% with a hardness of 190 HV1. The pearlite was spheroidised. The second had a more even and higher carbon content and was nearly 100% pearlite with a hardness of 320 HV1.

Two others came from the Roman site at Gestingthorpe, Essex.[20] One (No. 95) had an oval eye and was of Type 1 construction with a cutting edge of tempered martensite giving a hardness of 515 HV. The other (No. 31) had an edge with equal amounts of ferrite and pearlite. The latter had a fine granular distribution. The hardness was 210 HV1.

The asymetrical axe from Wanborough (790281) was an attempt at Type 1, made by folding over a piece of wrought iron on which a layer of steel had been welded.[19] It seems that some decarburisation occurred during heating before the final welding, leaving a thin line of ferrite between the two high carbon layers. The steel core was P+F with a hardness of 206 - 269 HV. Outer layers were ferrite with a hardness of 156 HV.

The 9th-10th century carpenter's axe from Winchester again shows a steel insert. But this has been hardened to 390 HV giving tempered martensite. The outer layers were ferrite with hardnesses in the range 117-150 (Fig. 25). The tool has been made by the expected technique of folding over, with a mandrel through the eye.

The Saxo-Norman axe from the River Thames at Kempsford, Wilts.,[36] first came to our attention as a radiograph in an article by M. Corfield (Fig. 26). This clearly shows a Type 2 (welded-on) steel edge which is unusual as most early axes have Type 1 edges. This type is a carpenter's axe and is shown in the Bayeux tapestry being used by Norman shipwrights. Outwardly it resembles the 13th century axe found at Winchester.[24]

The conservation report reads as follows:

'An iron axe head with a blade 25cm in length and 4.5cm width (widening to c. 5cm towards the middle where the blade meets the hafting 'neck'). In cross section the axe blade is triangular, narrowing from approx. 1cm across the back to a fine cutting edge. Clearly visible on the X-ray, and slightly visible to the eye, is

## Table C. Axe-heads

| Site | Site/Lab.Ref.No. | Date | Structure/Hardness HV Edge | Structure/Hardness HV Back | Type |
|---|---|---|---|---|---|
| Catsgore, Som. | F-14 | R-B | P/320 | | 3 |
| " " | F-176 | R-B | F+P/190 | | 1 |
| Gestingthorpe | 95 | R-B | TM/515 | | 1 |
| " " | 95a | R-B | P+F/210 | | 3 |
| Wanborough | 790281 | R-B | P+F/206 | F/156 | 1 |
| Kempsford, Glos (SU160967) | | S/N | TM/483 | F/87-113 | 2 |
| Winchester | BS6262/6317 | 9-10thC. | TM/390 | F/117 | 1 |
| Chingley, Kent | Fig. 32/61 | 16-17thC. | TM/391 | F/133 | 1 |

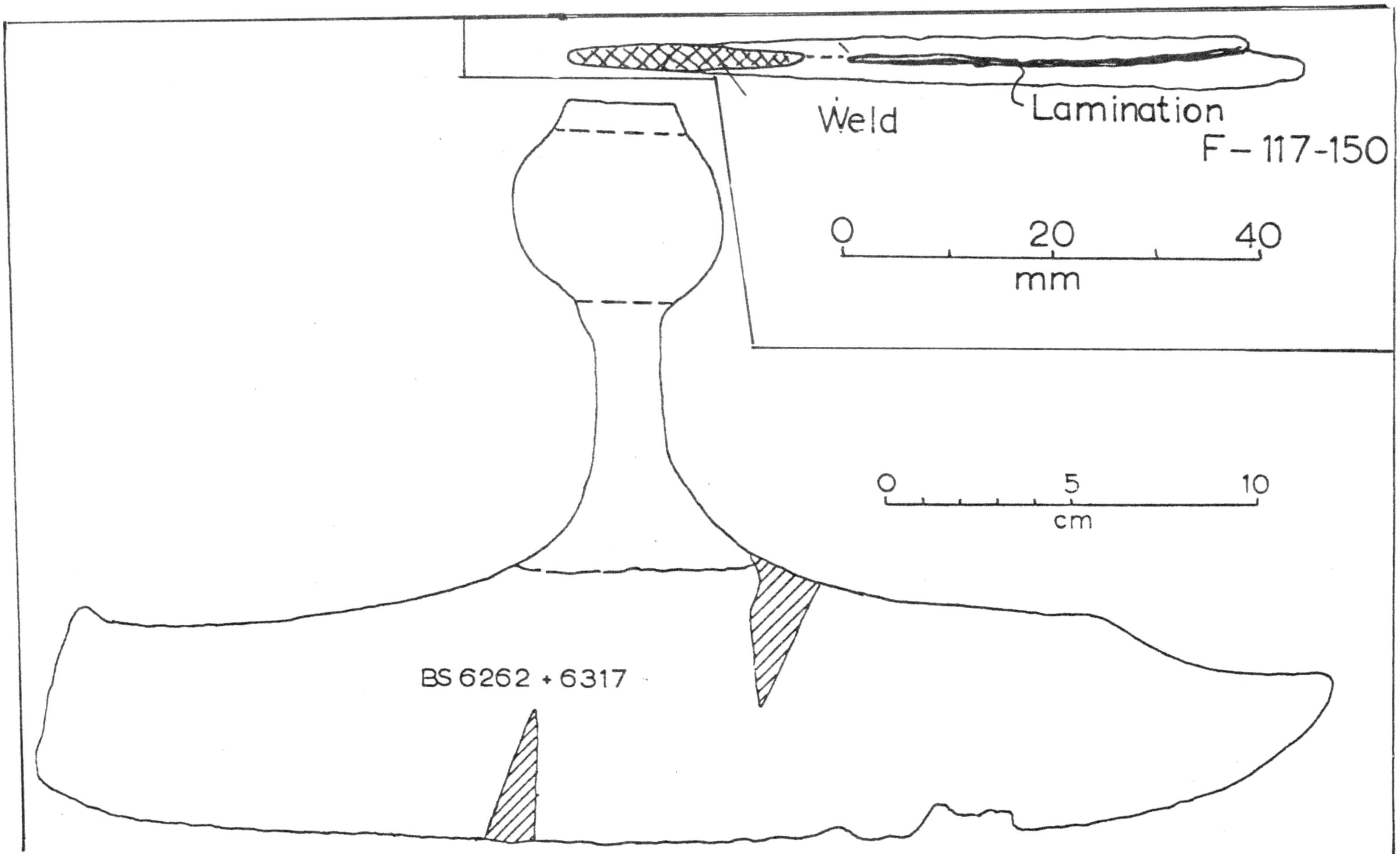


Fig. 25. Profile and section of Carpenter's axe from Winchester of the 9-10th century showing hardened steel insert. (BS 6262).

a line running the length of the blade approx. 2cm from the cutting edge. This line is the join between the harder iron for a good cutting edge and the iron used to make up the rest of the axe body. The axe was hafted at right-angles to the cutting edge, and fragments of wood are still visible inside the hole - these particles were consolidated with Polymethacrylic Ester in toluene before the stabilisation of the iron in lithium hydroxide. The axe terminates in a square head (3cm sq.) above the shaft hole.

The axe was in good state of preservation since the corrosion had not caused the metal to deteriorate seriously at all.'

Two 'Vee'-shaped sections were removed from the blade. Section A was cut from the cutting edge inwards (Fig. 26), while Section B was cut from the side near the opposite end. In the unetched state it is clearly a wrought iron with slag lines parallel to the length of section. It is very well forged with few cavities. Upon etching, it quickly shows a tempered martensite welded edge of Type 2 form. The original weld has a thick white line of resiual ferrite across which there has been quite a lot of diffusion to give a ferrite+pearlite structure (Fig. 27).

It would appear that the 20mm wide steel edge piece has been formed by carburising its end for a distance of 14mm and side welding this onto the axe as shown. The junction between the carburised length and the rest is very diffuse with no weld. The carbon content of the low carbon half of the edge is about 0.3%, tapering away to pure large grained ferrite near the other surface. The structure of this metal is spheroidal pearlite and martensite showing that the whole edge has been quenched, as might be expected.

The ferrite has nitride or carbide content and the hardness is 110 HV which indicates a low phosphorus content. The cutting edge is tempered martensite with a hardness of 483 HV1. The quenching temperature was probably in the region of 750°C, sufficient to fully harden the high carbon parts but not the low carbon areas.

The second section (see Fig. 27) was an oblique (or taper) section taken near the side in order to elucidate the problem of the ending of the corroded weld-line short of the sides of the blade and at the same time clarify the welding technique. The section is shown in Fig. 27 and it is clear that the steel was not intended to reach the side. The limits of the piece of steel are clearly shown by slag and rust-filled cavities, and the original weld is shown by a white line of residual ferrite. In this area the carburisation of the original piece of steel did not extend to the full thickness so that the weld was one of iron-to-iron than iron-to-steel, as in Section A. Some diffusion of carbon has taken place across the original weld but the hardness of the coarse-grained ferrite was in the range 87-113 HV1 which agrees with the hardness of 110 in Section A. However, the hardness of the steel is now only 250 HV which suggests a somewhat lower carbon content or slower quenching rate.

Clearly this axe has been made by steeling a piece of low phosphorus wrought iron. The edge had been prepared for the welding of a piece of steel as shown in Fig. 27.

780123B

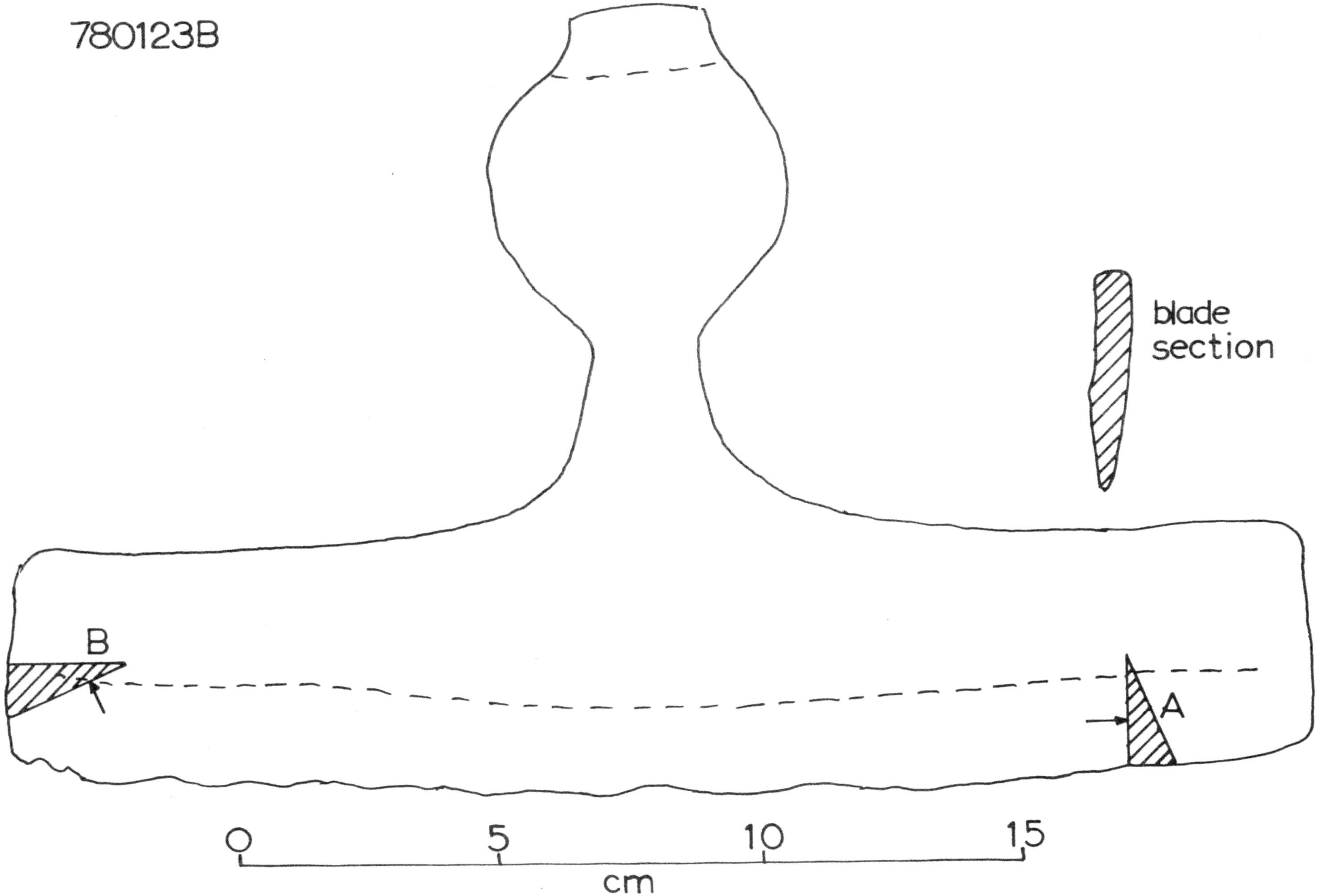


Fig. 26. Profile of carpenter's axe-head from Kempsford, Glos. Probably Saxo-Norman.

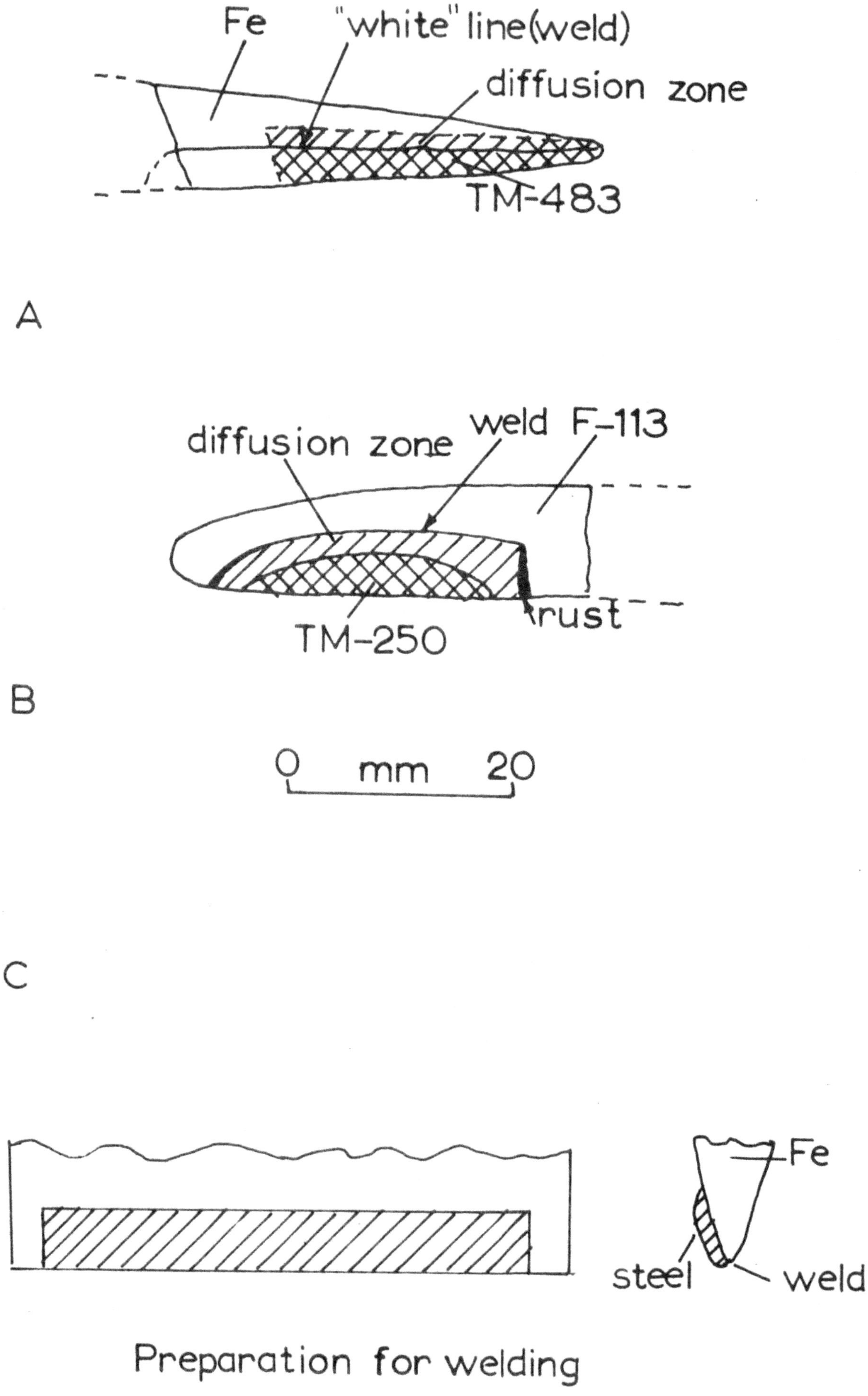


Fig. 27. Sections through cutting edge of above axe showing how steel was applied by "welding-on".

The steel was made by carburising a piece of iron to a depth varying from 2 to 3mm. The depth was uneven and the carburisation did not extend to the inside edge. (It probably did not extend to the outside edge either but his was ground away before use). The carburisation would require at least 6 hours at 950°C, and it would be difficult to get consistent results over a length of 23cms in a simple blacksmith's forge. The welding was reasonable but one would not expect a good weld at the scarf along the line shown on the radiograph. But the quality of the iron-steel weld at mid-thickness is good and shows that the smith achieved a high enough temperature (1200°C) and a clean surface free of scale. The hardness of Section A shows that the degree of carburisation and the quenching conditions were good enough to produce a cutting edge of adequate hardness.

An axe was also found on the 16th-17th century site at Chingley, Kent.[27] This was a perfect example of a Type 1 structure with a high carbon layer sandwiched between two layers of wrought iron. Marked As enrichment is present at the welds and there is slight diffusion of carbon from the steel into the iron. The steel has a tempered martensite structure with a hardness of 391 HV at the cutting edge and 376 at the back. The iron sides have a hardness of about 175. (Table C).

The 18th century site at Lymm Slitting Mill, Cheshire, produced an axe-head which was again made with the aid of a steel insert. This was very symmetrical with the steel insert positioned exactly along the centre line (Fig. 28). The steel was in the form of tempered martensite and had a hardness of 530-550 HV, with low carbon iron surrounding it with a hardness of 133 HV.

(d) Scythes, Sickles and Bill-hooks

This group of agricultural tools might be thought of as being of sufficient importance to make out of good quality steel. However, like the Silchester shears, it would appear that scythes, for example, were made only of rather poor wrought iron until more recent times (Table D). The Romano-British scythe from Poundbury[21] had been well-forged to give the distinctive reinforced back of the blade from a wrought iron with a hardness of from 102-129 HV. The edge was badly rusted but it appeared to have been wrapped with a thin layer of carbon steel giving a Type 4 structure with a hardness of 220 HV. But the later 16-17th century scythe from Chingley,[27] although well-forged like that from Poundbury consisted merely of wrought iron with a hardness of 119 HV. One wonders if this was hammer-hardened on the edge using a field anvil and that his region has corroded away. In contrast, a scythe of the 19th-20th century was examined and found to be wholly of tempered martensite with a hardness of 425 on the reinforced back and 590 HV at the cutting edge.

The sickles do not seem to be any more consistent. Two from the Romano-British level at Wanborough[19] show different structures. One consists solely of a medium carbon steel with a ferrite+pearlite structure with a hardness of 171 HV; perhaps this was hammer-hardened on the edge. The other is double sandwich structure of Type 1a. It consists of two ferrite + pearlite layers (256 HV) sandwiched between ferritic layers (132 HV) Fig. 29). Being typically Roman, the steel has not been quench-hardened but there are signs of cold working on the cutting edge. Another sickle of the Roman period comes from Brancaster. This is of Type 1a with a layer of ferrite+pearlite with a hardness of 230 HV welded to ferrite with

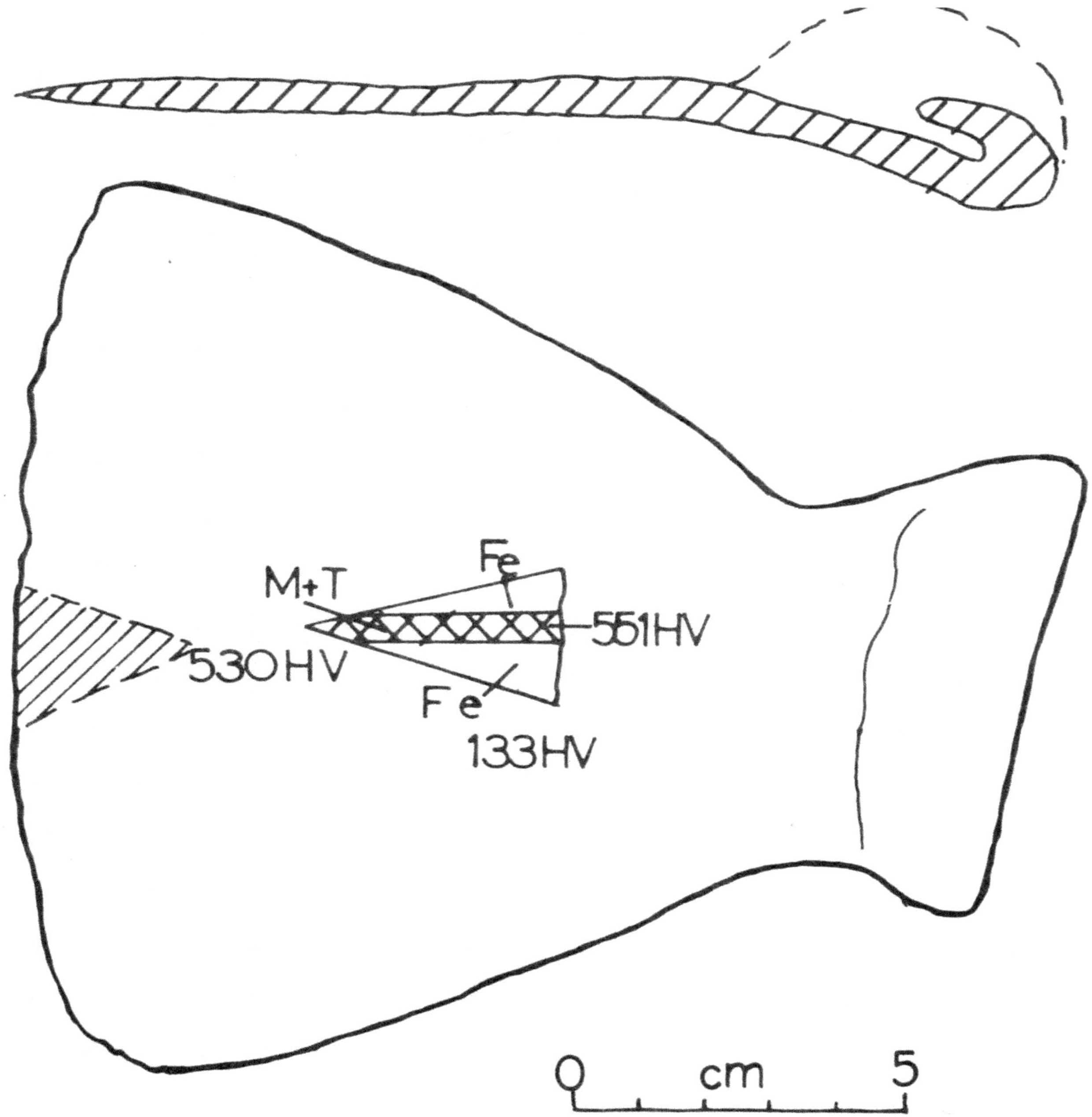


Fig. 28. 18th century axe-head from Lymm, Cheshire, showing hardened steel insert.

Table D. Scythes, Sickles and Bill-hooks.

| Site | Site/Lab.Ref.No. | Date | Structure/Hardness HV Edge | Structure/Hardness HV Back | Type |
|---|---|---|---|---|---|
| **Scythes** | | | | | |
| Poundbury, Dorset | 82001 | Late R-B | P+F/220 | F/102 | 4 |
| Chingley, Kent | Fig. 29/13 | 16-17th C. | F/119 | | 0 |
| French | | 19-20th C. | TM/590-425 | | 3 |
| **Sickles** | | | | | |
| Wanborough | 690337 | R-B | F+P/256 | F/132 | 1a |
| " | 800060 | R-B | F+P/171 | - | 0 |
| Brancaster, Norfolk | 773492 | R-B | F+P/230 | F/166 | 1a |
| Winchester | BS2602 | 13th C. | SC/ | F/87 | 2? |
| " | ACS358 | 14th C. | TM/350 | TM/370 | 3 |
| **Bill-hook** | | | | | |
| Gestingthorpe | 31 | R-B | P/270 | F/- | 1 |

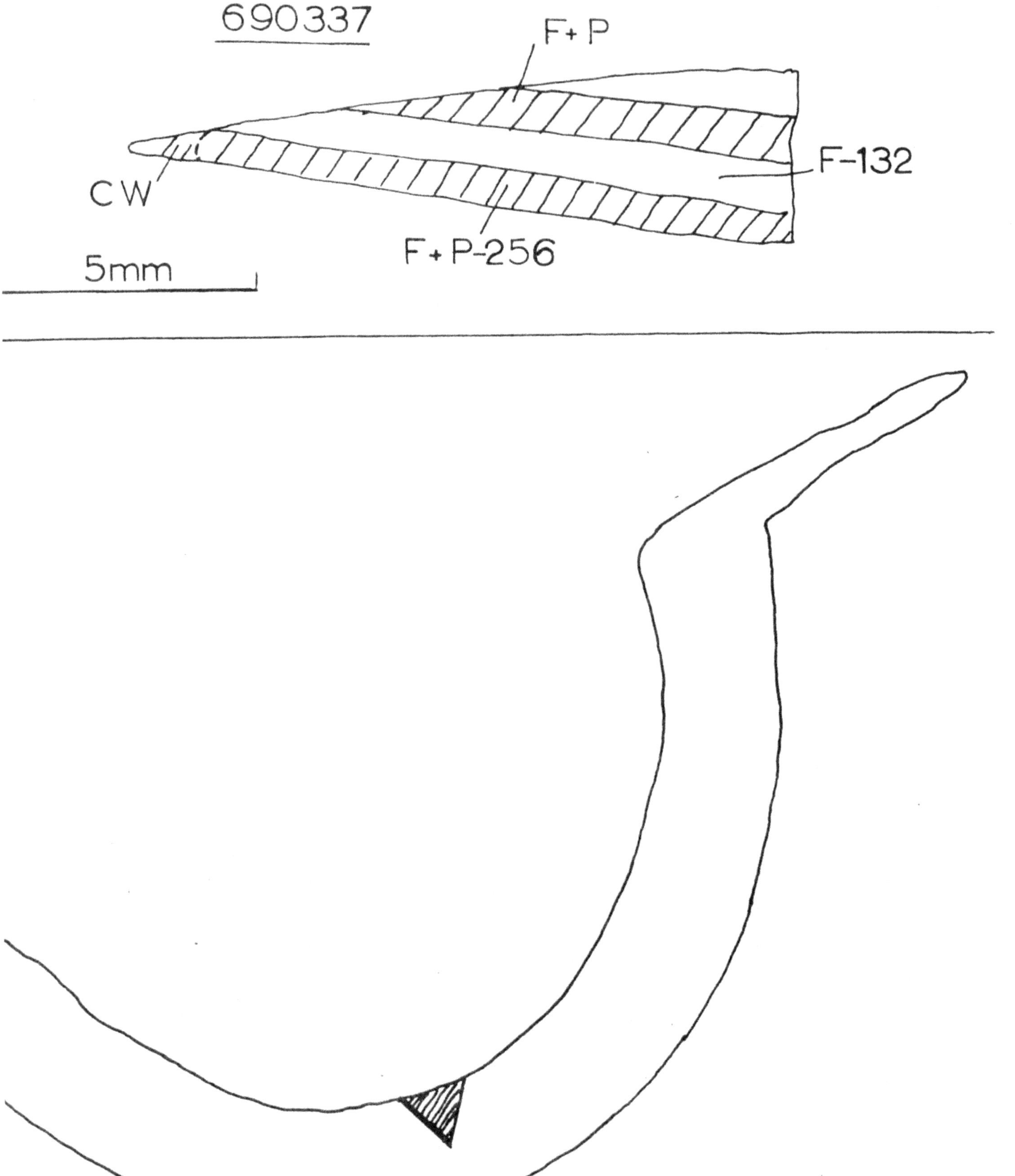


Fig. 29. Outline and section of a sickle from Romano-British Wanborough, Wilts.

a hardness in the range 143-166. This was welded in such a way that the steel projected far enough to form the cutting edge (Fig. 30).

Two medieval sickles were examined from Winchester.[24] The 13th cent. example was low carbon ferrite with a hardness of 87 HV. But the edge had a small amount of carbide (sorbite) which, with some hammer hardening, might have been fairly efficient. The other, of the 14th cent., was a piece of homogeneous carbon steel which had been quenched and tempered to give a hardness of 350-370 HV throughout.

The billhook came from the Romano-British site of Gestingthorpe and was of Type 1 structure with a carbon steel central layer, mostly 100% pearlite and with a hardness of 270 HV, sandwiched between two ferrite layers (Fig. 30).

### (e) Wood-cutting Chisels and Gouges

#### Chisels

Chisels may be solid (often with mushroomed heads), socketed, or tanged like those of today. The bolster, as in the case of the knives, does not seem to have been introduced before the 16th century. Chisels may be narrow, for mortices, or splayed. We have examined 7 Roman chisels from Britain and most of these are low carbon steel or wrought iron. In these the hardnesses vary from 124 to 250 HV depending on the degree of cold work and carbon content.

Two were Type 1 chisels. One from Ware[18] consisted of ferrite and pearlite (C = 0.4%) with a hardness of 206 HV welded to iron with a carbon content of 0.1%, but the other, a socketed chisel from Wanborough,[19] was far superior with a layer of tempered martensite (HV = 414) between two layers of wrought iron (Fig. 31). By the 16-17th century we find the steel has been welded to one side (Type 1(a) and the hardness has increased to 710 HV (Table E). The 18th century chisels from Lymm are of modern design but are still of Type 1 consisting of a thin layer of steel with a hardness of 690 HV welded to wrought iron (Fig. 32).

Homogeneous steel chisels occur from the 14th century onwards with hardnesses of 400 HV (Winchester). The modern chisel usually has a hardness somewhat higher than this - in the range 450 - 600 HV.

The most sophisticated of the gouges is that from Romano-British levels at Ware.[18] Here a Type 2 edge has been applied to a Type 1 blade (Fig. 31). The former consists of tempered martensite with a hardness of 320 HV. The quenching has resulted in ferrite and tempered martensite remaining on the outside of the back due to insufficient temperature for the low carbon content of this region, and ferrite and pearlite remaining on the inside due to too low a cooling rate in the centre. We doubt whether the structure of the back was intentional. The gouge from Saxon or later levels at Poundbury is a two layer structure with the softer on the inside which would not be the right way round for a gouge where the outside would be ground away to form the cutting edge (Fig. 31). A gouge from Chingley stands in marked contrast with the chisel and one suspects that it is not a tool (Table E). But the Roman gouge from the Fortress baths at Caerleon has the cutting edge welded on to the inside so that it will be used correctly when sharpened (Fig. 31D).

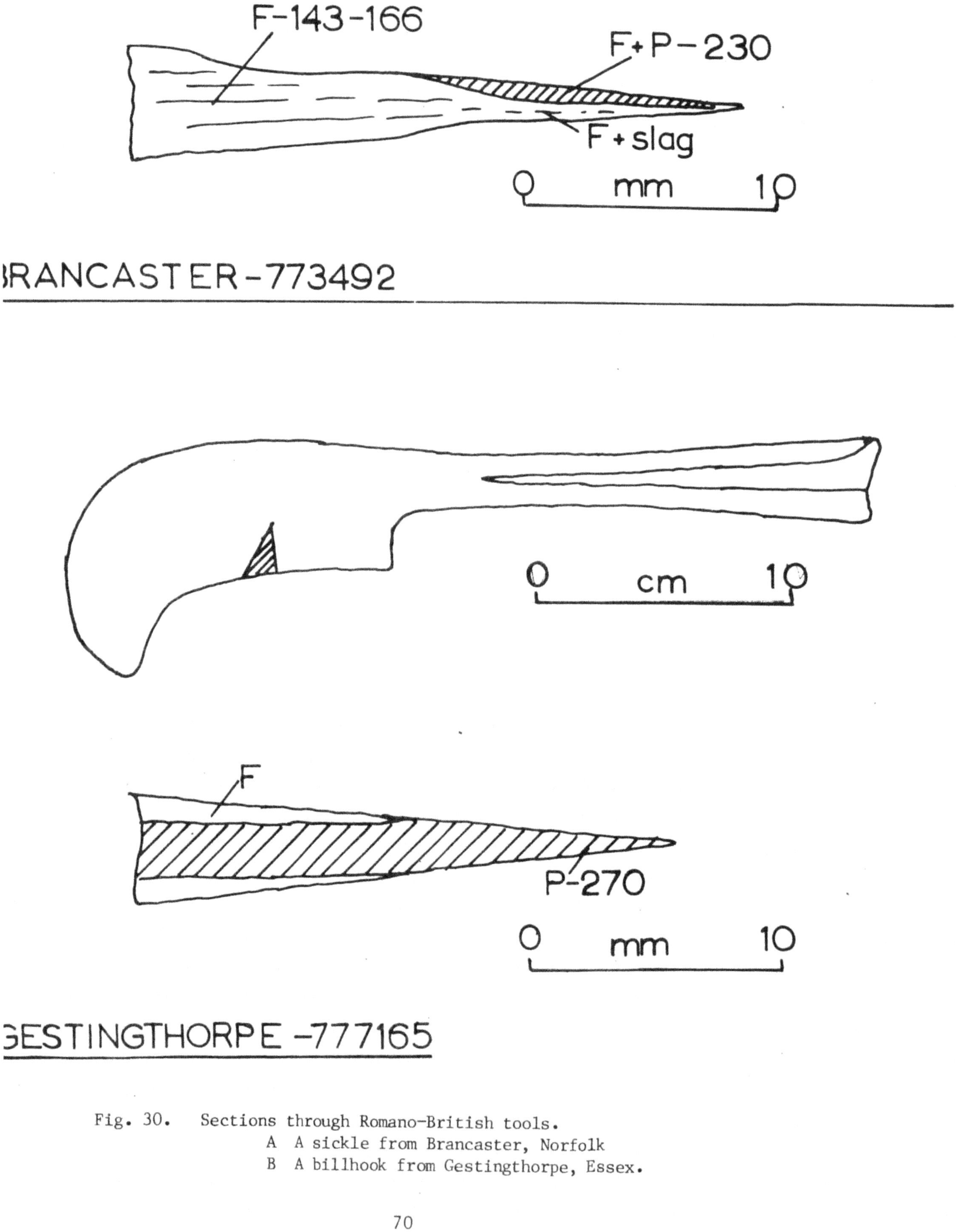


Fig. 30. Sections through Romano-British tools.
A A sickle from Brancaster, Norfolk
B A billhook from Gestingthorpe, Essex.

Table E. Wood-cutting Gouges and Chisels.

| Site | Site/Lab. Ref. No. | Date | Structure/Hardness HV Edge | Structure/Hardness HV Back | Type |
|---|---|---|---|---|---|
| **Chisels** | | | | | |
| Ware, Herts | 301 | R-B | F/206 | - | 1 |
| Wanborough (Socketed) | 800052 | R-B | TM/414 | F/210 | 1 |
| Gestingthorpe (Solid) | 117 | R-B | F/124 | | 0 |
| Gestingthorpe (Tanged) | 95a/57 | R-B | F/250 | | 0 |
| Chelmsford | 41 | R-B | F/137 | | 0 |
| Winchester | WP1607 | + 14th Cent. | TM/400 | | 3 |
| Chingley (Socketed) | 7S | 16th-17th C | M+T/710 | F/139 | 1a |
| Lymm, Ches. | LY1 | 18th Cent. | TM/690 | F/172 | 1a |
| **Gouges** | | | | | |
| Caerleon, baths. | - | R-B | S/329 | F/120 | 1 |
| Ware, Herts | 2074 | R-B | TM/320 | F+TM/305 | 2 |
| Poundbury, Dorset | Fe 62 | + Saxon | F/177 - 146 | | 0 |
| Chingley, Kent | 32/64 | 16-17th C. | - | F/112 | 0 |

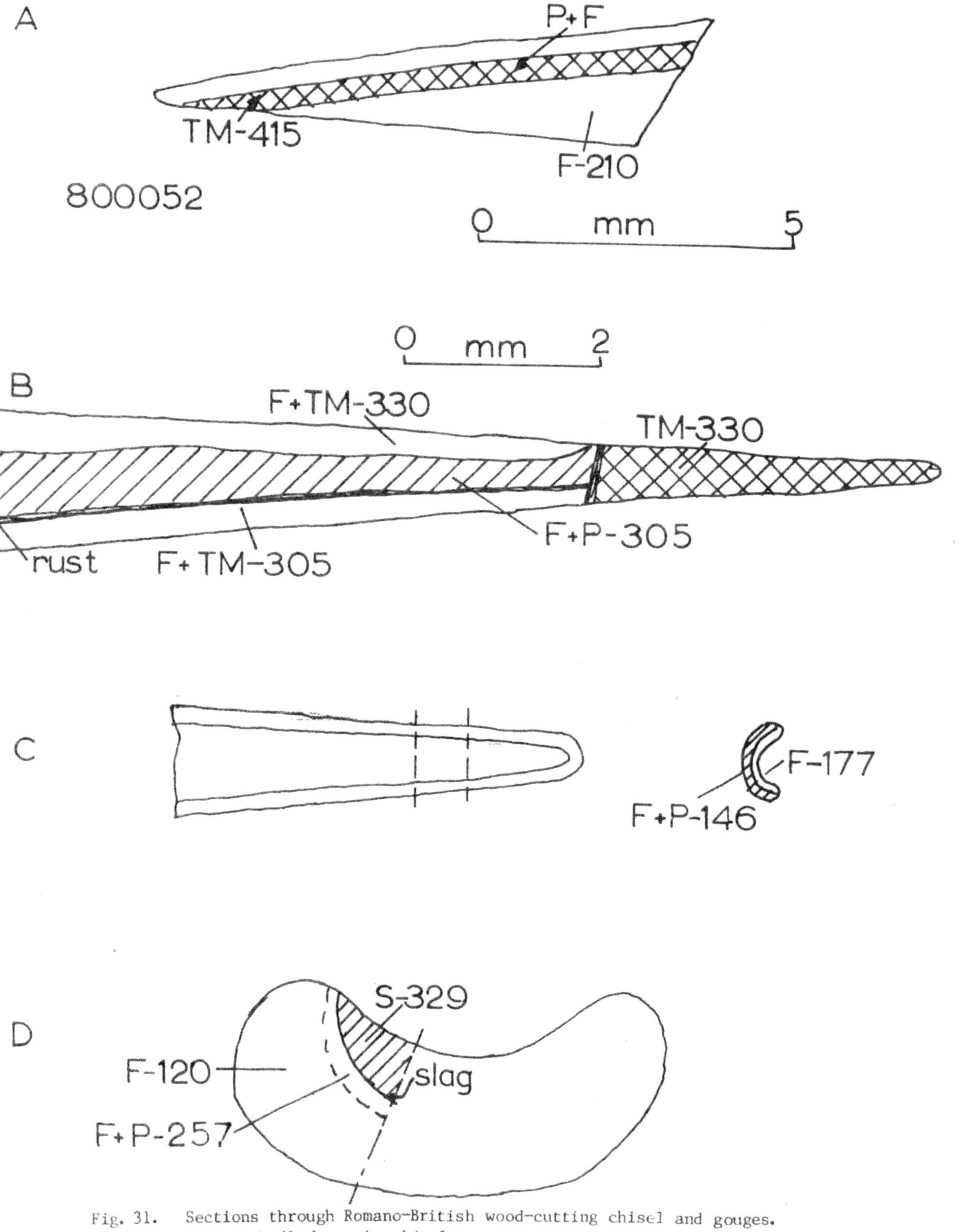


Fig. 31. Sections through Romano-British wood-cutting chisel and gouges.
A Wanborough, chisel
B Ware, gouge
C Poundbury, gouge
D Caerleon, "

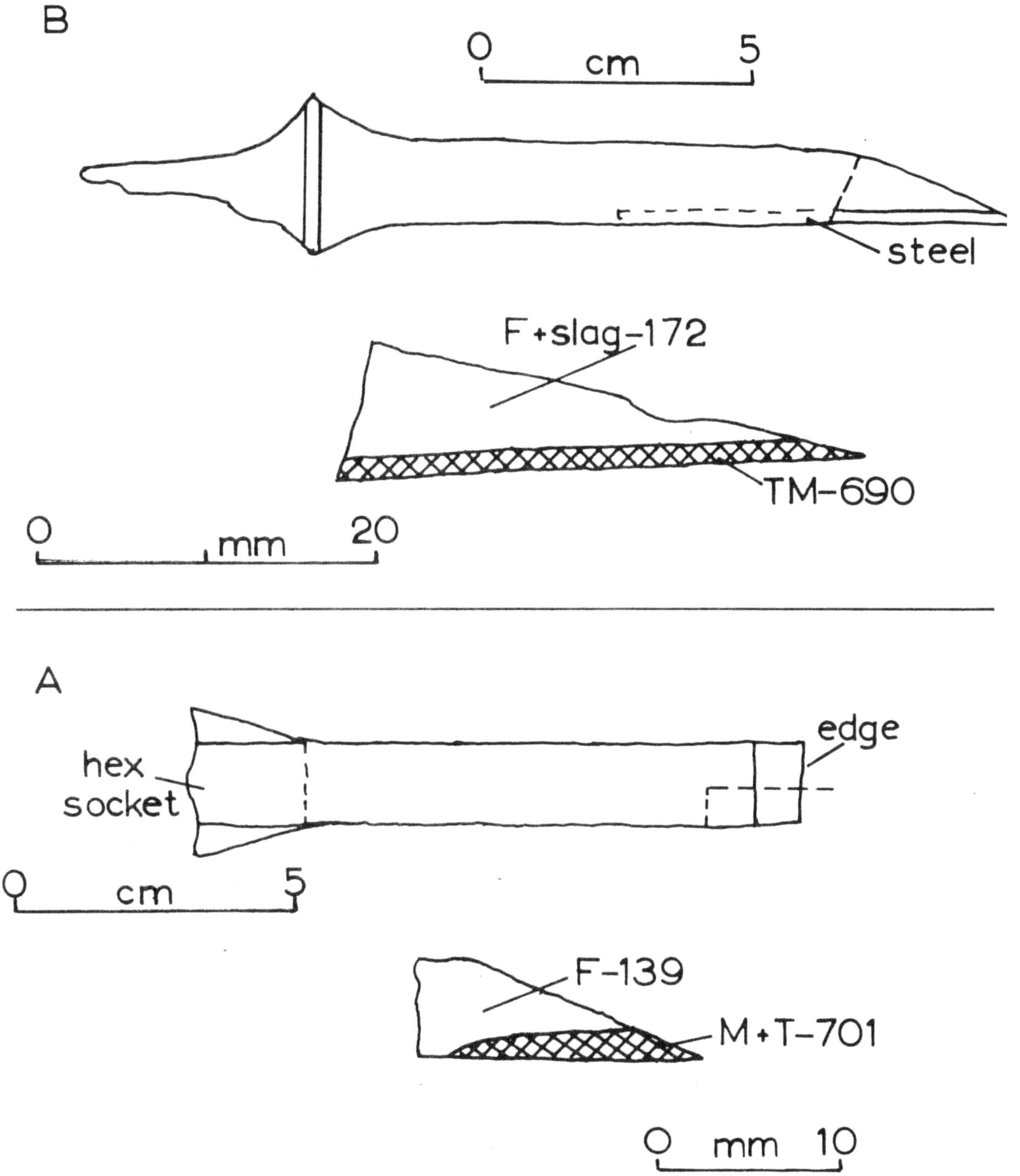


Fig. 32. Sections through wood-cutting chisels.
A A socketed chisel of the 16-17th century from Chingley, Kent.
B A tanged and bolstered chisel of the 18th century from Lymm, Cheshire.

(f) Ploughshares and Spadeshoes

In Roman and later times the metal parts of a plough consisted of a coulter which was a vertical bar or knife which was intended to cut the sod, and a share which sheared the ground horizontally and turned the sod over. Unfortunately we have not been able to examine a coulter. A share from the Romano-British site at Gestingthorpe was found to consist only of high phosphorus ferrite with a hardness of 230 HV, while another from the nearby site of Upper Yeldham also consisted of ferrite with a somewhat lower phosphorus content with a hardness of 109-145 (Table F).

Spade edges or shoes were applied to wooden spades to render them wear resisting like ploughshares. The 9th century example from Winchester (CG1690) could have been intentionally carburised, as the slaggy wrought iron with a hardness of 141 has had its hardness increased on the edge by carburising to give a mainly pearlitic structure with a hardness of 210 HV (Fig. 33). This is one of the few examples of the post Roman era where we seem to have hardening by carburising rather than by welding-on of steel.

A mason's trowel from an 11th century level at Winchester was found to be no more than wrought iron with a hardness of 127 HV.

## 5.3 Metal and Stone Working Tools

(g) Hand Hammers

One would expect large masses or iron, like hammers, to be made of bloomery iron with its variable but low carbon content. So it was not surprising to find that the Romano-British hammer head from Ware was no more than wrought iron consisting of coarse grained ferrite and slag with a hardness of 190 HV which suggested that it contained some phosphorus or was affected by cold work (Table G). Gestingthorpe[20] produced a hammer with an elongated eye, a slightly domed face and an almost pointed pein. Perhaps this was a tiler's hammer. (A similar hammer was found at the Lunt[39]). The pointed pein was ferrite with a hardness of 261. Most of the hardness is due to the phosphorus content and the carburising was not intentional.

The 17th century hammers from Winchester were very different. One had a chisel pein (Fig. 34), the head of which had been steel-faced. The steel layer was 3.5mm thick and curved slightly. The chisel-pein end was sectioned longitudinally and showed that it had been made by the insertion of a piece of steel (Fig. 34). The steel and its iron surround had mushroomed somewhat during use. The steel consisted of a quenched martensite+troostite with a hardness of 430 HV1 while the iron was ferritic with a hardness of 164. The flat head had been made by welding on a 10mm thick piece of steel. The weld was good at the edges but full of oxide and cavitites in the centre. But, since the head was used under compression, lack of adhesion in this area would not be detrimental. The head had been quench-hardened to give a mixture of martensite and troostite. The hardness in the martensitic areas was 633-644 HV5 and in the troostitic areas 341. The mean hardness, therefore, would be about 490 which is only slightly less than that of the modern hammer (550 HV). The metal that made up the bulk of the tool was ferrite which had a hardness of 174-178 under the head. This rather high hardness could be explained by the cold deformation under impact which would be expected of metal in such a

Table F. Spade-shoes and ploughshares

| Site | Site/Lab. Ref. No. | Date | Structure/Hardness HV Edge | Back | Type |
|---|---|---|---|---|---|
| **Shares** | | | | | |
| Gestingthorpe | 116 | R-B | F/230 | | 0 |
| Upper Yeldham | 118/794638 | R-B | F+P/145 | F/109 | 0 |
| Winchester | BS3394 | 17th C. | M+T/430 | F/164 | Steeled |
| Winchester | BS3766 | 18th C. | M+T/745 | F/205 | Steeled |
| **Spade-shoe** | | | | | |
| Winchester | CG1690 | 9th Cent. | P+F/210 | S/C | |

Table G. Hammer-heads

| Site | Site/Lab. Ref. No. | Date | Structure/Hardness HV Edge | Back | Type |
|---|---|---|---|---|---|
| Ware, Herts | 1623 | R-B | F/190 | - | 0 |
| Gestingthorpe | 69 | R-B | F+P/261 | F/205 | 0 |
| Winchester | BS3394 | 17th C. | M+T/430 | F/164 | Steeled |
| Winchester (flat head) | BS3766 | 18th C. | M+T/745 | F/205 | Steeled |

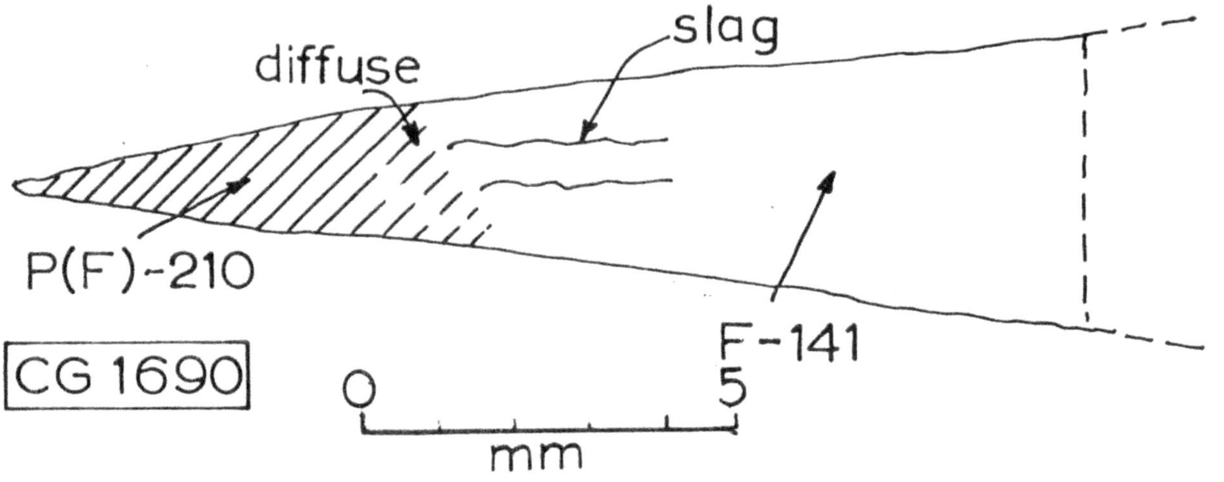


Fig. 33. A carburized spade-shoe from the 9th century, Winchester. (CG 1690).

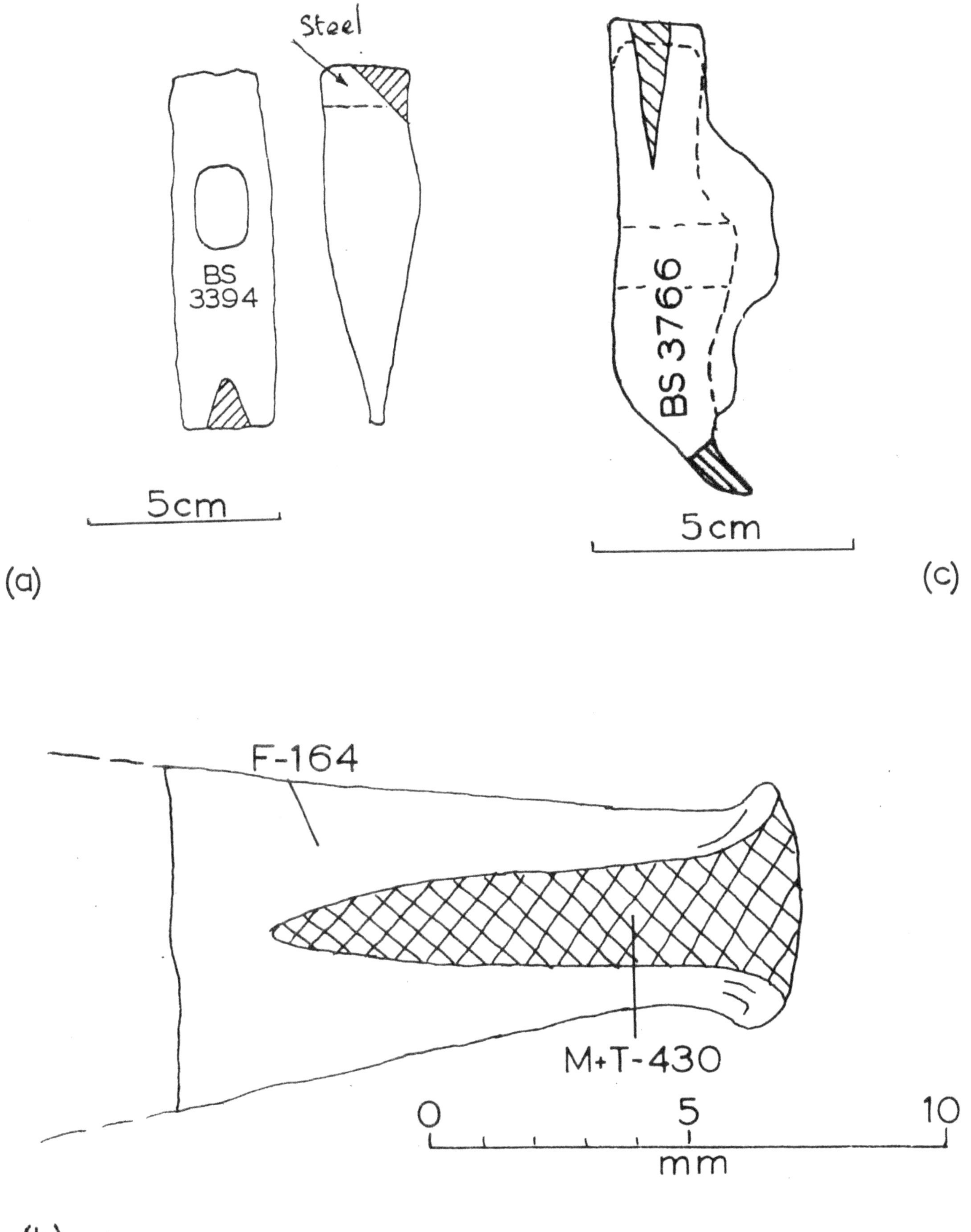


Fig. 34. Two hammer heads from Winchester showing evidence of steeling and heat treatment (BS 3394 and BS 3766). (a) shows chisel-pein head and (b) detail of same. (c) shows a claw hammer with steeled face but wrought iron claw. 17-18th century.

position. There were signs of Neumann lamellae - a common symptom of such deformation in early iron.

The second hammer from Winchester was a claw hammer (Fig. 34). The 'claw' was sectioned across its length and was found to consist of ferrite with a great deal of slag and a hardness of 131 HV. This would not be a very effective implement.

The flat head was also sectioned and was found to have been made in exactly the same way as the first. The steel layer was 3.5mm thick and curved slightly over the sides of the head. The steel seemed to have been made of two or three layers. The structure was martensite and troostite showing that it had been heat-treated by quenching. The hardness varied from 745 HV1 in the martensitic areas, through 400 in the troostite+martensite areas, to 330 in the completely troostitic areas. The mean hardness is 490 as in the first.

(h) Masons Chisels, Smith's Chisels etc. Punches and Wedges

This is a confusing group to write about. One would normally expect all but the wedge to be made of hardened steel. As they have to cut hard materials one would expect the edges to be short and well supported by metal behind, unlike the slender wood-cutting chisel.

A chisel-shaped end is obvious, but it may be the end of a round, square or even hexagonal bar. A smith's chisel may cut red-hot metal in which case it will be held by the smith and hit by the hammer of the striker (who may be the smith). In this case it will be called a 'set'.

A cold chisel will trim cold metal, make incisions or marks on cold metal and may be similar to a mason's chisel. Both will be hit with steel faced hammers and may have mushroomed heads.

A punch may make a hole in hot or cold metal. A drift is a tapered tool used to open out a hole. A wedge is merely a tapered piece of metal which has not been intentionally hardened in any way.

Naturally, one may be readily transformed into another by forging and re-heat treatment. But to be of any use, hot-working tools must be chilled at intervals during use or they will soon lose their hardness. For this purpose the smith will have a water 'bosh'. Naturally it is possible for tools to have been superficially hardened accidentally by such means in early times, although it is difficult to believe that a smith did not know the art of quench-hardening from the earliest of times. His main difficulty was getting homogeneous steel to make his quenching operation really effective. The results of our examination underline this problem.

Chisels

Starting with chisel-edged tools shown in Table H, we have a group from Roman times which vary enormously, from the Chelmsford chisel which is no more than pure ferrite to one of the Gestingthorpe chisels which attains a hardness of 440 HV in certain areas (Fig. 35). But the smith had the usual problem of getting a homogeneous piece of steel, and his lack of success in this respect, was the cause of the variation in properties in spite of the

## Table G. Mason's Chisels, Smith's Chisels, Punches, Wedges, Drifts and Sets.

| Site | Site/Lab. Ref. No. | Date | Structure/Hardness HV Edge | Structure/Hardness HV Back | Type |
|---|---|---|---|---|---|
| Chisels | | | | | |
| Brancaster (Mason's) | 773529 | R-B | F+P/224 | F/178 | 3 |
| Chelmsford | CHS1/41 | R-B | F/137 | | 0 |
| Gestingthorpe | G/122 | R-B | F+M/440 | F/177 | 1(a) |
| " | G/65 | R-B | F+P/230 | P/240 | 3 |
| Wanborough (cold) | 790779a | R-B | TM/800 | F/182 | 4 |
| " | 790239 | R-B | F/153 | | 0 |
| " | 771356 | R-B | F/140 | | 0 |
| Winchester | WP1607 | +14th C. | TM/400 | | 3 |
| " | BS2631 | 13thC. | F/140 | | 0 |
| " (twisted shank) | BS5866 | None | F+P/186 | | 0 |
| Punches | | | | | |
| Brancaster | 773725 | R-B | F+P/305 ——— | 224 | 3 |
| Chelmsford | CHS72/47 | R-B | F+P/181 | | 3 |
| Catsgore, Som. | F 505 | R-B | F+P/250 - 126 | | S/C+ |
| Gestingthorpe | 60/105 | R-B | F+P/290(CW) | | 0 |
| " " | 34/93 | R-B | - | F+P/153 | - |
| " " | 66/ | R-B | - | SC/193 | 3 |
| Ware, Herts | 1710 | R-B | M/385 | F/240 | 3 |
| " " | 97 | R-B | F/150 | F/150 | 0 |
| Drifts, Wedges | | | | | |
| Gussage AS | 283 | Pre-R IA | TM/532-473 | TM/353-401 | 3 |

Table G. Continued

| Site | Site/Lab. Ref. No. | Date | Structure/Hardness HV Edge | Back | Type |
|---|---|---|---|---|---|
| Winchester | ACS511 | Late R-B | F/131 | | 0 |
| " " | WP4218 | Late R-B | M/660 | M/570 | 3 |
| " " | ACN216 | 10th-11thC | F/167 | | 0 |
| " " | CY80 | | F+M/590 | F+P/224 | 2 |
| Chingley, Kent | Fig. 31/56/19 | 16-17thC | F202/265 | | 0 |
| " " | " 31/54/17 | 16-17thC | M+T/7226 | P/329 | 1(a) |
| " " | " 31/55/13 | 16-17thC | F/146 | | 0 |

+S/C = Surface carburised.

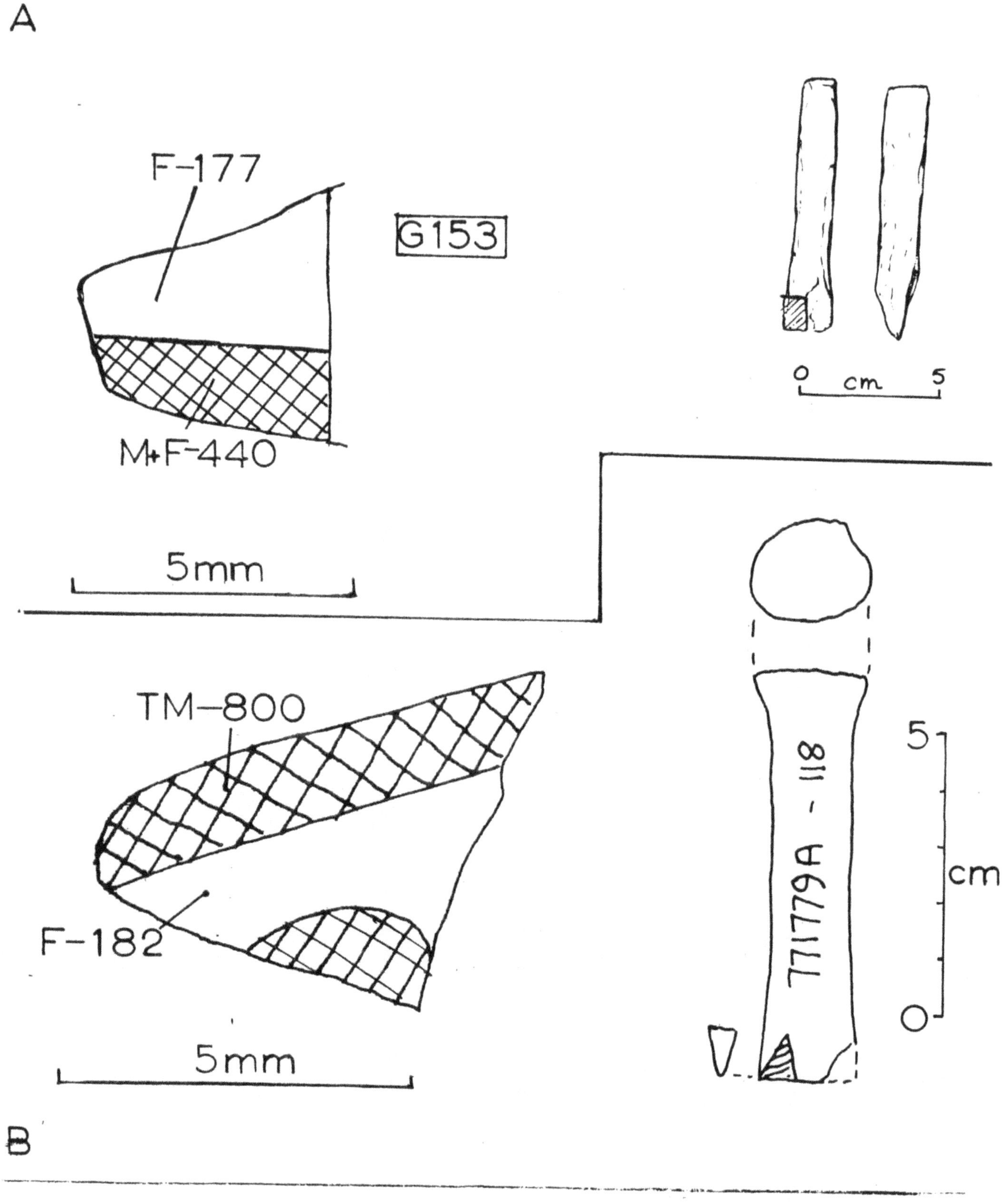


Fig. 35. Romano-British cold chisels from Gestingthorpe and Wanborough. (B).

application of quenching. The smith at Wanborough did much better as far as the edge was concerned and achieved a hardness of 800 HV (Fig. 35).

The medieval chisels do not seem to be any more consistent and the maximum we get is a hardness of 400 HV for a homogeneous martensitic steel (WP. 1607).

When we come to look at the Roman punches we must remember that they may have been used for marking or decorating non-ferrous metals and therefore high hardness was not necessary. We find most of them have hardnesses in the region of 290–400 HV and that they are essentially low to medium carbon steel. One from Ware was quenched, another from Catsgore, appeared to have been surface carburised.

Romano-British

Wanborough

771779A - Cold Chisel

This is fairly free of slag. But the edge consists of a layer of steel welded to iron with another layer of steel sandwiching the iron. This has been quenched from a high temperature coverting the steel at the tip into martensite. Further back the central areas seem to be ferritic. The white lines showing high As welds seem to be full of single phase slag. Much of the martensite is dark showing evidence of (? self) tempering. The hardness of the martensite is 800 HV; that of the ferritic material away from the edge is only 182 HV (see sketch, Fig. 35).

790239 - Cold Chisel

Mostly fine-grained ferrite with slag. The hardness is 153 HV.

790356 - Cold Chisel

Fine-grained ferrite with single phase slag. Also contains a central band of very coarse ferrite. The hardness is 140 HV.

Brancaster.

773529

A square-sectioned, flat-edged chisel with an expanded cutting edge. A segment was removed about 2.5 cm from the cutting edge. Made of piled pieces of variable grain size and carbon content, the latter varying from 0 to 0.6% C. The hardness varies from 178 HV (all ferrite) to 224 HV (ferrite and pearlite). The pearlite is lamellar indicating slow cooling though 700°C.

Chelmsford

CHS/1 (41)

Chisel-shaped, but merely fine grained ferrite with some grain-boundary cementite and a hardness of 137 HV. A poor tool for a chisel.

Gestingthorpe

153 - Smith's Set or Chisel

The edge of this chisel consists of two pieces of metal joined along the centre. One has a high carbon content and the other hardly any. The carbon of the former varies from 0.2 to 0.6% and has a Widmanstätten distribution of ferrite and martensite. Clearly it has been fairly rapidly cooled from a very high temperature (1100°C) and, at a temperature of about 800°C when the ferrite was separating from the austenite, was quenched in water to give martensite with a hardness of 440 HV. The low carbon half of the metal has a hardness of 177 HV1 and as the carbon content is nearly zero probably contains some phosphorus (Fig. 35A).

154 - Smith's Chisel

The edge of this slender blade has been made by folding in such a way as to enclose some charcoal in the centre of the fold. The edge consists of ferrite and pearlite which has been cold worked to give a hardness of 230 HV1. The carbon content in this region is low, about 0.3% but near the centre rises to 0.8%, giving pearlite with fine grain boundary ferrite. The hardness in this region is 240 HV1 reflecting the higher carbon content and absence of work hardening.

Medieval

Winchester

WP 1607. Chisel (?) - 14th Cent. or later

An etched cross section shows a dark, tempered martensite with large grain size and some inclusions. The hardness is 400 HV1. This is a well-quenched and tempered homogeneous carbon seel.

WP 1607 - Chisel (?) - 14th Cent. or Later

An etched cross-section shows a dark, tempered martensite with large grain size and some inclusions. The hardness is 400 HV1. This is a well-quenched and tempered homogeneous carbon steel.

BS 2631 - Flat Bar or Chisel - 13th Cent.

A longitudinal section from one end was found to consist of ferrite+slag with some carbides but the carbon content does not exceed about 0.1%. The hardness was 140 HV1. Clearly this is nothing more than a piece of wrought iron.

BS 5866 - 'Twisted Chisel', no date

Consists of ferrite and sorbitic pearlite with a carbon content of 0.2 - 0.5%. The hardness is 186 HV. Twisting was quite a common form of decoration from Roman up to modern times.

BS 2631 - Flat Bar or Chisel - 13th Cent.

A longitudinal section from one end was found to consist of ferrite+slag with some carbides but the carbon content does not exceed about 0.1%. The hardness was 140 HV1. Clearly this is nothing more than a piece of wrought iron.

WP. 1138 - Mushroomed Cold Chisel - 16th Cent.

The cross section shows a steel edge in which the steeled part does not quite reach the tip. In both parts the structure is ferrite+pearlite but there is about 0.6% C in the steel portion (181 HV) and somewhat less elsewhere.

Punches (Romano-British)

Brancaster

773725

A rectangular sectioned punch or wedge. A piece was removed from the sharp end and found to be steel varying from 0.1% C on one surface, to 0.7% C in the middle. Made by piling, with decarburised zones between. The steel had not been hardened as it consisted predominantly of ferrite and pearlite. The hardness of the low carbon areas was 224, and of the high, 305 HV.

Chelmsford

CHS72 (47)

A piece was removed from the end. This had a square cross section. It consisted of ferrite with a varying amount of spheroidised pearlite in the range 0 to 0.5% C with a mean carbon content of about 0.4%. The structure was fairly homogeneous and the hardness was 181 HV. No heat treatment had been given.

Catsgore

F.505 Cat. 1062

A punch or drift. A piece was removed from the tip of the tool. It proved hard to cut but on examination was found to be predominantly fine-grained ferrite with a hardness of 126 HV1; however, extensive areas of the surface were carburised up to 0.7% C. These areas consisted of fine lamellar pearlite with a hardness up to 250 HV1 although the actual tip of the tool was only 126 HV. It must be remembered that a considerable amount of the tool had been lost by rusting and what was examined may only represent the border line between a heavily carburised surface and the ferritic core. It was clear that there had been no attempt to form the tip of the punch by welding on steel.

Gestingthorpe

60 - Smith's Punch

A rectangular piece was removed from the tip. This had been formed by folding over and forming a poor medial weld line. The structure was ferrite and pearlite with a carbon content of 0.1 - 0.2% and the pearlite was spheroidal showing that it had been held for a short time at 600 - 700°C. The ferrite grains are heavily elongated indicating that it has been cold worked. The hardness is 290 HV1.

34 - Punch or Chisel

The head is a homogeneous mixture of ferrite and pearlite (0.1 - 0.2%). The hardness is 153 HV. Spheroidal pearlite. Tip not examined.

66 - Smith's Punch

A piece was removed from the side. This consists of 100% sorbite with a hardness of 193 HV1. The carbon content is about 0.6% and the evenly diffused structure must have been formed by heating at about 700°C for an appreciable time followed by slow cooling.

Ware

97 - Punch

A piece was removed from the point and found to be very slaggy wrought iron with a ferritic structure and a hardness of 150 HV1. This would not make a very good punch.

1710 - Punch

A better one. An attempt had been made to harden this by quenching but the carbon content was not high enough for most of the metal to respond to quench-hardening. Most of the structure is fine-grained ferrite, giving hardnesses up to 140 HV1, but there are occasional small areas of martensite one of which has a hardness of 385 HV.

Drifts and Wedges

We have several high grade tools which can be classified as chisels or drifts, and some edged tools of lower quality which are probably no more than wedges. Amongst the better tools is one from the pre-Roman site of Gussage All Saints. A longitudinal section was cut through the edge of this tool (Fig. 36). The structure was found to be martensite and martensite+troostite with hardnesses varying from 473 (M+T) to 532 (M). It is clear that this tool was intentionally made, either for use as a chisel or drift. A cross section through the middle of the tool showed it to be very heterogeneous with a 'lap' across the centre of the section. One surface is light after etching and has a hardness of 193 HV with a ferrite+pearlite structure. The pearlite was not resolvable suggesting fairly fast cooling from about 700°C. Towards the centre the structure is mainly tempered martensite with a hardness of 353 HV5. Further, towards the other side across the lap, the structure is the same with a hardness of 401 HV5. Some areas are lighter with a hardness of 208 HV. It is clear

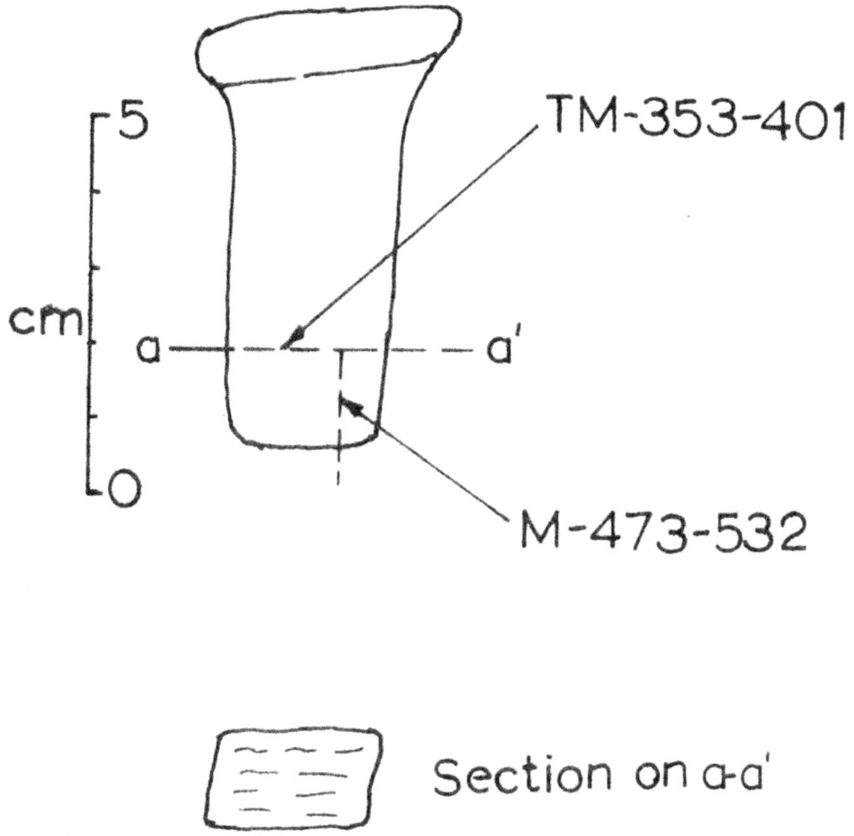


Fig. 36. Section through chisel or drift from a pre-Roman Iron Age level at Gussage All Saints, Dorset.

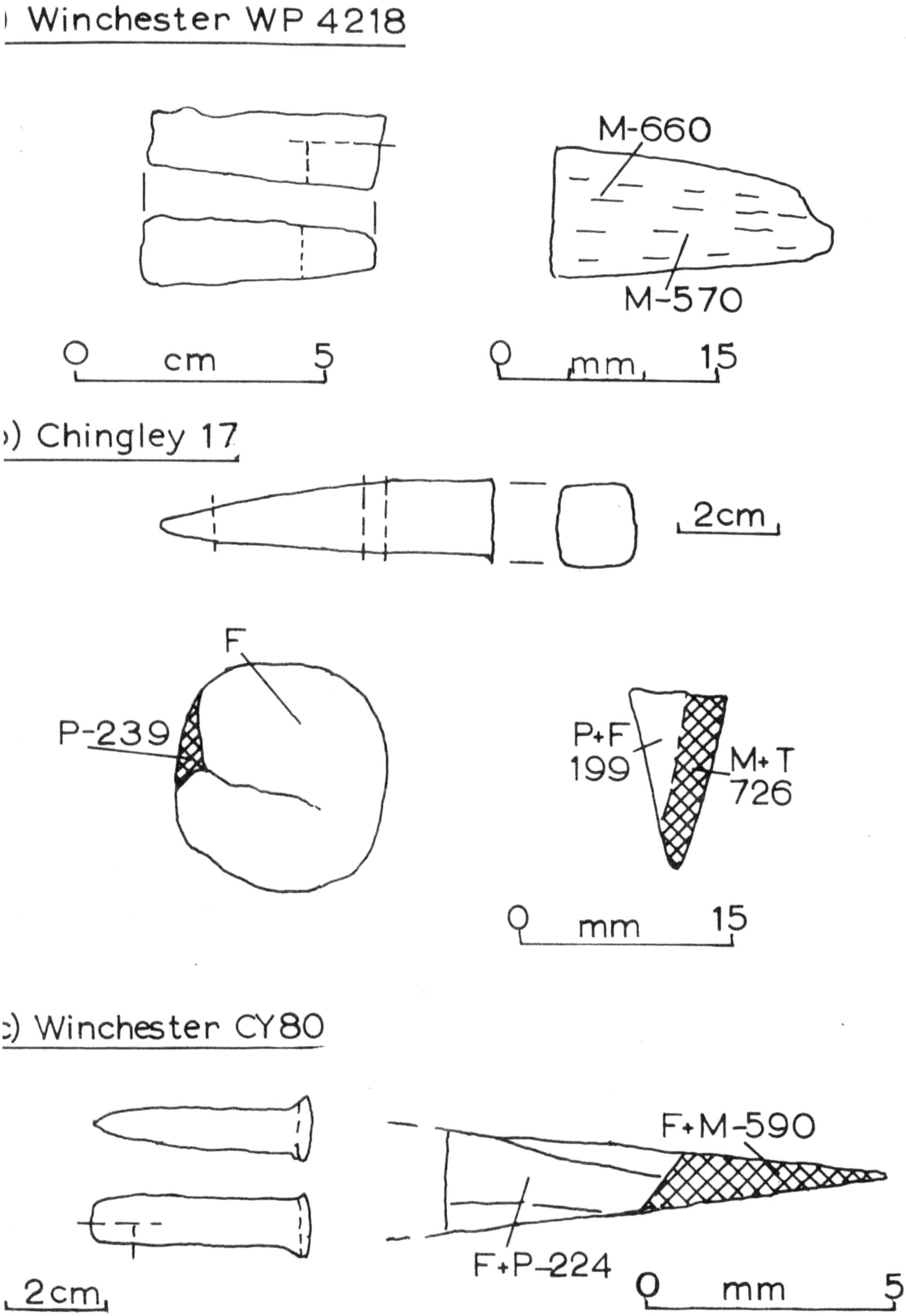


Fig. 37. Sections through chisels and drifts from Winchester and Chingley.

that this steel was of uneven carbon content and was quenched and tempered.

The late Roman tool from Winchester (WP 4218) is a good piece of carbon steel, fully hardened to martensite and therefore seems too good to be a wedge (Fig. 37).

Chingley (16th - 17th Cent.)

Two drifts were examined; one is circular in cross section and the other square. The latter (No. 17) is wholly ferritic with a hardness of 146 and the tip has not been hardened in any way. The other (No. 13) has been steeled by having a strip of carbon steel welded on to one side for the whole length of the drift so that it takes up most of the area of the point (Fig. 37). The carbon content of the steel must be about 0.7 - 0.8%, and that in the body of the drift has a pearlitic structure with a hardness of 239. But at this point it has been quench-hardened to give a structure of martensite + troostite with a hardness of 726. The low carbon body of the drift had a ferrite + pearlite structure with a Widmanstätten distribution which shows that it has been heated at some time to high temperature and cooled fairly fast, probably when the tip was hardened. This tool shows a good understanding of the technique of heat treatment and would have been a very useful tool for making the rivet holes in the scissors, for example.

No. 19 is a 2.4 cm dia. wrought iron bar with a blunt end and a mushroomed head which has been deformed by hammering. A piece was taken from the head and found to be heavily deformed ferrite with a twinned structure and a hardness of 202 - 265. The material was 100% ferrite and the high hardness reflects the deformation. But in order to get deformation twins (Neumann lamellae) by hammering, the iron usually has to have a high P or $N_2$ content.

Finally, we have the undated tool from Winchester (CY 80) which has a welded-on steel tip (Type 2) and is similar to the above with a tip hardness of 590 HV (Fig. 37). Some of the rest are probably wedges and their structure accidental.

(i) Saw Blades and Files

The saw-teeth were from short pieces of saw blade found on the Roman site of Wanborough (Table J). Considering that their hardness was only in the range 470 - 210 HV they must have been designed for wood-cutting. One is the result of welding carbon steel to mild steel (Fig. 38) and in typical Roman style there has been no attempt at quench-hardening. The second, in constrast, is a homogeneous high carbon steel fully quenched and lightly tempered to give a hardness of 470 HV. The third ferritic with enough phosphorus to give a hardness of 210 HV.

It would seem that the smiths took their saw blades seriously and tried to produce good blades within the limits of their technology.

780169 - Saw Tooth

This shows stringers down the centre delineating a weld line between iron and steel. The steel is once again 0.6% C with a pearlitic structure. The other side is ferritic. The whole blade has been heated to a high enough temperature to give a marked Widmanstätten structure. The low

## Table J. Saw-blades and Files.

| Site | Site/Lab. No. | Date | Structure/Hardness HV Edge | Core | Type |
|---|---|---|---|---|---|
| **Saw Teeth** | | | | | |
| Wanborough, Wilts | 780169 | R-B | P+F/245 | F/142 | 1(a) |
| " " | 790274 | R-B | F/210 | | 0 |
| " " | 435242 | R-B | TM/470 | | 3 |
| **Files** | | | | | |
| Winchester | Rake (?) | U/S | TM/690 | | |
| Winchester | LBSi40 | U/S | TM/1100 | | |
| Wanborough | 790887 | -1750 | TM/535 | F+TM/275 | |

## Table K. Awls.

| Site | Site/Lab. No. | Date | Structure/Hardness HV Edge | Core | Type |
|---|---|---|---|---|---|
| Chelmsford, Essex | CH2.215/13 | R-B | F+P/165 | | 0 |
| " " " | CH2.214/ | R-B | F/146 | | 0 |
| Gestingthorpe | 62/106 | R-B | F+P/124 | | 0 |
| Ramsbury, Wilts. | 12 | Sax. | F+P/205 | | 0 |
| " " | 9 | Sax. | B/370 | | 1 |

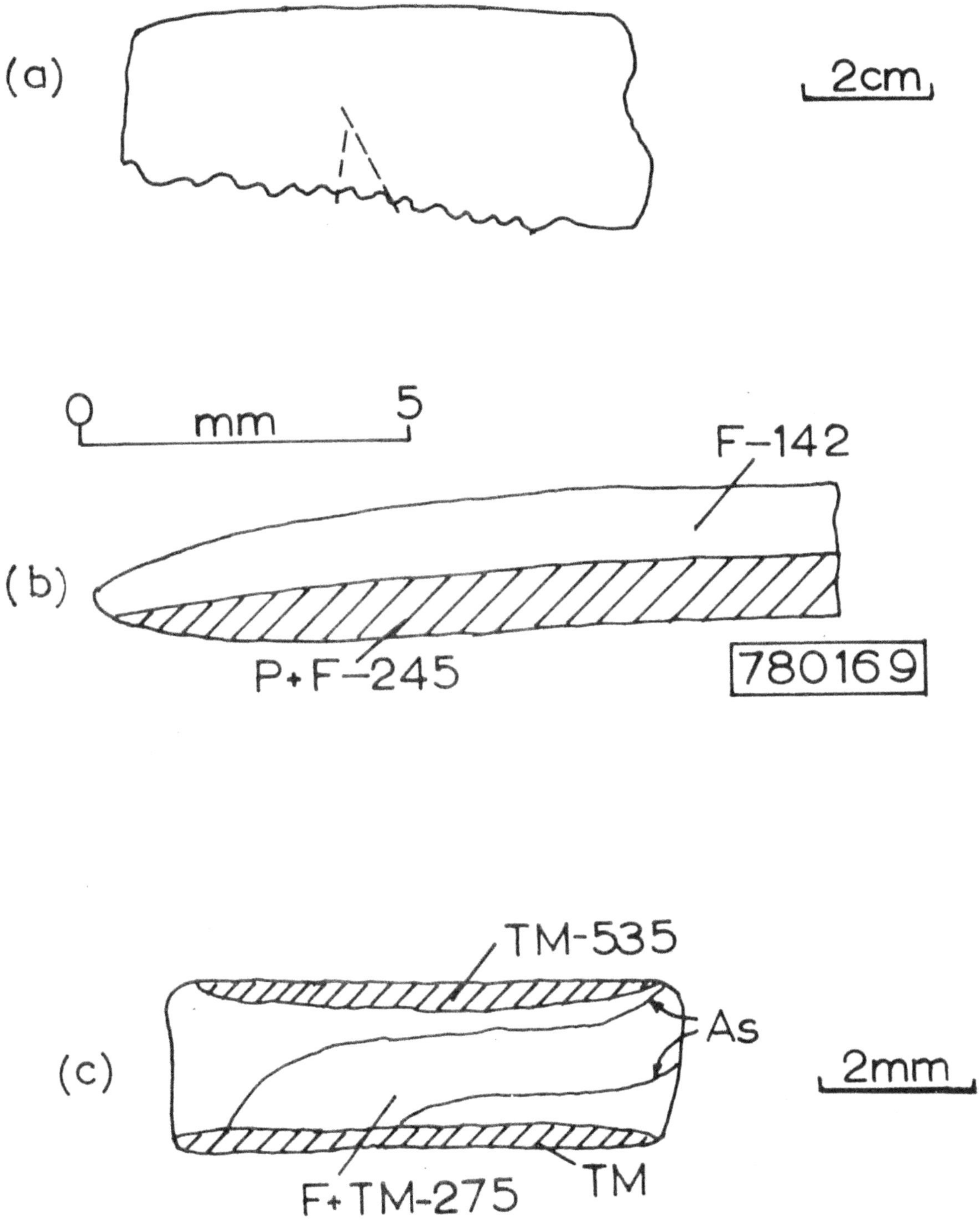


Fig. 38. Section through a Romano-British saw-blade from Wanborough and a piled steel file from Wanborough.

carbon side contains about 0.2% C and has a hardness of 143 HV, while that of the high carbon ferrite+pearlite is 245 HV (Fig. 39).

435242 - Saw Tooth

A clean bit of steel with some carbide inclusions. Like the knife 684039, this was resistant to etching but re-etching gave a dark structure of light, granular, carbides in a dark matrix of lightly tempered martensite. The hardness was 470 HV1.

790274 - Saw Tooth

This consists entirely of ferrite of medium grain size. The hardness at the edge is 210 HV which indicates a considerable phosphorus content.

The only files examined were from Wanborough and Winchester and they are all unstratified (Table J). That from the Roman site at Wanborough clearly pre-dates Huntsman (c. 1750) as it is made of carburised wrought iron, folded over to give 3 layers of high carbon steel which has finally been hardened after cutting the teeth (Fig. 39). The resulting hardness was 535 HV1, adequate for working wrought iron.

The other two files were from Winchester and were probably more recent being homogeneous high carbon steel with hardnesses in the range 960 - 1100 HV1.

## 5.4 Boring Tools

### (j) Awls, Augers and Boring Tools

Awls were one of the first metallic tools to be made by man. They are used for boring holes into wood, leather or bone but more rarely metal. They are of square or circular section and the later ones normally have a bolster to prevent them being pushed up into a bone or wooden handle.

One of the Roman awls from Chelmsford is square-sectioned, of ferrite, with a hardness of 146 HV. The other contains some carbon which raises the hardness to 165 HV1. One from Gestingthorpe has a pyramidal head and a tapering stem and was mostly wrought iron with a hardness of 124 HV (Table K).

Two awls were examined from the Saxon site of Ramsbury. The first (9) has a circular and a square end. Presumably the square end fitted into the handle. The other (12) was more or less rectangular in section.

9. Awl: A section was cut through the middle. The section showed a diagonally-placed layer of pearlite enclosed by ferrite-and-pearlite with a widmanstätten structure. The dark etching pearlite or bainite, which was not resolvable and therefore not very slow-cooled, had a hardness of 370 HV. The high-carbon central zone shows evidence of slag lines on one border, but the lower-carbon zones are unlikely to be products of decarburisation. This tool would seem to be an example of intentional steeling.

12. Awl: This showed ferrite with areas of ferrite-and-pearlite, with much fine slag. The pearlite was not resolvable, which shows relatively fast cooling. The hardness of the ferrite is 205 HV, which suggests a fairly

high phosphorus content.

On the whole we cannot see why an awl needs to be harder than this for the work it has to do. Like the styli from Thistleton and elsewhere, a high phosphorus wrought iron with some work-hardening at the tip is quite adequate and superior to the modern bradawl which often seems to lack hardness.

The spoon auger used for boring holes in timber for the insertion of wooden dowels was known from the Roman period and was used throughout the Medieval period unchanged. The example shown in Fig. 39 comes from the River Ivel near Letchsworth, Herts. It is basically a piece of low phosphorus (0.091%P) wrought bloomery iron onto which carbon steel has been welded. Both edges have had steel applied so it could be used clockwise or anti-clockwise. The hardness of the martensitic edge is 460 HV, but the other edge is somewhat softer and was probably carburised by accident. A hardness of 460 is quite adequate for a wood-working tool and there is no doubt that this auger would be very efficient.

One of the most interesting boring tools recently found is the cannon boring bar from the 17th century site of Chiddingly, Sussex[41]. It is 3.55 m long, weighs 85.72 kg and has an expanded end into which 4 bits are inserted making the outer diameter 14 cm. These bits project about 15 mm forwards and sideways (Fig. 40). Each bit is made of a plate of wrought iron, steeled on one surface (Fig. 41) with a 2.5 mm piece of carbon steel. These bits were wedged into place with shims and could be taken out and sharpened or replaced.

The hardness of the ferritic component was 172 HV1. The steel varied from 713 (maximum) to 326 near the weld, where it had a martensite+troostite structure (Fig. 42). The outer surface was fully martensitic and it would appear that the quenching rate was not sufficient to give 100% martensite over the whole 2.5 mm thickness. As the martensite was not tempered the steel must have been kept cool during use, perhaps by means of a water spray over the cutters.

This bit would have been used for cleaning out the bore of the iron cannon which had been cast with the aid of a core. After the middle of the 18th century all iron cannon were made by casting solid, and boring the bore. This required a very much more sophisticated boring machine.

## 5.5 Discussion

The two most striking conclusions to come out of this work are, first, that the majority of tools are made by the welding of iron to steel and secondly that until the Migration period it was not general practice to heat-treat steeled or carburised tools to get the best properties out of the steel component. We appreciate that steeling by carburising is a very slow process but it is clear from continental work by Pleiner and others that surface carburising was practised a great deal in the pre-Roman and Roman periods (see for example Table 7). Continental practice shows also that heat-treatment was not generally used until the Migration period.

If we turn our attention to the cutlery situation in Britain we see that steeling was in vogue in the immediate pre-Roman period and that it was possible to make good knives. This seems to contrast with the small amount

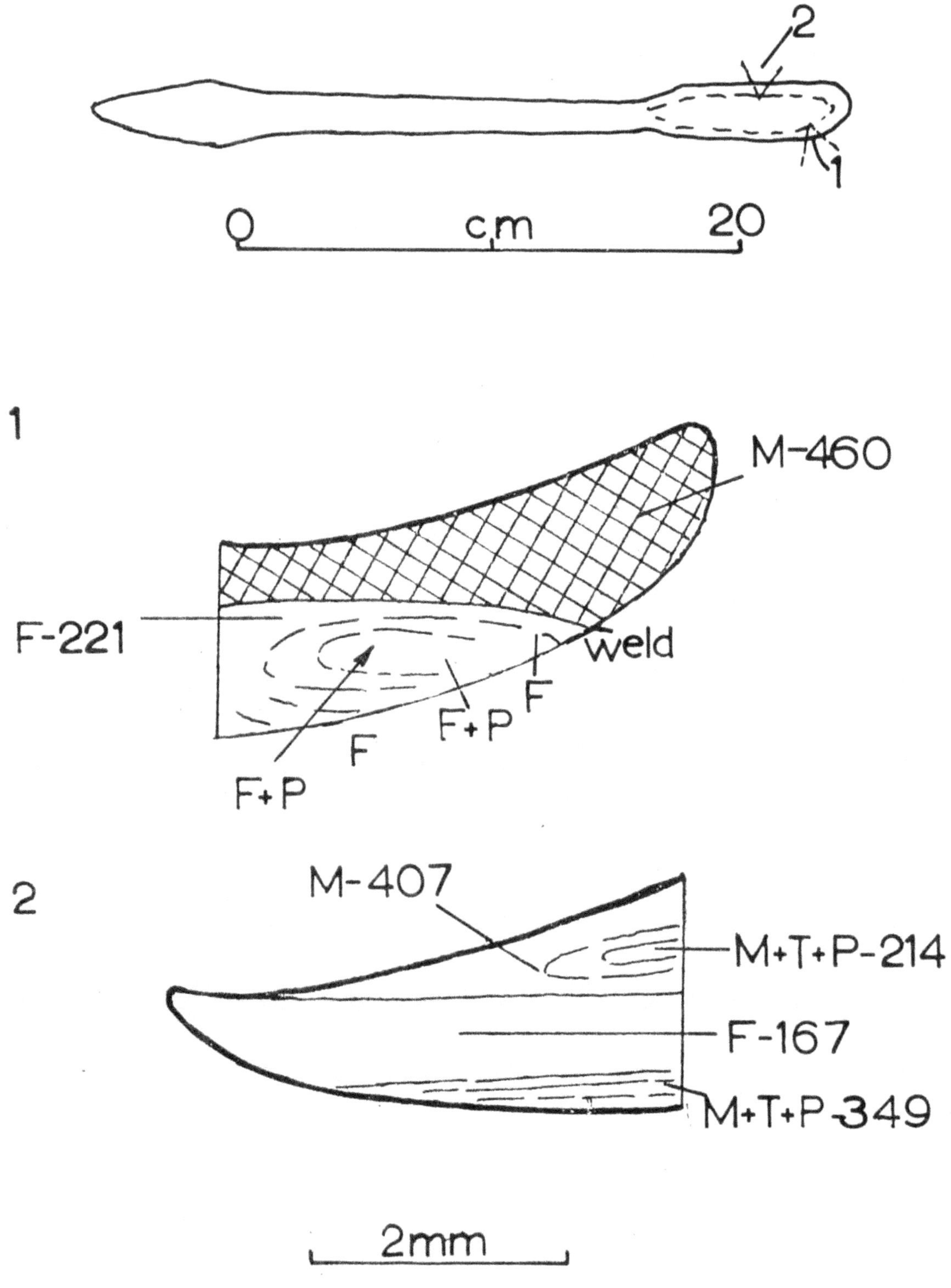


Fig. 39. Roman or medieval spoon-bit from Letchworth, Herts. showing steeled cutting edge.

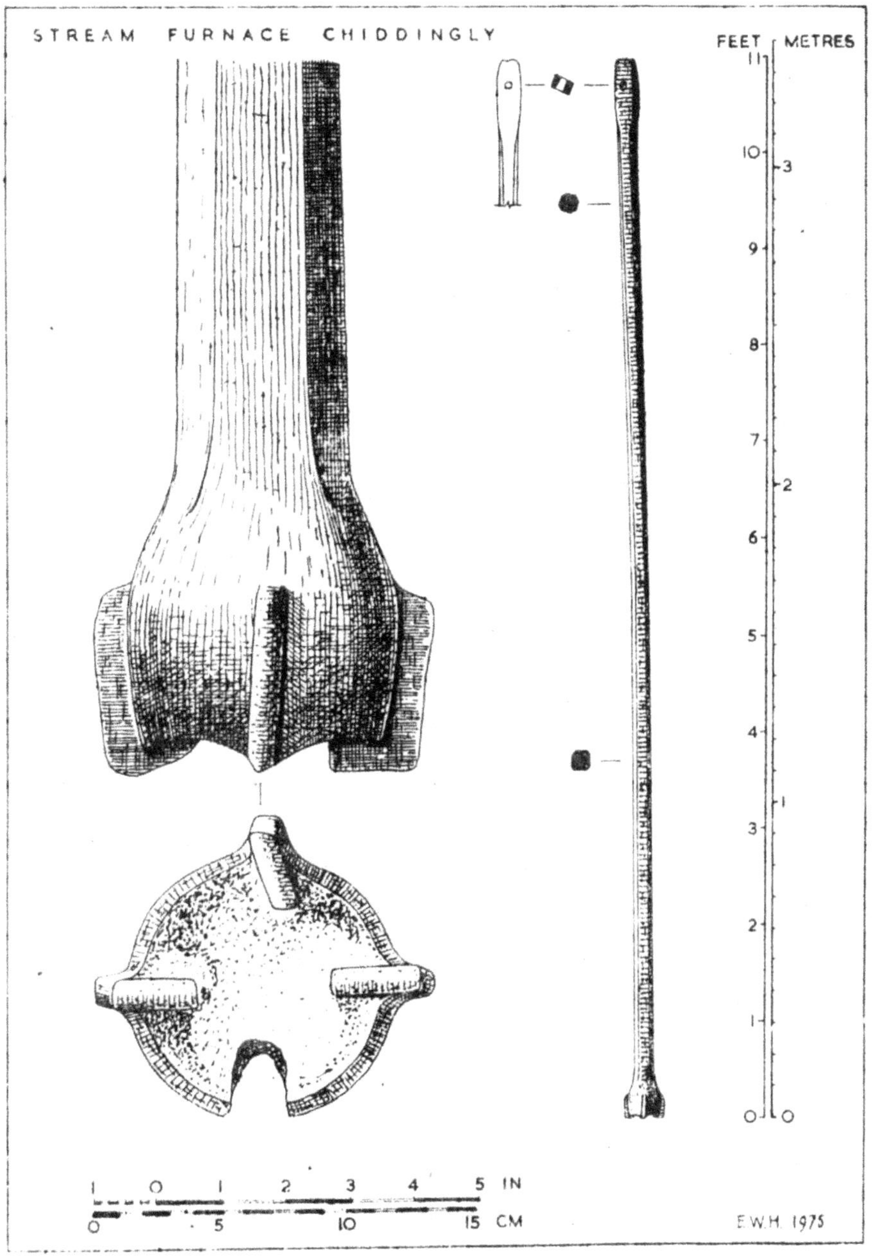


Fig. 40. Cannon-boring bar and bit from Chiddingly, Sussex (After Butler and Tebbutt[41]).

Hardness (2.5kg)

| | |
|---|---|
| Top | 847 |
| | 762 |
| | 464 |
| | 339 |
| | 150 |
| | 119 |
| | 146 |
| | 148 |
| | 159 |
| | 172 |
| | 126 |
| | 146 |
| | 155 |
| | 145 |
| Bottom | 162 |

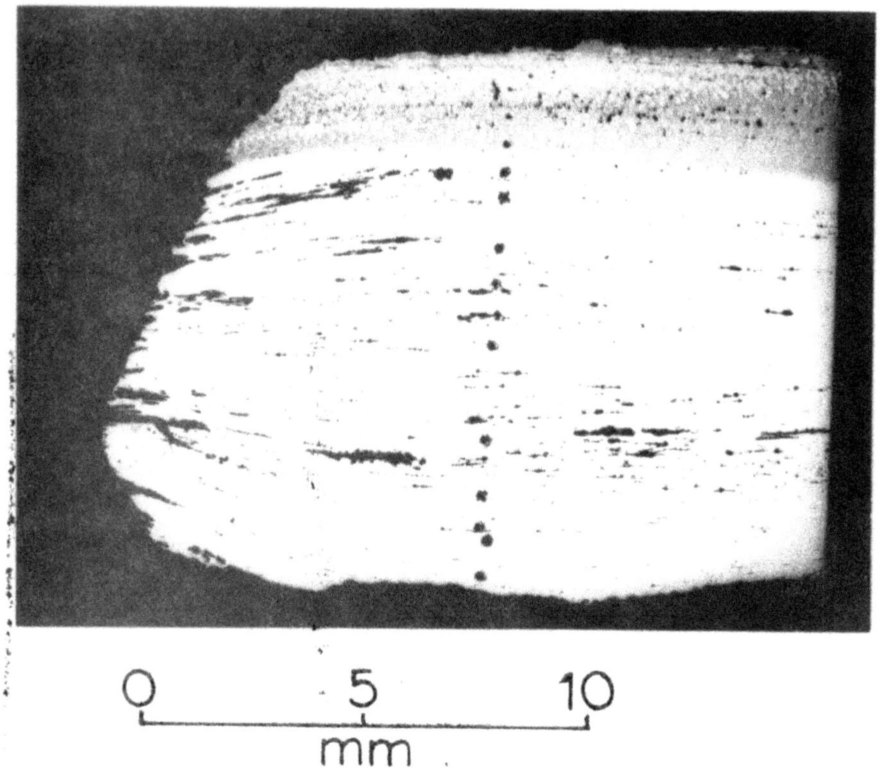


Fig. 41. Section through one of the steeled inserts in the head of the above, with hardness measurements. (Courtesy of E.M. Trent).

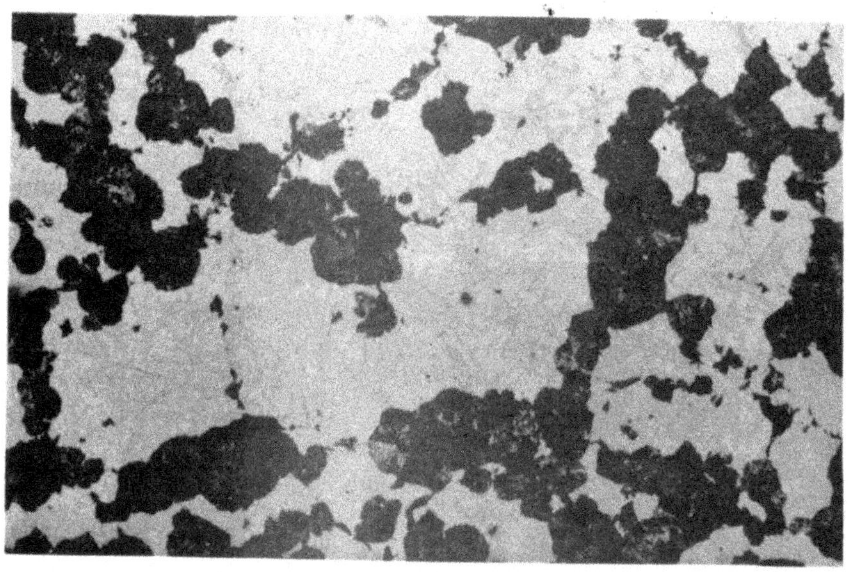

Fig. 42. Structure near weld of cannon-boring bit. Troostite and martensite X 325.

Table 7

Metallography of edge tools;

Summary of Pleiner's results on Czechoslovakian material (Ref. 42).

| No. | Site | Date | Object | Structure | Micro-hardness HV (50g) | Type RFT. |
|---|---|---|---|---|---|---|
| 10 | Lovosice | Hallstatt | Knife | F+P | | - |
| 12 | " | " | Shears | F+P | 150-208 | - |
| 14 | Stradonice | " | Knife | F+P | 168-225 | - |
| 15 | " | " | " | FO+P (Mart) | - | - |
| 16 | " | " | Shears | P(+F) | 301 | Carb? |
| 17 | " | " | Axe | Mart. | | 3 |
| 18 | " | " | (same) | " | | 3 |
| 19 | " | " | Hammer | Mart/Sorb | 212/669 | 3 |
| 20 | " | " | socketed chisel | F+P(+P) | 144/223 | Carb? |
| 21 | " | " | Saw | F+P | 165/297 | Carb? |
| 22 | Letky | " | Saw | F(+P) | 185/304 | " |
| 24 | Stradonice | " | Fireshovel | F+Sph.P | 143/223 | Piled |
| 25 | Lhotice | " | Scythe | P(+F) | 186/451 | 1 |
| 26/27 | Stradonice | " | Spadeshoe | F | 111/251 | - |
| 31 | " | Roman | " | F | 139/166 | - |
| 32 | " | " | Shears | F | 139/152 | - |
| 33 | " | " | " | F+P+N | 176/254 | - |
| 34 | " | " | Knife | F | - | - |
| 35 | Belec | " | Axe | F+P | 156/227 | Carb. |
| 36 | Pecky | " | " | F+P | 118/157 | " |
| 37 | Tisice | " | Razor | F+P | 113/157 | " |
| 42 | Vrdy | Burgwall | Axe | P(+F) | 123/151 | Carb. |
| 43 | Bilichov | " | " | Mart. | 231/463 | " |
| 44 | Praha/Ruzyne | " | " | P+F | 172/221 | " |
| 47 | Klucov | " | Ploughshare | F(CW) | 102/301 | 0 |
| 48 | Vinavice | " | " | F(F+P) | 191 | - |
| 49 | Zerotin | " | Scythe | F(+steel) | 131/205 | 2 |

Table 7 (continued)

| No. | Site | Date | Object | Structure | Micro-hardness HV (50g) | Type RFT. |
|---|---|---|---|---|---|---|
| 50 | Letky | Burgwall | Sickle | F+P | – | 3 |
| 51 | Libusin | " | Knife 14 | Mart | 251/304 | 2 |
| 52 | " | " | " 16 | F+P | 204/239 | 2 |
| 53 | Caslav-Hradek | " | " 23 | M+F | 208/215 | 2 |
| 54 | Letky | " | " 26 | P+F | 185/235 | 2 |
| 55 | Libusin | " | " 13 | F+P | 171/236 | 2 |
| 56 | Caslav-Hradek | " | Awl | F+P | 198/217 | – |
| 57 | Levy-Hradec | " | smiths-hammer | Mart | 412 | 2? |
| 58 | Caslav-Hradek | " | claw-hammer (Chisel pein) | P(+F) | – | – |
| 59 | " " | " | Shears | F+P+ | – | 3 |
| 60 | " " | " | " | P(+F) | 269/359 | 2 |
| 61 | " " | " | " | smiths-chisel | 227/269 | piled |
| 64 | Nejdek | " | Axe | F+P | | 1 |

N.B. The hardness measurements were made with a load of 50g using the Hanemann system.

of evidence available from Greece and the Near East where it can be shown that homogeneous steel was being used from about 1000 BC onwards.[44] It is not known whether this was made by intentional carburising of iron or by the selection of the high carbon parts of the heterogeneous bloom. So little remains in many cases that it is possible that what we find is the occasional high carbon parts of a very heterogeneous blade.

When we come to look at the Roman scene in Britain one is appalled at what the Romano-British customer was willing to accept. Whereas we would expect a stainless steel knife to have a hardness of at least 550 and a kitchen knife to have one of 650 or more, the Romano-British cook would apparently be happy with one as low as 133 HV made of wrought iron (Table 8). It is true that it was within the power of users at Wanborough to demand a good knife with a hardness of 720 HV, but the results given for several Roman sites, including Wanborough, show that he rarely got it.

In many cases also, the Roman smith gave the user a piece of steel but for some reason was unwilling or unable to heat-treat it. This seems to be a waste of good steel (Table 8).

This changes with the arrival of the Migration people so that we see at Poundbury and Ramsbury knives with hardnesses of 615, 520, 553 and 880 HV, like those that we would expect today. This becomes normal in the full medieval period and went on until the 19th century when consistent carbon steel made the welding of iron to steel unnecessary.

The results of the stratified examples which mainly come from the 10th to the 14th centuries have been summarised in Table A3. These have been compared with results obtained on later sites such as those from Goltho, Barton Blount, Holyoak and Chingley. Thus we have a reasonable representation from the 10th - 18th centuries. The results clearly show that the commonest group was the steel-edged Type 2, which became the mainstay of the 12th - 14th centuries. In this connection it is worth noting the comment repeated in Lloyd's book on the cutlery industry[45] and dated to 1582:- 'This argument cuts like a Leadenhall knife where, as they say in common speech, if one pours on steel with a ladle, another comes and wipes it off with a feather'. It would appear that this quotation refers to steeling which got the bad reputation it deserved and which was gradually replaced by more homogeneous steel blades from the 14th century onwards.

So far, the only comparable series of knives examined in the British Isles comes from Greencastle in Northern Ireland.[46] These all come from 13th century levels. The results are summarised in Table 9 and show a similar variation in quality to those from similar levels at Winchester. The results of the metallurgical examination of edge tools from North-East Europe and Scandinavia shows that the welding technique was popular from the Migration period onwards. What is outstanding is the small number of examples of knives made by surface carburising. Out of the entire Russian collection of 47 examples reported by Kolchin only 2 appear to have been carburised.

The techniques used in the Medieval period are not unlike those used in the pre-Roman and Roman periods except that the ferritic knife whether phosphorus-containing or not, has virtually disappeared. The sandwich structures are now better developed and there are at least four main types, with the Type 1 having many sub-varieties. Comparison with Kolchin's work on the material from Medieval Novgorod[47,48] (Table 10) and other Russian

Table 8

Comparison of metallurgical levels achieved on various Romano-British sites

| Site | Numbers of artifacts | | |
| --- | --- | --- | --- |
| | Total | Carburised | Hardened |
| Gestingthorpe | 31 | 11 | 2 |
| Thistleton R.V. | 13 | 3 | 0 |
| Ware | 12 | 5 | 2 |
| Catsgore | 11 | 6 | 1 |
| Colchester | 12 | 2 | 0 |
| Brancaster | 8 | 5 | 0 |
| Wanborough | 25 | 19 | 8 |

Table 9. Metallography of Medieval Iron Artifacts from Greencastle.

(After Scott[33])

| No. | Date | Edge Structure | HV | Back/Core Structure | HV | Remarks Type |
|---|---|---|---|---|---|---|
| 13 | Late 13th C. | Mart. | 351 | P+F | 116 | 2 |
| 3b | " | | 141 | P+F | 173 | 1 |
| 14 | " | F+P+ | - | F+P | 107 | 1 |
| 39a | " | M+T | 421 | Fe+N | 97 | 2 |
| 39b | " | M | 362 | F+P | 119 | 2 |
| 57/42 | " | F+P+ | - | F+P | 114 | 3 |
| E44/67 | " | P+Fe+C | 152 | F+P | 118 | Carb? |
| 70 | " | M+T | - | Fe | - | 4 |
| GC/C5 | " | M | 468 | Fe | 100-127 | 2 |

Out of 9 knives, the edges of 5 were found to be made of heat-treated carbon steel with hardnesses greater than 300 HV.

Table 10.

Russian Knives (from Kolchin)[47]

| Date Century | TYPE 1 | 2 | 3 | 4 |
|---|---|---|---|---|
| 9th-10th | 8 | 2 | 0 | 0 |
| 10th | 5 | 1 | 0 | 0 |
| 10th-11th | 2 | 3 | 0 | 2 |
| 10th-12th | 1 | 2 | 0 | 0 |
| 11th | 1 | 0 | 0 | 0 |
| 11th-12th | 0 | 6 | 0 | 2 |
| 12th | 1 | 0 | 0 | 0 |
| 12th-13th | 1 | 2 | 0 | 0 |
| 13th | 1 | 1 | 0 | 0 |
| 13th-14th | 0 | 1 | 0 | 0 |

sites shows that Type 1 was an early type there also but it did not get replaced by Type 2 to the same extent as at Winchester, for example. Nevertheless, the similarity between the material from Russia, and Poland about which we know so much from the work by Piaskowski, shows a remarkable uniformity of technique across Europe. Given this uniformity across so wide an area it would seem that we are not going to be able to discern the work of individual smiths unless the blades contain makers' marks which can be related to individual smiths like the sword-makers.

Scissors were not introduced until the Byzantine period so they do not appear in the Roman or pre-Roman periods in Britain. In fact in Britain we have none from Winchester or any other of the well-known Medieval sites, and those from Salisbury could be anywhere in the period 14th to mid-19th centuries. Those from Chingley are more closely dated to the 16th - 17th centuries. Before the Medieval period shears were the main shearing tool and their use continued for certain purposes until recent times.

The ferritic blade from Roman Silchester is of interest not only because it shows the ultimate in the use of high phosphorus iron in the Roman period, but because it shows the presence of Neumann lamellae (deformation twins) which prove that the blade was severely cold-worked to get the high hardness of 282 - 290 HV.

Axes are not as common in the Iron Age as they were in the Bronze Age. Perhaps this is because they lasted longer and when no longer fit for further work were re-used by the smith and were lost to the archaeological record. The vast majority of Bronze Age axes are late and no doubt were in process of being re-used when the Iron Age supervened. In our work we have not looked at any from the pre-Roman Iron Age but reference in this respect should be made to the work of Scott[33] who examined several Irish axes of the common pre-Roman socketed type and found to be made from a single sheet or iron from which a flap projected and was used to form the cutting edge. The flap appears to have been intentionally carburised on one surface so that when folded it consisted of ferrite and pearlite with 0.4%C in the centre of the section, surrounded by ferrite at the surface. The carburised area was exposed after sharpening the edge.

Romano-British axes were of the shaft-hole type and judging by the pieces removed from the cutting edges were either Type 1 or homogeneous carbon steel. Hardnesses as high as 515 HV were achieved but lower hardnesses in the range 200-300 seem to be acceptable. Naturally these were not quench-hardened and agree with the axe from Silchester examined by Coghlan[52] which has a hardness in the range 118-210. It is debatable whether hardnesses as high as 515 were necessary for a tool used for cutting wood and a lot depends on how it was intended to be sharpened.[53] If this was to be done by cold-working on a field anvil then a low hardness would be quite effective. African axes today are not quench-hardened although many are made of steel of European origin. A Sudanese axe was made from a piece of high carbon steel. The edge showed a very fine pearlite-ferrite structure with a hardness of 360-400 HV. The area around the shaft hole was 309. A Nigerian socketed axe was made from a 0.5% carbon steel again with pearlite and grain-boundary ferrite. This had a hardness of 257-274 on the edge and 178 on the socket.

It is therefore a surprise to find that the British medieval and post-medieval axes have been quench-hardened to give a hardness in the range 390 to 550 HV. Three out of four have been made in Type 1 with a piece of steel carefully placed between the piece of iron that was wrapped round a mandrel to form the shaft hole. The Kempsford axe was an application of the Type 2 weld, but was very carefully made in spite of the use of an easier and cheaper way of making a steel edge. Comparison with continental axes show much the same technique. Salin[54] gives us an axe of the 5th century from Trémont on the Meuse which has a carbon steel insert of Type 1 giving a maximum hardness of 437 HV which has been obtained by quenching and tempering. Haithabu in Viking Schlesvig has produced axes of piled steel of varying carbon content so arranged that the highest carbon content was in the centre.[55] But this was not quench-hardened and the hardness was in the range 121 to 199; not a very efficient use of a carbon steel.

The distinguishing feature of the scythe is the reinforcement at the back which is necessary to support such a long blade. Since this implement needs no greater hardness than a shear blade it is not surprising that the late Roman and 17th century blades are much the same, being of low hardness (119-220). This contrasts with a modern blade of homogeneous carbon steel which has an edge hardness of 525. But in the 19th century a good scythe - a Crown scythe - was made by welding in a piece of carbon steel between two pieces of wrought iron, i.e. Type 1 blade[3] (Fig. 43). This was forged down to give the reinforcement or 'set' shown in the figure. The blade would be quenched and tempered which would harden the carbon steel only and give a good impact-resistant blade. Such blades were made in England until the 1960s.

Sickles are not such a great problem and, as Table D shows, they were made in all types giving hardnesses in the range 87-370 HV. Billhooks were probably made by knife-smiths using the same techniques.

Wood-cutting chisels and gouges do not need to be very hard but many have been made using the Type 1 technique with a piece of steel welded onto one side. Some consist of homogeneous carbon steel, and hardnesses in the range 400 to 710 are obtained.

As Dr. Rees has shown,[56] the Roman plough is a complicated piece of equipment consisting of a coulter to cut the sod, a share to undermine it and a mould board to turn it. Unfortunately we have not examined the coulter which one would expect to be the hardest of the three. But a Roman coulter was found at Stanton Low and examined by the United Steel Cos Ltd. in 1960. Clearly one part of this was made of wrought iron, with 0.08% Ni and 0.01% Cu and some slag. The structure was that of ferrite but only one piece was examined and it is possible that a piece of steel was welded to it. Pleiner has examined European coulters and there is strong evidence for steel.[42,57] As for shares, we know that these were originally of wood and are not surprised to find that the two Roman ones from Essex are merely wrought iron with hardnesses in the range 109-230. The spade shoe from medieval Winchester seems to have been made by carburising but the hardness is only 210.

A Roman mason's chisel from Chesterholm was examined by Smythe.[58] While it contained some martensite the carbon content of the tip was variable and the hardness varied from 464-579 HV. On the whole this is fairly typical of the better ones that we have examined such as those from Gestingthorpe and Wanborough. Viking period chisels from Helgö in Sweden

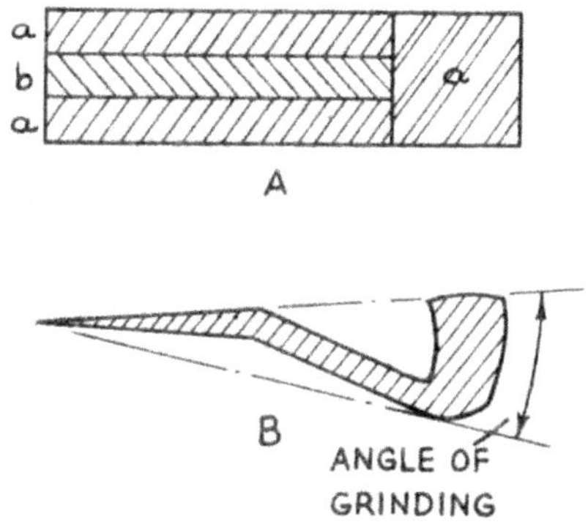


*Making a crown scythe.*
*A.—the unforged blade (a) wrought iron or mild steel plates sandwich the cast tool steel plate (b).*
*B.—Section through the forged blade showing the 'set'.*

Fig. 43. Detail of crown scythe (After ref. 3).

are not very different[59] (Table 11).

It is a pity that none of the files examined are stratified but we do know that fully martensitic files were being used in Europe in the Roman period[60] and that microhardnesses of 700–1000 HV were possible. The file from Wanborough with a hardness of only 535 HV1 would not be a very efficient tool. But the hardness of those from Winchester would be well above the Roman Bavarian tool if it had been measured on a microhardness scale.

Two other files have been found in the British Isles. One comes from the pre-Roman Iron Age site of Gussage All Saints and has not been metallographically examined. It is 162 mm long and has 10–12 cuts/cm on one of its faces. The other is also from the pre-Roman Iron Age and comes from Fiskerton, Lincs.[61] Unfortunately it has been almost completely corroded and it now consists of a martensitic or cementitic envelope from which the contents have been removed, presumably because they were anodic to the casing.

The results of the Roman saw-teeth show that good saws could be made in the Roman period. But one from Oxyrhynchus had a hardness of only 127–140 and as Coghlan[52] noted 'would not remain sharp for very long when cutting even soft wood'.

Pleiner has examined at least three Roman spoon augers.[42] As would be expected from British evidence none of the bits were quench-hardened but they were made of steel which in one case was hypereutectoid (more than 0.8% C) with microhardness reaching 800 HV. But the fact that the Letchworth auger was not only welded but quench-hardened argues for a post-Roman date.

Table 11

Summary of results obtained by Modin and Pleiner[59] on the metallography of edge tools from Helgö V (1), 1978)
Viking period.

| | | |
|---|---|---|
| Knife | | Type 1, sandwich. 200-1400 HV; Martensite with As lines. |
| Knife | | Wholly martensitic; Type 3; 600 HV piled; homogenous with As lines. |
| Arrowhead | 1 | 0.15-0.2% C; HV = 200-210; ferrite with g.b. pearlite. |
| " | 2 | F+P and phos. sorbitic. Possibly carburised. HV 220-535. |
| " | 3 | Ferrite and F+P; about 0.2% C; HV 180-400. |
| " | 4 | wrought iron mostly; F+P; HV 200-600. |
| Chisel | 1 | Piled Fe and slag with welds; Fe + pearlite; HV 200-250. pearlite sph. not heat treated. |
| " | 2 | Type 1, carburized at end and then sandwiched between Fe; (HV 250-270). Tip sorbite; HV 400-500. |
| " | 3 | homogenous martensitic; HV 600-900; As lines (400 HV). |
| " | 4 | " " HV 600-800. |
| " | 5 | Martensite + troostite; 0.4% C, HV 300-700. |
| Punch | | Pearlite and ferrite, 0.6% C; HV 180-400. |

Hardness HV; Load = 30 g.

## Part 2

# 6. Results of the Examination of Edged Weapons.

## 6.1 (1) Arrowheads and Spearheads.

Two types of arrowheads predominate - a small broad head and a thin dagger type - the long bodkin (Jones).[62] Effectiveness is measured by the delivery of the maximum amount of energy on the smallest possible area. Mail armour is easily penetrated but plate armour 3 mm thick is protective. It takes a penetration of 11 mm to the thorax to cause fatal wounding through 2 mm plate. By 1405 it was recognised that arrowheads needed accurate heat-treatment and hardnesses of 400 HV were attained; but before 1480 few arrowheads exceeded the range 120-220 HV. Eventually, apart from helmets, plate armour was found to be useless against the later firearms and was discarded.

For civil purposes, i.e. the penetration of flesh, iron is all that is required for the sort of game likely to be met within Britain. If edges lose their hardness due to contact with bone or the ground they can be repaired and sharpened. Spearheads and arrows for military use on the other hand, require a harder edge and it would seem that this is what we have in the case of the medieval arrow and the Anglo-Saxon spearheads. Wrought bloomery iron has a hardness in the range 100-150 HV, while a carburised ferrite + pearlite structure would have hardnesses in the range of 200-300 HV. It would appear that the former was adequate for the penetration of shields but that harder edges would be needed for certain types of body armour.

Only three arrowheads were examined. One from the Romano-British site of Wanborough which was sectioned on the shank just below the head and was found to have been made of a tempered martensitic steel with a hardness of 390 HV. It would seem that the steel head had been welded to a ferritic shank. Another Roman arrowhead came from Brancaster. This was entirely ferrite, but with small areas of carbide reaching about 0.2%C and having a hardness of 156 HV. The third arrowhead was from Lambourn, Berks, and was dated to the 13th century.[63] This consisted of steel barbs welded to a ferritic core with a hardness of 128 HV. An attempt had been made to harden the barbs but, as they only contained low carbon the hardness of the tempered martensite was only 240 HV. (Fig. 44). (Table L).

Spearheads from the Lunt, Baginton, Coventry. These came from a Roman fort dated to the period 60-76 AD. A piece 11 mm long was removed from the blade of that shown in Fig. 46 (1) and consisted almost entirely of coarse ferrite and slag with a small amount of intergranular carbide at one end. The hardness was 123 HV1 which suggests that the ferrite has a relatively low phosphorus content.

A piece 7 mm long was removed from the blade of the second spearhead from this site. (Fig. 46 (2)), and consisted again of coarse ferrite with some slag but with needles of carbide or nitride. There are also small areas of fine ferrite + pearlite indicating a carbon content of about 0.1-0.2%. At one end of the specimen the carbon content rose to 0.4% with fine irresolvable pearlite possibly bainite. The hardness of the ferrite was 175 HV1 suggesting an appreciable phosphorus content.

Table L. Arrowheads and Spearheads.

| Object | Site | Site/Lab Ref.No. | Date AD. | Structure/Hardness HV Edge | Back/Core | Type | Ref. |
|---|---|---|---|---|---|---|---|
| Arrowhead | Wanborough | 790279 | R-B | TM/390 | - | - | |
| " | Brancaster | 773685 | R-B | F/156 | - | 0 | |
| " | Lambourn | S 466(1) | 13th C. | TM/240 | F/123 | - | Coghlan[63] |
| Spearhead | Thames, Kempsford | 790311 S 38 | c 600BC | F+P/149 | F/112 | 0 | |
| " | Wanborough | 790277 | R-B | F/105 | - | - | |
| " | " | 771769 | R-B | F+P/168 | - | 0 | |
| " | Gestingthorpe | 8/10 | R-B | F/83 | - | 0 | |
| " | " " | 92 | R-B | F/133 | - | 0 | |
| " | " " | 95/98 | R-B | F/99 | - | 0 | |
| " | Brancaster | 773642 | R-B | F/140 | - | 0 | |
| " | The Lunt, Coventry | (1) | 60-76AD | F/123 | - | 0 | Hobley[39] |
| " | " | (2) | " | F/175 | - | 0 | " |
| " | Thames, Kempsford | 790309 (S 35) | 6-11th C. | F+P/172 - F/116 | | 0 | |
| " | " " | S 36 | 10-11th C. | TM/496-272 | F+P/277-108 | PW | |
| " | " " | S 33 | 9-10th C. | P+F/269 | F+P/234-F/122 | PW | |
| " | Winchester | BS 3433 | 13th Cent. | P/276 | F/178 | 1 | |

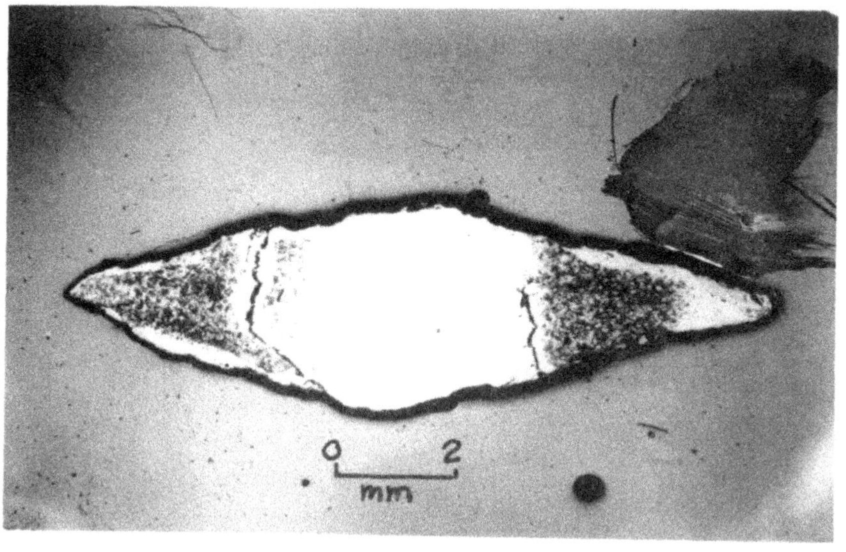


Fig. 44. Medieval arrowhead from Lambourn, Berks. After (Coghlan and Tylecote). (63).

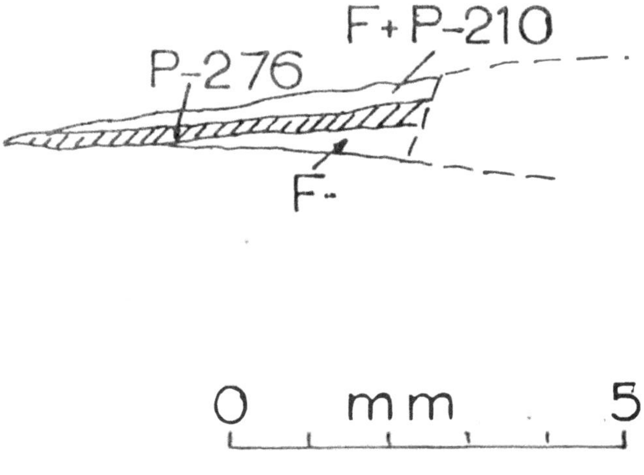


Fig. 45. Section through arrowhead from 14th century Winchester (BS 3433).

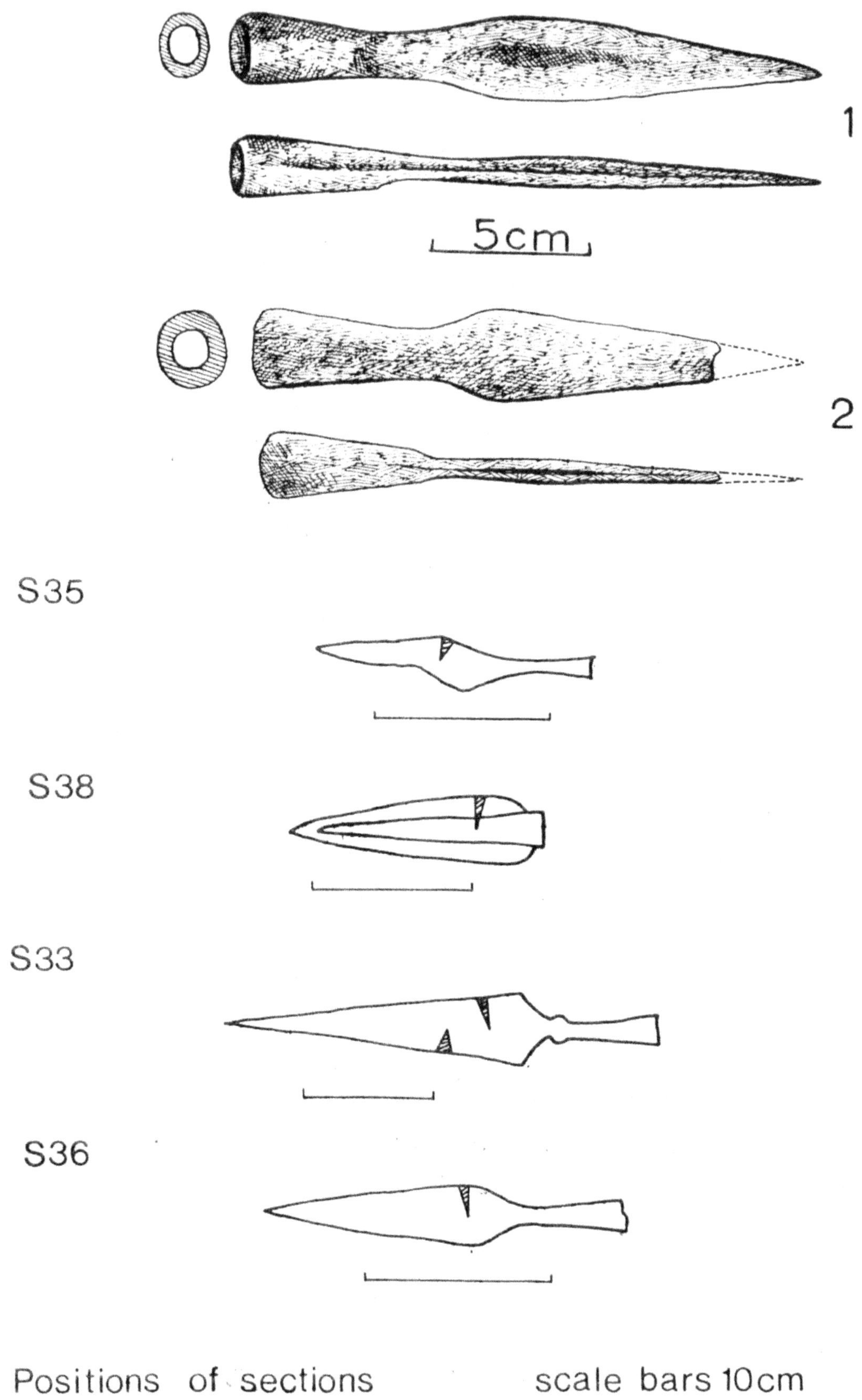


Fig. 46. Outlines of spearheads.

Two spearheads came from Wanborough but these were poor quality with low hardnesses. One was ferrite with a hardness of 105 HV, and the other ferrite + pearlite with a hardness of 168. Both were dated to the 4th century. Three came from Gestingthorpe and all consisted of ferrite with hardnesses in the range 83-133 HV. Another Romano-British example came from Brancaster. It was square-headed with a fluted tang. A segment was removed from one corner of the head, 4 cm from the tip. It was all ferrite with some slag and a little carbide in the grain boundaries giving a hardness of 140 HV.

Winchester has provided a 13th century socketed arrowhead or spearhead, BS 3433. This has been made of a sandwich construction with a high carbon core which outcrops at the edge. The hardness of this is 278 and it appears to be ferrite and pearlite. (Fig. 45).

S 35. A spearhead from the Thames (Table L), was found in the river Thames at the old ford at Kempsford in Gloucestershire, measured 15.9 cm long and 2.7 cm wide. The surface had mostly corroded away but corrosion was not deep, and the majority of the weapon survived with a good metallic core, and as the corrosion had been fairly even, the shape survived quite well. Its type and date are uncertain, but it probably falls in the period 6 - 11th century. In outline it resembled the Swanton series K, (Swanton, 1974, p. 23), although the profile was quite different, being convex in section (Fig. 46).

A narrow V-shaped section was cut from near the widest part extending to just over half way across the blade. Unetched, the section showed the slag content to be generally fairly low, mostly present as small flattened ribbons and spots with a few larger and one particularly massive irregularly shaped inclusion. The slag inclusion tended mostly to congregate along lines along the main axis of the section, giving a somewhat streaky or layered appearance.

When etched in nital (Fig. 47), a central line of slag spots became more prominent[64] being highlighted by a narrow but distinct line of pearlite which appeared superimposed upon it. This most probably marked the position of a weld dividing the blade into two halves along the main axis of the section. The macrograph in Fig. 47 shows this and the accompanying diagram shows a simplified version of macrograph. The two halves of the blade section, (a) and (b) in the diagram, showed up mainly as fairly even, grey pearlitic zones representing a carbon content of about 0.2-0.3%. In somewhat darker areas in the upper right and lower left parts of the macrograph the carbon content was rather higher, probably about 0.3-0.4%. Conversely in the upper left and lower right hand areas of the macrograph the pearlitic zones faded out altogether to leave localised areas of large grained ferrite. The pearlite occurred as fairly well spheroidised lamellae in a small-medium grain size matrix of ferrite.

The streaky effect created by the parallel broken lines of slag across the section suggested that the metal of both parts (a) and (b) may have been subjected to piling or a process of forging out, folding and forging out again until this kind of layered effect was produced.

The overall, fairly even, carbon distribution across the section may have been achieved by the carburisation of the two pieces (a) and (b) before these two pieces were welded together as part of the final forging of the

(a)

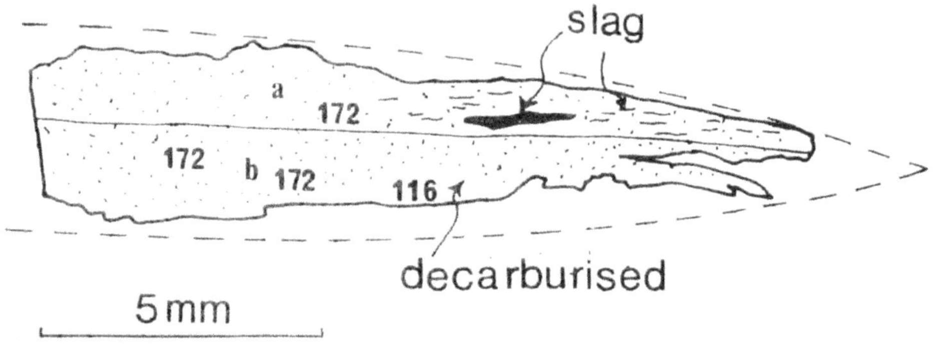


(b)

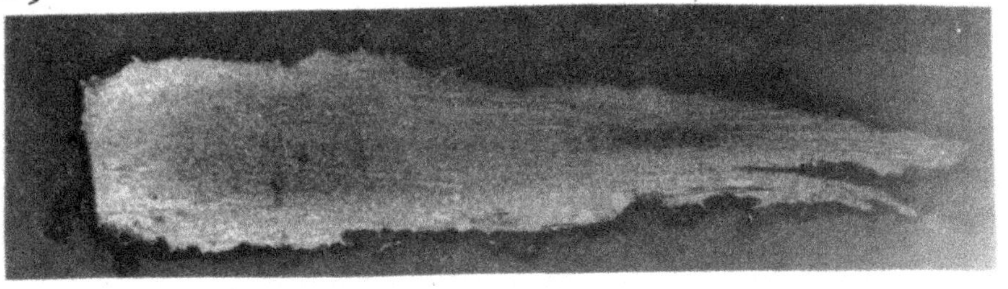

(c)

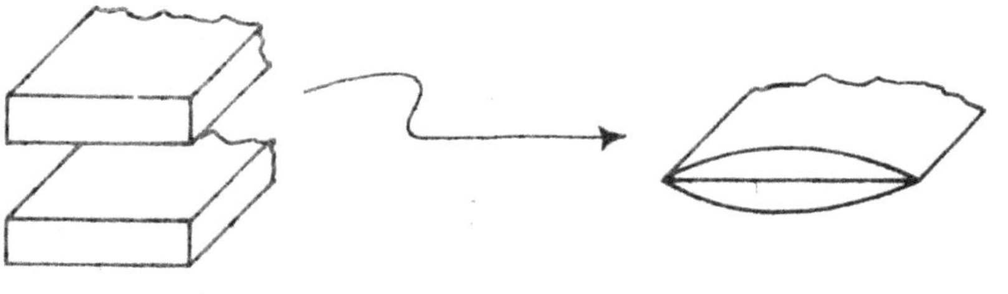

Fig. 47. Section through a small spearhead from Kempsford, Glos. No. 35.

spearhead. The two ferritic areas where the carbon content fades out altogether might then be explained as localised areas where carbon absorption did not take place during the carburising process or possibly just as areas which were decarburised during the final forging of the spearhead. The latter is probably less likely because a more general surface decarburisation might have been expected in this case. The third possibility is that (a) and (b) were at one time both parts of the same bar of steel which did include smaller areas of ferrite. This possibility too is probably less likely in this case because a process of folding and forging out several times would have tended to distribute the ferrite areas more evenly or at least flatten them out more.

Four hardness readings were taken, one in part (a) where the carbon content was at its highest, approximately 0.3-0.4% carbon, and three readings from another high carbon area to be decarburised area where only ferrite was visible. The results were as follows; part (a) near the centre, 172 HV; part (b) near the centre 172 HV and (b) nearer the outside of the blade where the carbon content was somewhat lower, 172 HV; part (b) near the tip of the cutting edge where only large grained ferrite was visible, 116 HV.

The great similarity between parts (a) and (b) does suggest strongly that they were at one stage part of the same bar of metal that was either cut or folded then forged out to produce the spearhead. The reconstruction diagram of Fig. 47 illustrates this. The overall quality was quite good, the weapon mostly consisting of a low carbon steel of a fairly low slag content. The spheroidised nature of the pearlite of the steel suggests that the spearhead was subjected to fairly prolonged heating at a little below the lower critical point, possibly about 650 - 700°C; long enough for the pearlite lamellae to partially ball-up or spheroidise. This procedure of subcritical annealing may have been done deliberately to toughen the metal after final forging of the spearhead.

S 38, Another socketed spearhead (Table L) found in the River Thames at the old ford at Kempsford, Gloucestershire. It measured 15.7 cm in length by 3.9 cm in width at the widest part of the wings of the blade (Fig. 46). The holes to locate an iron securing peg lay on opposite sides of the socket near the mouth and the corroded remains of the peg survived in situ. The spearhead was very similar in shape and size to Late Bronze Age socketed spearheads and therefore may be an early Iron Age copy of a Late Bronze Age type of spearhead i.e. a transitional Bronze Age/Iron Age type, possibly of about 600 B.C. The full extent of corrosion was difficult to gauge but it appeared to be mainly confined to fairly heavy surface pitting. A section extending to just over half-way across the socket was cut from near the widest winged part of the spearhead. When the section was viewed unetched both sides of the socket showed a fairly low slag content with the shorter surviving part containing proportionally more than the other. In each case the slag was present as small spots and ribbons which tended to be segregated into lines which gave a flow pattern to the metal of the socket, on either side, which was aligned along each arm of the section and joined along a line which coincided with the main axis of the section from the centre of the weapon to the tip of the wing or flange at the edge.

When etched with nital this flow pattern effect was emphasised (Fig. 48). It could now been seen that the section consisted of two halves, (a) and (b) (Fig. 48) which joined at a weld line along the main axis of the

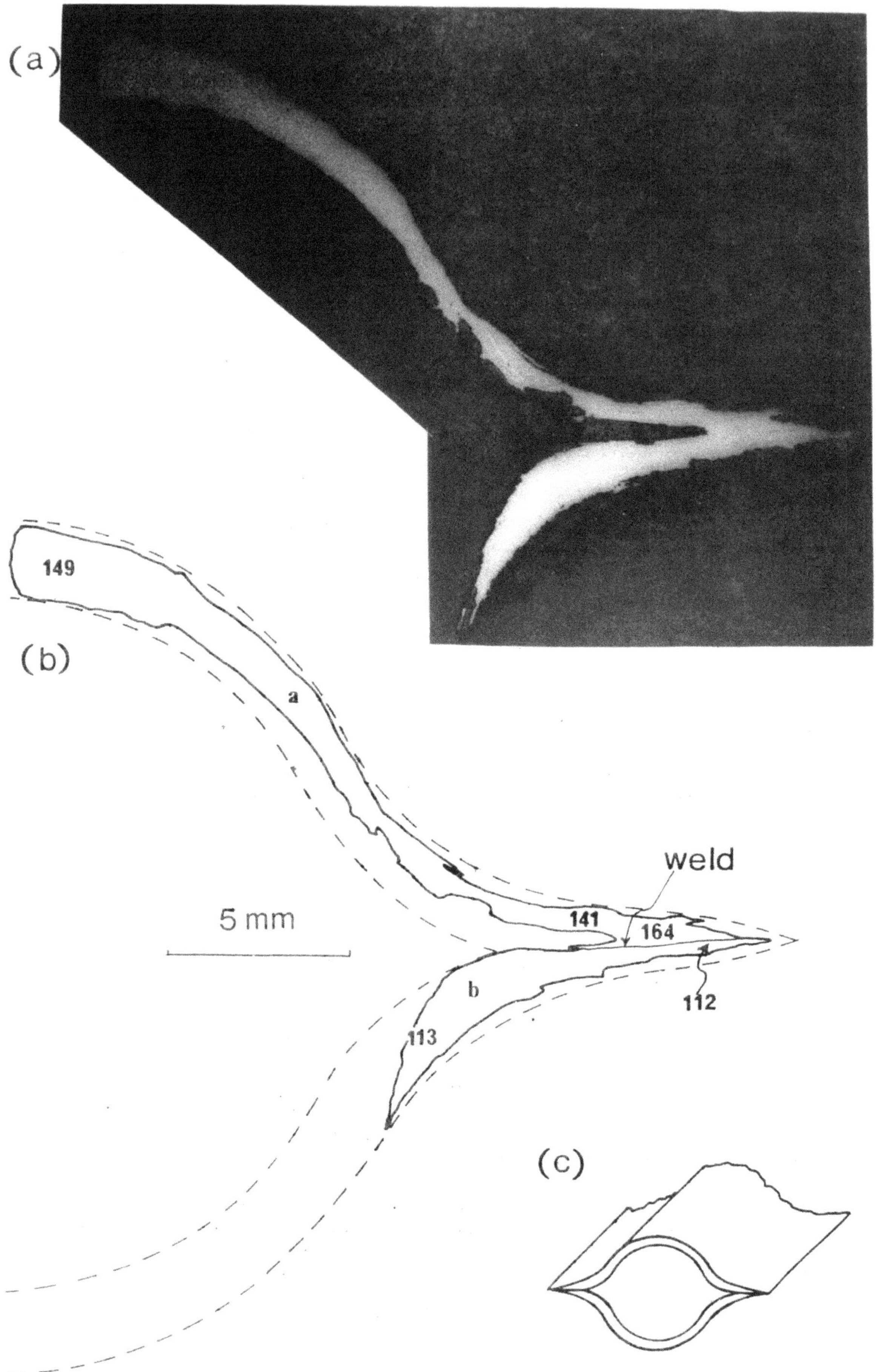


Fig. 48. Section through remains of socketed spearhead from Kempsford, No. 38.

section. This weld line was marked by a white line that coincided in places with a broken line of flattened slag ribbons.

Area (a) proved to be mainly a low carbon steel of about 0.2% carbon which consisted of a fairly even distribution of fairly small grained lamellar pearlite over most of the socketed part, while over the wing or edge part the pearlite became rather banded, interspersed with narrow bands of larger grained ferrite. The carbon content of (a) was less, approximately 0.1% in this wing part. Area (b) consisted of alternate narrow bands of large grained ferrite and small grained lamellar pearlite resulting in an overall carbon content of (b) of about 0.1%. Short rod-like nitride or carbide needles were also visible in places in the ferrite bands particularly along the innermost band, i.e. next to the socket void, of area (b).

Hardness values of 141 and 149 HV were produced from either end (in section) of the fairly low carbon iron, part (a). Corresponding tests from either end of the very low carbon iron, part (b) gave hardness values of 112 and 113 HV. The ferrite line marking the weld between (a) and (b) gave a hardness value of 164 HV. The white line may represent a band of localised arsenic enrichment at the weld junction.

The low slag content of the two halves of the blade showed that the initial forging work on the bloom after smelting was done well and the banded nature of the slag present may also reflect that the metal was forged out, folded, and forged out again (piled) several times before the final assembly and forging of the spearhead. If this is correct then the alternate ferrite and pearlite bands observed in areas (a) and (b) may indicate that surface carburisation may also have taken place at each stage and the rather higher carbon (0.2%) and more homogeneous part of area (a) might have been the result of variable reducing/oxidising conditions in different parts of the smith's hearth.

S 36 Another fairly small spearhead (Table L), came from the same site as the last. The surface had been extensively, but quite evenly, pitted by corrosion, although the metallic core survived quite well. The surface corrosion meant that the original profile was difficult to gauge but it appeared to be roughly convex with some tendency towards a central rib. (Its dimensions were 20 cm long by 3.2 cm wide (max) (Fig. 46)). Its outline, size and profile were similar to Petersen Type G spearheads[65] although the socket was proportionately longer in this case. (Petersen, 1919). It would appear to belong to the Late-Saxon period, possibly to the 10th or 11th centuries.

A transverse V-shaped section extending to just over half-way across the section was cut from the wide part of the spearhead in such a way that one edge (the one to be examined) was cut obliquely to the edge of the weapon but at right angles to what appeared to be the main weld lines visible on the radiograph.

When viewed unetched the slag content was seen to be quite low. Several broken lines of mostly small slag ribbons and spots could be seen running across the section from side to side suggesting that there were welds present running in this direction dividing the section possibly into several parts although it was not possible to determine how many at this stage. An irregular patch of partially flattened slag ribbons and spots, aligned this time along the main axis of the section, occurred centrally near the tip of the cutting edge.

When etched with nital the structure became much clearer as can been seen in the macrograph of Fig. 49. The diagram of Fig. 49 shows a simplified version of what is visible in the macrograph. It was possible to see that the section was divided into four main areas which are marked (a) - (d) in Fig. 49. A possible fifth area is also shown which divides what would be a much larger area (d) otherwise containing very little slag, into three parts - d, e and f as shown on the diagram and macrograph. Part l occupies the more or less exact centre of the spearhead and two rather discontinuous and irregular lines of slag inclusions, indicated as broken lines on Fig. 49, mark the probable position of welds between these central pieces, d, e and f. Traces of these welds could be seen on the radiograph.

The cutting edge part (a) gave a rather uneven structure. The area at the surviving tip of the cutting edge etched lightly and had a distinctly acicular martensitic appearance. Away from the tip this soon gave way to pale etching Widmanstätten-looking ferrite and pearlite, the pearlite fraction not being optically resolvable. This would appear to represent a quenched low carbon steel probably not more than 0.2-0.3% carbon, if that. Where the central patch of slag inclusions occurs, that particular area consisted almost entirely of ferrite (apart from the slag). The very dark etched zone which can be seen on the macrograph to occupy most of the inner area of (a) consisted of a central very dark acicular looking area of tempered martensite. This was mixed with a certain amount of unresolved pearlite, and away from this central area the structure gradually changed into a paler zone of pearlite with an increasing proportion of ferrite.

The carbon content of the dark etched area (a) appears to be much higher than at the tip of the cutting edge and may be nearer 0.4%-0.5%. Between the cutting edge tip and dark etched area the carbon content decreased to well under 0.1% where the central area of slag inclusions occurs. Although the appearance has been exaggerated somewhat by the dark etching effects of the heat treatment, this part (a) was clearly of an uneven composition. A varying carbon distribution like this is not to be expected from a piece of wrought iron that has been carburised so this piece (a) may have been a piece of specifically produced bloomery steel.

The weld line between cutting edge piece (a) and the first of the inner pieces (b) was very clearly marked by an almost continuous line of slag ribbons which was highlighted by a diffuse white line along its length. Part (b) appears to have been a twisted and possibly composite strip - the folded appearance being outlined in places by a few slag spots. The main area consisted of a fairly even fine grain low carbon distribution of pearlite in ferrite, the carbon content being probably mostly less than 0.1%. The smaller two areas, outlined by the slag spots, appeared to have a rather more patchy distribution of pearlite in ferrite and with a generally larger grain size and slightly higher carbon content, possibly about 0.1%. The weld boundary between (b) and (c) was less easy to see but was marked by a few slag stringers and spots and one deep corrosion pit. Part (c) showed a banded structure.

The bands consisted of a fairly fine-grained distribution of pearlite in ferrite very similar to the main area of part (b) and with a similar carbon content of about 0.1% or less.

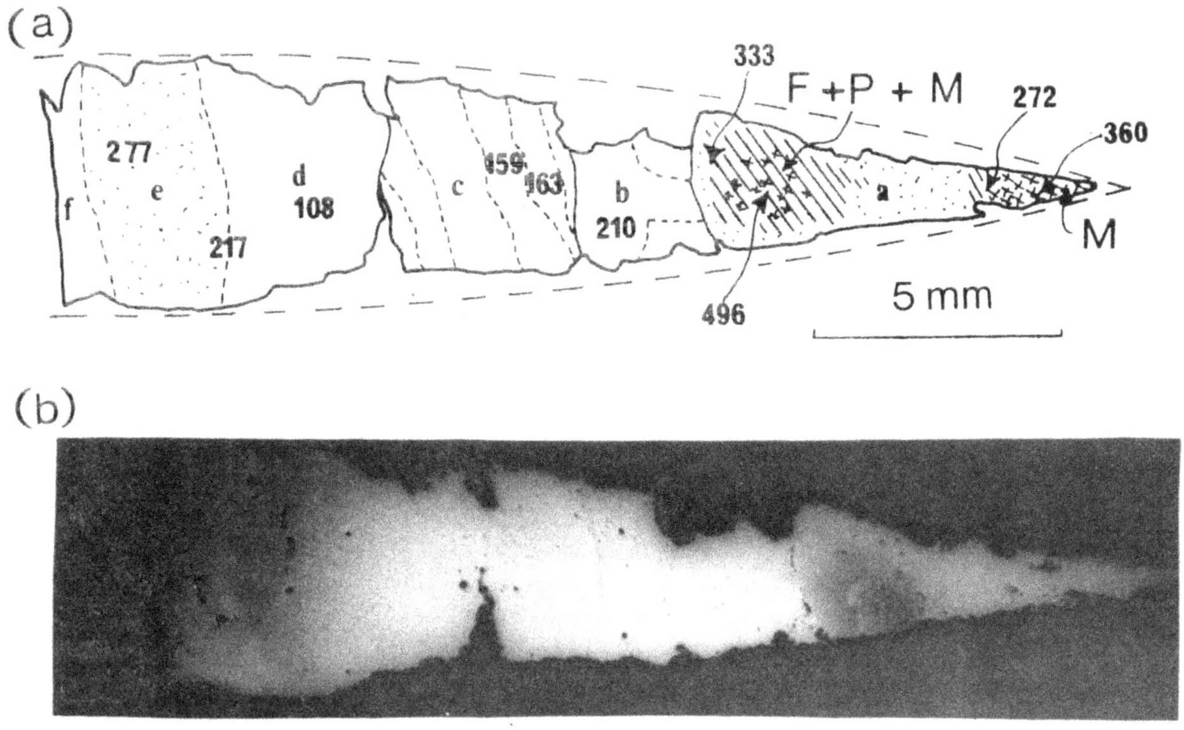


Fig. 49. Section through a spearhead from the Thames at Kempsford, No. 36.

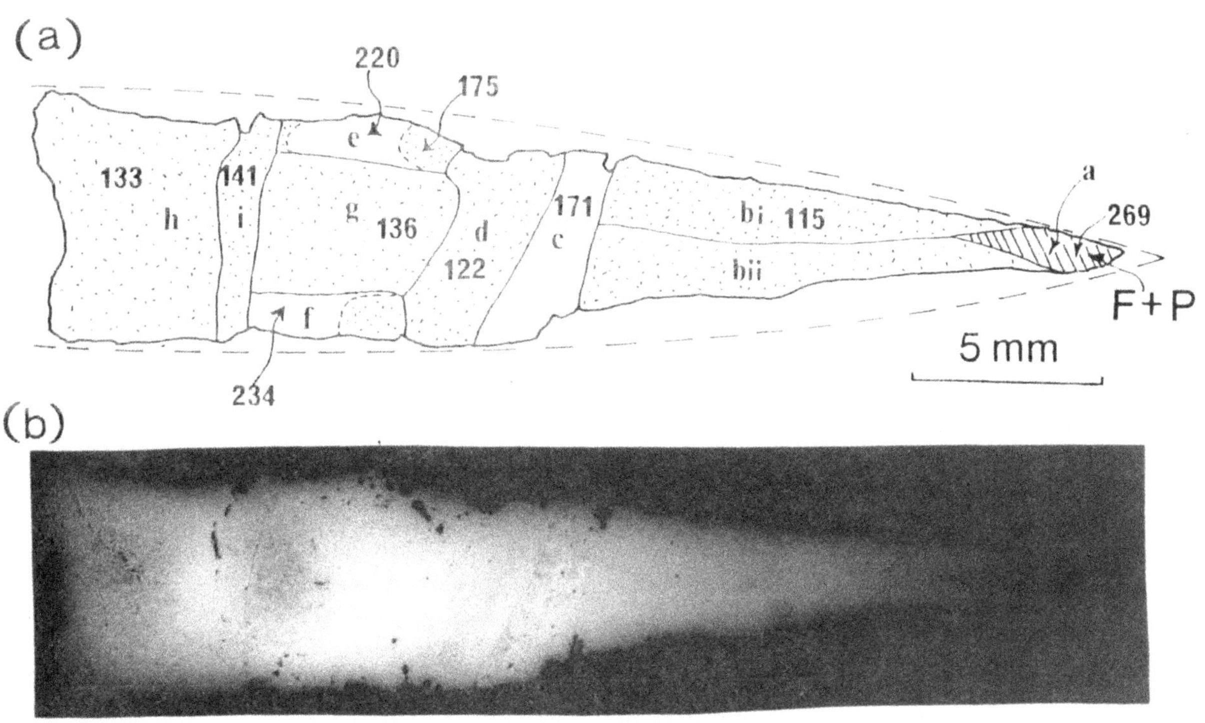


Fig. 50. Section through a pattern-welded spearhead from the Thames at Kempsford, No. 33

(c)

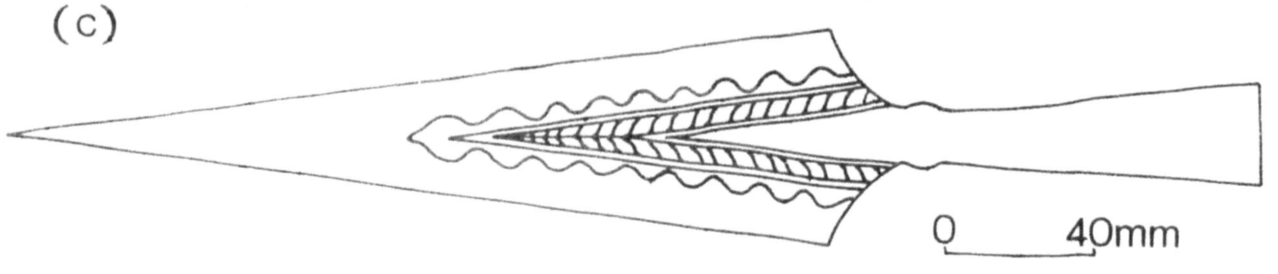


(d)

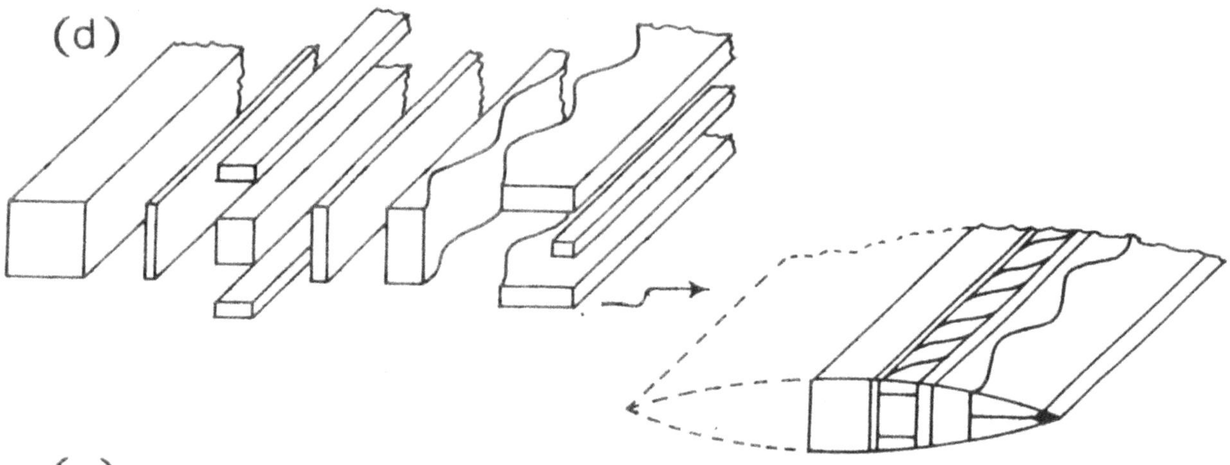

(e)

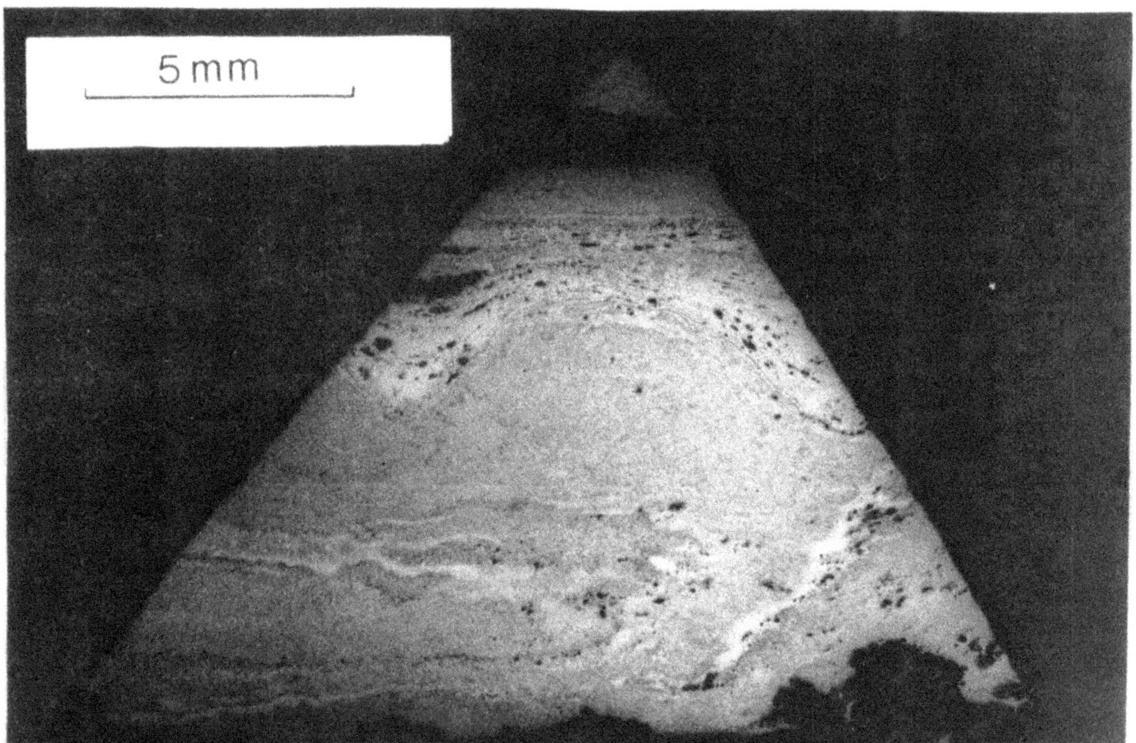


Fig. 50 (continued) Section from same spearhead to show 'wavy' weld + surface detail drawn from X-ray and three-dimensional view.

The boundary between the composite strip (c) and core piece (d) was fairly clearly marked by a few slag spots and ribbons and also by two very deep corrosion pits which have clearly followed the weld line between (c) and (d). As mentioned above this core part of the spearhead probably originally consisted of three parts, marked d, e and f on the diagram, Fig. 49a, separated by an irregular line of slag spots and ribbons. It would appear that these composite parts of (d) were forged together before the final assembly of the spearhead. Part (e) showed a fairly even distribution of fine grain pearlite and ferrite. The pearlite was mostly unresolvable (optically) amorphous looking, although somewhat granular with a few acicular patches possibly of martensite. The carbon content of the central part appeared to be about 0.3-0.4%.

Four hardness readings were taken on the edge part (a): firstly near the tip which showed a martensitic appearance, 360 HV; secondly a little further in where the martensitic area soon gave way to one of ferrite and pearlite (272 HV), thirdly further in again and in the centre of a very dark area tempered martensite (496 HV), and finally near the inner weld where a dense area of amorphous looking pearlite or grain boundary carbide of piece (b) gave 210 HV. One of the paler very low carbon bands of piece (c) gave 163 HV while an adjacent ferrite band with slightly more pearlite gave 159 HV. Of the central parts of the spearhead, the ferrite of the iron piece (d) away from the zone of carbon diffusion gave 108 HV although in the area heavily affected by carbon diffusion a hardness value of 217 HV was obtained. The small grain pearlite of central steel piece (e) gave a value of 277 HV.

There has been a great deal of carbon diffusion outwards into parts (d) and (f). The pearlite proportion gradually decreases away from (e) and the pearlite fades out altogether about half way across part (f) to give way to large grained ferrite.

It would appear that the core of the spearhead was originally made up as a sandwich of two pieces of wrought iron between which was forged a piece of medium carbon steel which may have contained about 0.4-0.5% C. This composite piece must then have been heated above the upper critical point - approximately 800°C for a fairly long time, possibly several hours in what may have been a deliberate attempt to produce a composite piece of a fairly homogeneous piece of steel.

<u>S 33. A spearhead found near the ford at Kempsford in Gloucestershire,</u> (Table L). Its wide blade (Fig. 46) closely resembled Petersen's Type H Viking spearheads (Petersen,[65] 1919) which would appear to date it to around 900 A.D. Its condition was fairly typical of river finds with fairly heavy surface corrosion pitting but with fairly good metal survival underneath. Length 33.1 cm by 5.6 cm at its widest point. Date approximately 9th to 10th cent. A.D.

A narrow wedge-shaped section was cut from the blade 4 cm down from the shoulder of one of the wings. A second section was cut from a point about 8 cm down from the shoulder on the other side of the spearheads. This was cut so as to get a surface view of the cutting edge and part of the pattern welded design. The transverse section unetched showed a fairly low slag content across the section except for a band in the centre which contained much slag, mostly small spots and ribbon-like inclusions of one or two darker phase constituents (probably fayalite and a glassy matrix) but with some of the larger inclusions showing a third phase (of wüstite) as a dendritic pattern.

When etched with nital the structure showed up clearly (Fig. 50) the section can be divided into 10 main parts, these being the different components of the spearhead. The weld lines were fairly clearly marked by 'white' lines which consisted of ferrite probably locally enriched by arsenic. These were superimposed by a few small two-phase slag inclusions of a pale and dark, probably wüstite and fayalite in a glassy matrix. Also superimposed were narrow linear segregations of pearlite.

The small tip part (a) appeared to be a strip of steel welded between the edge of the inner parts (bi) and (bii) which were of wrought iron. A fairly fine grain rather amorphous or granular looking distribution of pearlite with ferrite indicating a carbon content of about 0.5% over much of (a) but decreasing towards the welds where fairly extensive carbon diffusion had taken place into the adjacent ferrite of parts (bi) and (bii). The mainly large grain ferrite parts (bi) and (bii) each showed a slightly pearlite area next to the inner weld with part (c) indicating a slight degree of possibly accidental carburisation while the individual pieces were being forged. The two halves (bi) and (bii) were virtually identical even to the extent of partial carburisation suggesting that they were two halves of the same piece of wrought iron folded in half, welded together and forged out again during which time the steel edge piece (a) was welded in.

The weld line between edge parts (bi) and (bii) and the inner band (c) on X-ray gives rise to the sine wave pattern which forms the outer part of the pattern welded zone. The line is fairly clearly marked by small inclusions of slag and particularly by one large inclusion showing a dendritic pattern of wüstite against a darker matrix of fayalite or a glassy constituent which was trapped between the parts (b) and (c) during welding. The part (c) itself had a high slag content of varying inclusions mostly of a dark phase ? fayalite. Apart from its slag content part (c) consisted of mainly large grained ferrite.

The part (d) which had shown a fairly pronounced straight-grained pattern on X-ray showed up as a rather featureless band in section. It consisted of ferrite plus pearlite with a carbon content of between about 0.1 and 0.3%.

The central plain core piece (g) was not suspected from the X-ray and consisted of ferrite with a rather variable pearlite distribution very similar to part (d) giving a carbon content of about 0.1–0.3%. It would appear that parts (d) and (g) are both little more than low carbon iron; although patchy the carbon content is one of a low carbon steel. They are so similar that they probably came from the same piece of low carbon iron.

The surface parts (e) and (f) gave rise to the spiral or cork-screw like inner pattern in the welded zone visible on X-ray. They are very similar to each other and consisted of alternating bands of pale very large grained ferrite and darker grey small medium ferrite plus pearlite with a carbon content of 0.1%. These bands gave a gently folded appearance showing how a transverse view of the spiral structure looks. The spiral structure of parts (e) and (f) was achieved by the welding together and twisting of the paler and darker elements which show in the section.

The narrow band (i) shows up as a medium grain ferrite plus a little pearlite with a carbon content of no more than 0.1%, i.e. a low carbon iron.

The inner piece (h) showed a fairly even ferrite + pearlite distribution with a carbon content of about 0.2% but towards the edges this faded out suggesting that the piece may have been at one stage a fairly even piece of low carbon steel which became extensively carburised during the initial forging of the socket of the spearhead. This is uncertain and part (h) can probably be regarded as a piece of low carbon iron of a patchy composion.

The steel edge part (a) gave a hardness of 269 HV and the iron part (b), to which it had been welded gave a hardness of 115 HV. The pale high slag iron part (c) gave a hardness of 171 HV. The low carbon iron of parts (d) and (g) gave hardnesses of 122 and 136 HV. Hardnesses of 220 and 234 HV were obtained for the pale bands in each of the welded parts (e) and (f) and the low carbon iron of part (e) produced a hardness of 143 HV. Strip (i) gave 133 HV and 141 HV for the central piece (h) with its similar or slightly higher carbon content. The relatively high hardness values for the ferrite parts (c) and especially (e) and (f) are probably the result of a high phosphorus content.

The second etched section gave the structure shown in Fig. 50e. This shows how part of the surface of the weapon may have looked if it was polished and etched in a similar way. As shown in Fig. 50a parts (a) - (d) and a very small part of (e) or (f) are represented here. Most of the darker etched steel part (a) has corroded away but a small part survives. The wider edge part (b) shows an interesting 'watered' pattern a result of a variable but low carbon content with areas of paler larger grain ferrite which on forging out have given this very streaky appearance which is similar to that observable on some medieval and later Japanese sword blades. The mainly small grained clean and grey metal of part (h) contrasts sharply with the pale mainly large grain ferrite of (c) with its slag streaks, and the weld line giving the sine wave pattern shows up clearly. This pattern seems to have been achieved by the initial preparation of the edges of (b) and (c) that were destined to be welded together. These edges must have been given a corrugated profile during initial forging so that these two profiles would interlock and could then be forge welded together to give a single piece with parallel sides but which would then give this sine wave pattern on later surface preparation. The grey fairly small grained low carbon iron piece (d) occupies most of the rest of Fig. 50e and the dark band of corrosion runs roughly through the middle of it. A very small part of the inner pattern welded parts either (e) or (f) appears but it is too small an area to really show any detail. The method of construction is shown in Fig. 50 (c) and (d).

### (m) Scramasaxes (Seaxes) and Other Single Edged Weapons

A seax or scramasax is a single edged sword which was usually hilted in much the same way as a double-edged sword. Its section is more like that of a knife but it was usually decorated by pattern-welding or by grooves or lines and patterns of non-ferrous inlay or (occasionally) complex inscriptions. Scramasaxes are less common in Britain but were popular in Norway. Their length varies considerably so that some are no more than knives used as daggers.

S 50, Scramasax from Barham Down (Table M), (Maidstone Museum (no acc. No)), is a heavily corroded scramasax (Fig. 51), missing the tip of the blade and the end of the tang, apparently found in a grave at Barham Down in Kent in 1842 although there is no further record of this. It was covered in a heavy crust of iron corrosion products which have split into several large flakes. The approximate shape with its gently curving back and cutting edge could just be made out and this seemed similar to other examples from Anglo-Saxon graves and looks like Wheeler's Type 1, or 'Frankish type' of 6th-8th century date (Wheeler,[66] 1927, p. 178). It seems most likely that this blade came from an Early Anglo-Saxon burial and that it is roughly of a 5th-7th century A.D. date.

Two wedge shaped sections were cut from the blade, one from the back and one from the cutting edge. The sections were cut from fairly near the middle of the blade but were staggered a few centimetres apart so as not to weaken the metallic core too much but so as still to achieve an overall complete transverse section.

When viewed unetched much slag was visible in both sections. Some very large irregular three phase inclusions occurred in both sections and in these the central phase was a pale grey consituent (? wustite) in a dendritic formation in a medium and darker grey matrix (? a glassy constituent and fayalite). Some two phase, pale grey and darker grey inclusions were also visible. Most of the inclusions were smaller, ribbon-like and of one dark (? fayalite) phase or two phases, medium and darker grey (? fayalite and a glassy constituent). The slag content was much lower along a narrow margin on one side of the 'back' section and also nearer the tip of the 'cutting edge' section.

When etched with nital the areas of much lower slag content the margin of the back section and the area towards the tip of the cutting edge section, both rapidly went very dark (Fig. 52) and tended to stain blue. These darker etched areas (b) and (d) in Fig. 52 consisted of fine-grained pearlite with some ferrite. The pearlite was of a fairly well spheroidised, but still partly lamellar structure, and the overall carbon content appeared to vary between about 0.5 and 0.8%. The remaining areas of the section, those with the much higher slag content, etched slowly and are the pale areas (a) and (c), Fig. 52, in the back and cutting edge sections respectively. These consisted mainly of very large grained ferrite with a few patches of short rod-like nitride or carbide needles also visible.

There were no specific lines of slag inclusions which could be easily distinguished from the rest as representing weld lines between areas (a) and (b), (c) and (d) respectively, although in both cases it is quite clear that a piece of fairly high carbon steel, (c) and (d), has been welded to a piece of wrought iron, (a) and (c), of fairly low grade to judge from the high slag content. Fairly extensive carbon diffusion was also visible across the approximate weld positions.

The steel part of both sections gave in each case a hardness value of 248 HV. The very large grain ferrite core gave a hardness of 171 HV in both sections. The latter value is rather high for plain wrought iron, this and the very large grain size probably indicating the presence of phosphorus in this core piece.129

It seems most probable, despite corrosion having removed so much, that

Table M. Scramasaxes.

| No. | Site | Ref.No. | Date AD. | Structure/Hardness HV Edge | Back/Core | Type |
|---|---|---|---|---|---|---|
| S 50 | Barham Down, Kent | Maidstone Museum | 6-8th C. | P+F/248 | F/171 | 4 |
| S 42 | Leyton | 0.2275 | 8-10th C. | M/831 | F/157 | 1 |
| S 37 | R. Thames, Kempsford Glos. | Swindon | 9-10th C. | M/737 | F+P/236-174 | 1 |
| S 22 | R. Thames, Reading | 52.80 Reading | 9-11th C. | F/152 | F/184 | PW |
| S 9 | R. Thames, Hampton | A.27086 | 10-11th C. | | F/98-75 | |
| S 34 | R. Thames, Kempsford, Glos. | Swindon 790308 | 10-11th C. | P+F/204 | F/165 | PW |
| S 4 | Dorset | Dorset Mus. | 9-10th C. | TM/775 | F/152 | PW |

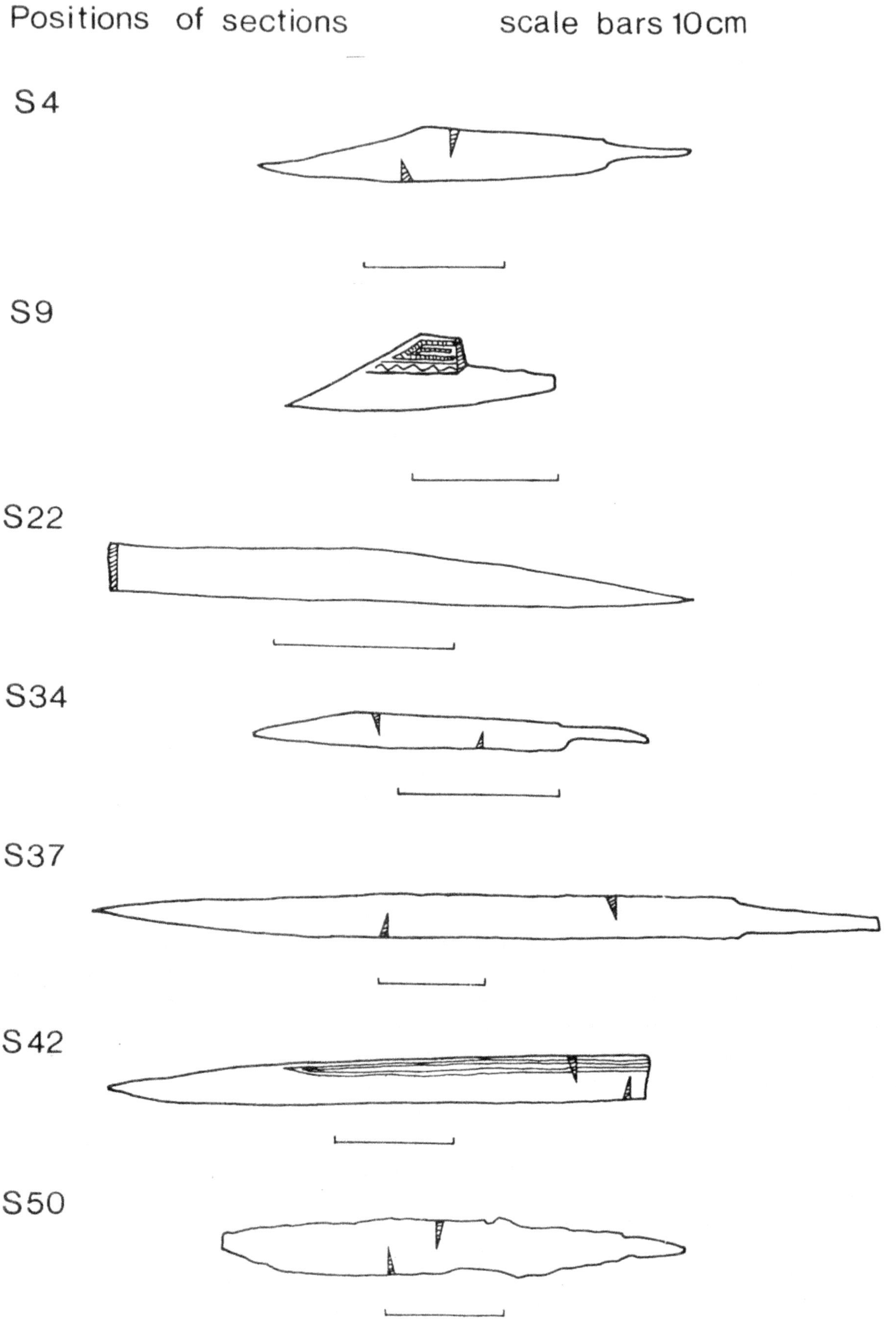


Fig. 51. Outlines of scramasaxes showing positions of sections.

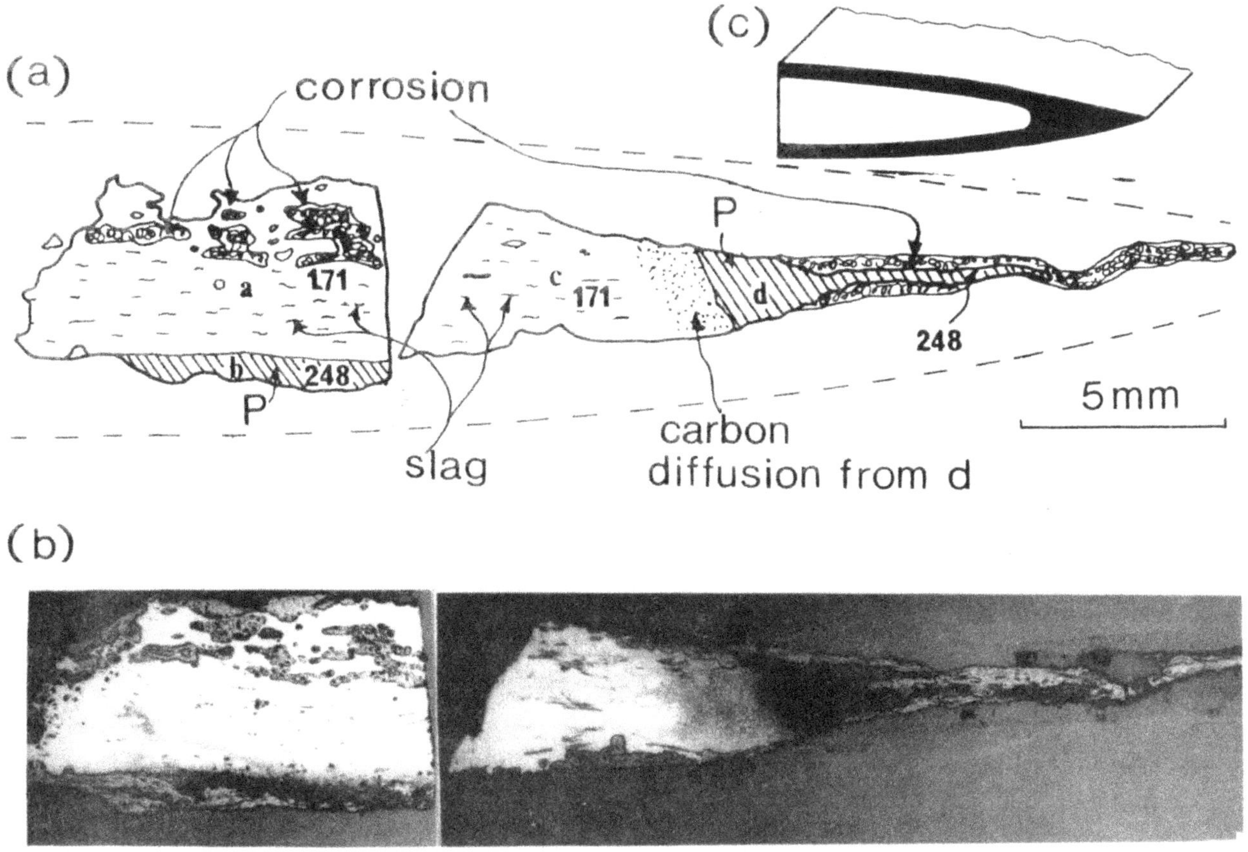


Fig. 52. Section through scramasax No. 50, from Barham Down, Kent.

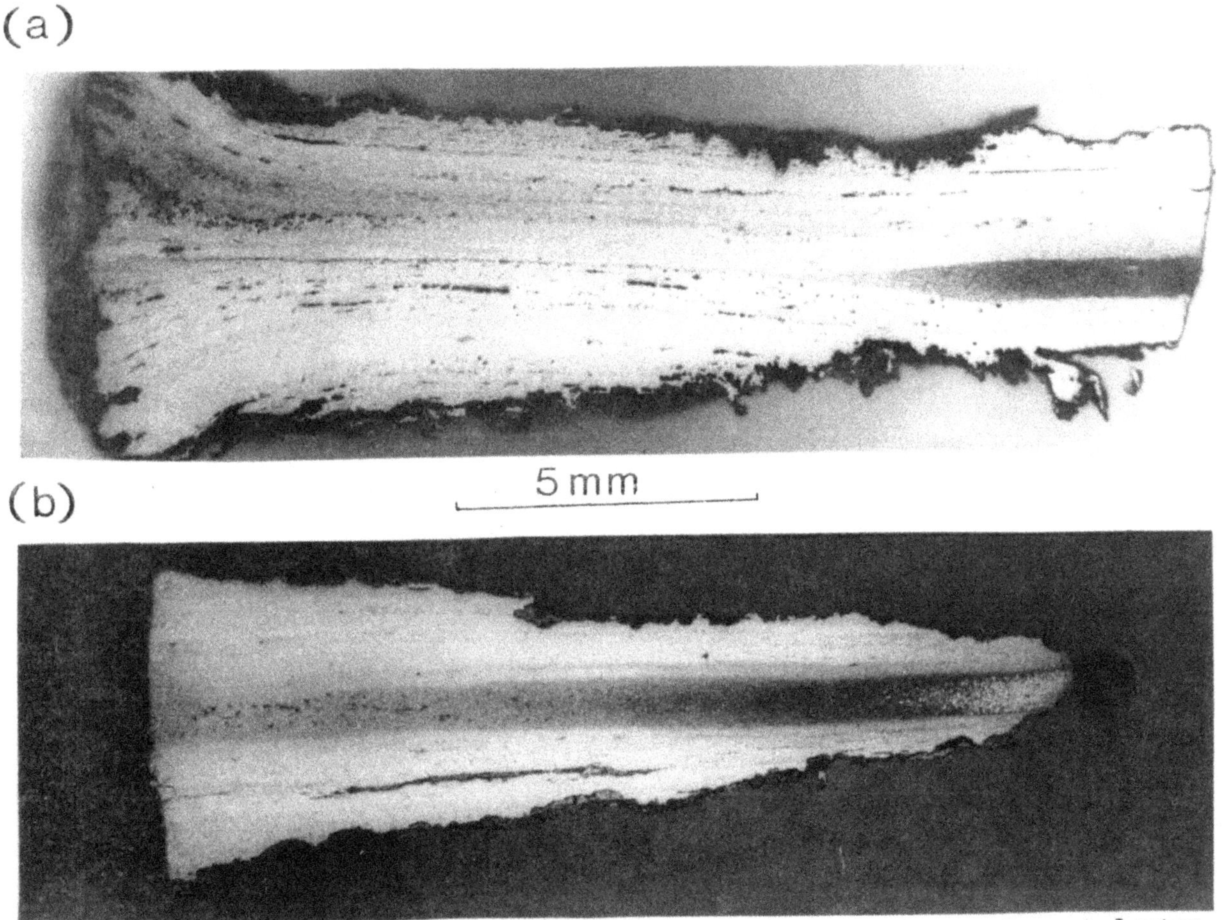


Fig. 53. Section through scramasax No. 42 from the River Lea at Leyton, Essex.

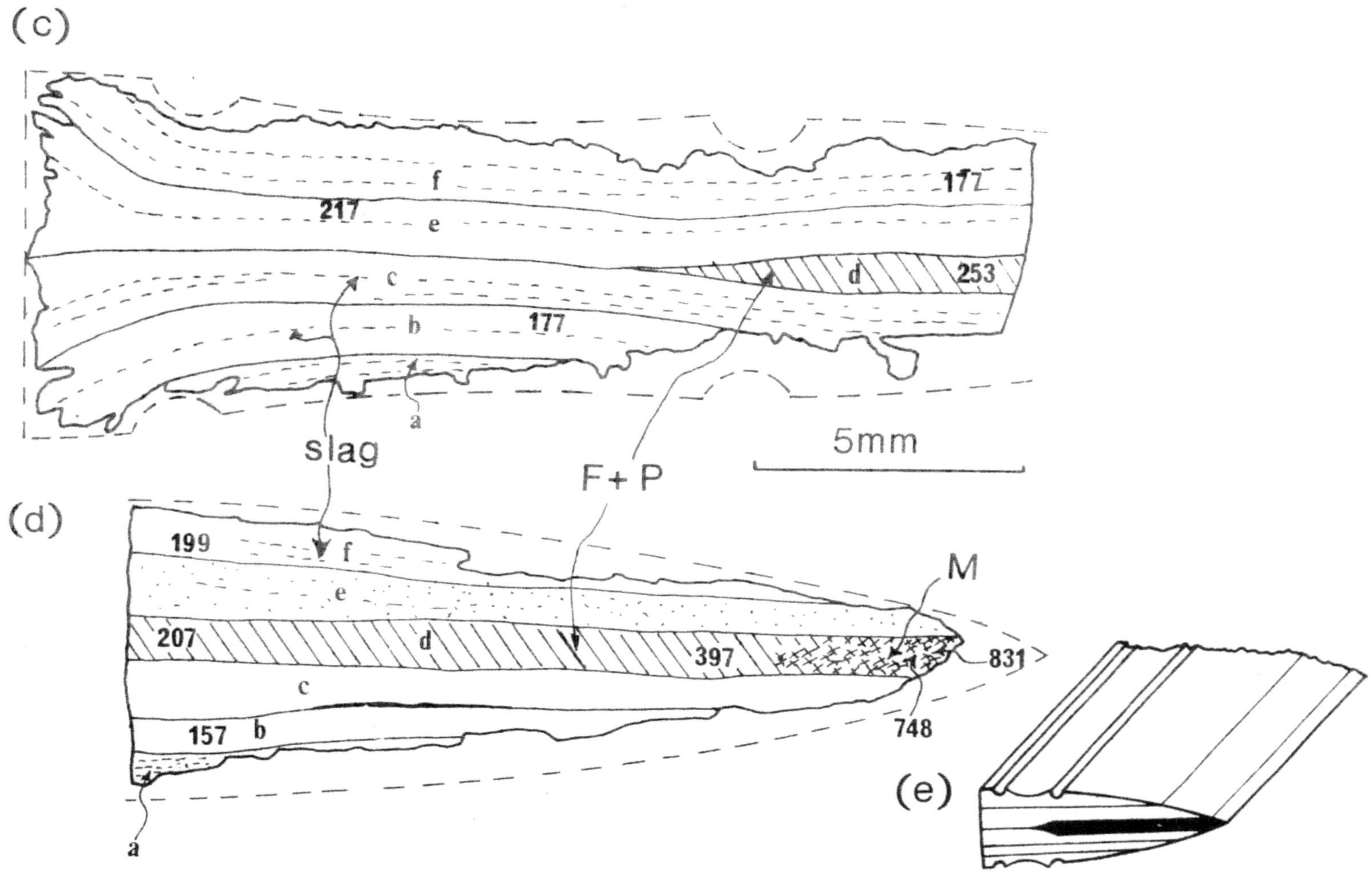


Fig. 53. (continued) Section through scramasax No. 42 from the River Lea at Leyton, Essex. + three-dimensional view.

areas (b) and (d) were parts of one piece of steel that formed a 'jacket' i.e. was a single piece folded and welded around a wrought iron core of which both (a) and (c) are representative areas. The high quality of the outer steel 'jacket' contrasted markedly with the wrought iron of the core. The finished blade appears to have been air-cooled, then subjected to a period of fairly prolonged annealing below the lower critical point of about 700°C, which was presumably done specifically to toughen the steel. The blade with its steel jacket and iron core is a clear example of the Type 4 blade (Fig. 1).

S 42, Scramasax (?) from the River Lea at Leyton (Table M), (Museum of London, 0.2275). The lower, or tip part, of a probable scramasax (Fig. 51) found in the present century. There was no indication that it had been bent or broken in antiquity but the break was not recent and the tang end may just have corroded away as a result of differential corrosion under water. The surviving piece of blade was much pitted by corrosion although the metallic survival of the core appeared to be quite good. It measured 64.2 cm in length by 3.7 cm in width and 0.85 cm at its thickest. Near the back of the blade, on either side, were two narrow grooves separated by a wide groove or fuller. The two narrow grooves ran parallel to the straighter part of the back and came together to form a point where the back of the blade curved towards the tip. The cutting edge was fairly straight, curving up a little towards the tip. In form this blade was closest to the Wheeler Type III or Hurbuck Type of scramasax, attributed to the 8th - 10th centuries (Wheeler,[66] 1927, p. 179).

Two V-shaped sections, each extending just over half-way across the width, were cut from either side of the blade so as to obtain a composite transverse section. The sections from the cutting edge and from the back were taken 1.5 cm and 9.5 cm respectively away from the broken end of the blade (Fig. 51).

Unetched, a varied slag content and distribution was visible, mostly as small spots and well flattened inclusions. These inclusions were grouped in fairly broad bands which ran along the main axis of both sections. All the slag visible was aligned in the same direction across the section. The slag content of these banded areas varied between very low and quite high.

When etched with nital the banded appearance of the sections became more pronounced and a quite distinctive 'flow' pattern showed up. (Fig. 53). A possible total of seven bands altogether were differentiated and these are marked (a) - (g) on the diagram of Fig. 53. Bands (a), (b) and (c) and (e), (f) and (g) can really be considered as the two largest components of the blade, as far as its final assembly is concerned. Each of these two components was composed largely of piled, or laminated, wrought iron. Between these two larger components a small strip of fairly homogeneous steel (d) in Fig. 53, was welded. This central steel strip shows up as the dark etched band. Towards the back of the blade this dark band can be seen to narrow, almost to a point, although it appears to continue as a narrow but distinct very dark grey line, marking here, a single weld line between the two piled wrought iron outer components of the blade. Near the back of the blade the bands of the outer piled pieces curve sharply outwards, which reflects some final hot forging at right angles to the main axis of the section, i.e. against the back of the blade.

The central steel strip (d) was probably of a fairly homogeneous carbon content of about 0.5 - 0.6%, although it gave a varying etched appearance from pale at the cutting edge tip, to very dark at the inner end. The tip end consisted mostly of martensite, which away from the tip end became progressively more mixed with a darker etched constituent which showed up, under higher magnification, as a feathery structure, probably of upper bainite (Fig. 53). The martensite only occurred near the cutting edge, and further away from this end the darker constituent became mixed with a gradually increasing proportion of ferrite, of an elongated somewhat acicular form. The darker constituent also varied and partly appeared to consist of poorly formed pearlite and granular or spheroidised carbide. Near the centre of the blade, however, the ferrite appeared rather more equiaxed in form. Towards the back of the blade the narrow, but distinctly visible grey line that the central band (d) tapered down to, consisted of ferrite in an acicular distribution with pearlite which appeared partially spheroidised.

The remaining parts of the blade all appear to have been wrought iron with little carbon, although after the final welding together, fairly extensive carbon diffusion occurred outwards from the central steel piece (d), into the adjacent iron piece (c) and (e). On either side of the central piece (d), the iron areas appear to consist of three bands. (a), (b) and (c) consisted of large grained ferrite and differed mainly in their slag content which was successively high, low and medium. The slag concentrations varied somewhat along the sections so that the content in band (c) was higher towards the tip of the cutting edge, while that of (b) increased towards the back of the blade where it was similar to that of (a), which had a consistently high slag content.

The bands (e), (f) and (g) varied in roughly the same way, (e) with a generally low slag content, while the slag content of (f) was similarly high to (a), and that of (g) fell approximately between the two. The band (f) differed somewhat from the others, in that it contained some carbon in the part towards the back of the blade. It was probably not much more than about 0.1%, i.e. little more than wrought iron, and may not have much significance.

Two hardness values, 831 and 748 HV, were obtained for the martensitic area near the tip of the central steel piece (d) and a third value of 397 HV, further in where a dark nodular constituent, probably troostite, showed up. A hardness value of 264 HV was obtained for part (d) near the inner end and a further value of 204 HV nearby, where it had become partly decarburised. From the iron parts that made up the rest of the blade, the following hardness values were obtained: (a) large grain ferrite - 157 HV, (b) medium grain ferrite - 177 HV, (e) ferrite and pearlite (probably diffused from (d)) - 217 HV, (f) large grain ferrite - 199 HV.

Before the final welding together of the blade components, the mainly wrought iron parts (a), (b), (c) and (e), (f) and (g) most probably already formed two single pieces between which the steel piece (d) was to be welded. They appeared to consist of individual pieces of iron of differing slag contents, which were welded or piled together.

After the final forging the blade was quenched, although a fully quenched or martensitic structure only occurred near the tip of the cutting edge, which means that the quenching was inefficient, possibly deliberately so. The acicular nature of the ferrite in the pearlite areas also indicated fairly fast cooling, but the somewhat spheroidised ferrite near the centre, may indicate that some self-annealing took place in the middle of the blade.

S 37, A long scramasax, found in the River Thames near the old ford at Kempsford in Gloucestershire, (Table M). The surface had been fairly evenly and lightly corroded, but under this the metal of the blade appeared to have survived well (Fig. 51). Its dimensions were: Length 77.5 cms, maximum width 4.2 cms, and maximum thickness 0.8 cms. On either side of the blade near the back was a wide groove or fuller. It is similar in form to the Wheeler Type III or 'Hurbuck' type of scramasax, which would place it within the Late-Saxon period, probably the 9th or 10th century.

Two V-shaped sections were cut from positions staggered along the blade, so as to enable a composite transverse section to be examined. When viewed unetched, the section across the back part of the blade showed a varying slag content, with two broad bands containing many partially flattened slag ribbons and spots running unevenly along the main axis of the section. These two bands were separated by a generally wider, fairly slag-free area, with a further slag free area to one side of the blade. Towards the cutting edge end of the back section, the slag band faded out and just the occasional flattened inclusion was seen here. The cutting edge section also showed a varying amount of slag, with very little in the areas along the sides, whereas the central zone along the main axis contained a slightly higher proportion of slag ribbons and spots.

When etched with nital the structure shown in the adjacent macrographs of Fig. 54 could be seen. This is also shown in simplified form in the diagram accompanying. These show that the blade consisted of a very clearly marked sandwich of three parts (b), (c) and (a). The weld position between the back (d) and the cutting edge components, was difficult to see as it was not marked by any discernible line of slag spots, although along part of its length it was marked by a narrow faint line of pearlite.

The back and the cutting edge were, however, of a markedly different composition. The back area (d) did appear to consist of one piece, although its structure, and hence its appearance, varied greatly. The two broad banded areas of high slag content noted above, consisted otherwise mostly of large grained ferrite. The adjacent low slag bands also consisted mostly of large grain ferrite. The macrograph of area (d) (Fig. 54) also shows two dense dark patches which represent apparent areas of quite high carbon steel in an otherwise almost mainly ferrite matrix. These showed a heat-treated microstructure of very dark etching, irresolvable pearlite, with some tendency towards an acicular martensitic appearance. The carbon content of these areas was roughly between 0.5 and 0.8% C.

There appears to have been considerable carbon diffusion outwards from the two steel areas within the back (a), which shows that this part of the blade was subjected to fairly prolonged heating. The banded nature of area (a) suggests that this back piece was first of all formed by the forging

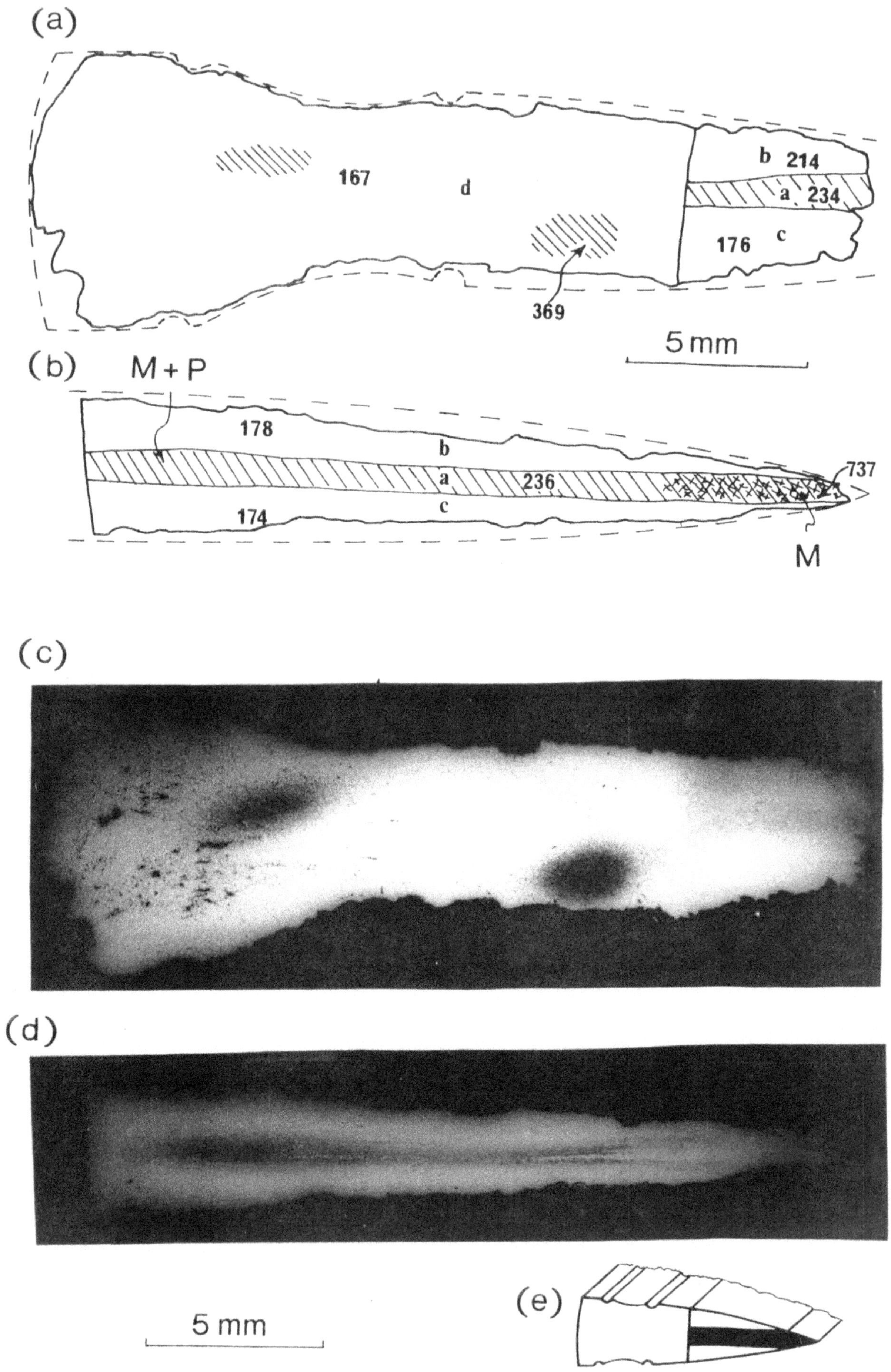


Fig. 54. Section through scramasax No. 37 from the Thames at Kempsford.

out, folding, and forging out again (piling) of one piece of wrought iron of greatly varying slag content or, more likely, a similar forging procedure involving two different pieces of wrought iron, one containing much slag, the other very little. The two steel areas are difficult to interpret for certain, but they are unlikely to represent steel wires welded into this back part of the blade, during any forging or piling procedure, as this would almost certainly have left some traces, probably in the form of entrapped slag of the weld junction, but none were observed. One possible explanation is that these areas, which both lay within bands of ferrite with a high slag content, are the product of uneven conditions within the bloomery furnace which produced this piece of wrought iron. The coarse nature and quantity of the slag in these bands suggests that the iron from this particular bloom was not subjected to careful or prolonged forging after smelting in order to remove slag. Accidentally produced particles of steel might survive as distinct areas within a piece of iron not subjected to much post-smelting forging as seems to have been the case here.

The cutting edge of this blade was of a very distinctive sandwich construction (Fig. 54d). The outer parts (b) and (c) were similar to each other, consisting mainly of fairly large grained ferrite, with small amounts of pearlite visible at the grain boundaries in places and with a quite low slag content. However, the boundary, in each case, with the central area (a) was marked by a wide diffuse zone of pearlite which faded out from the central area. This clearly represented diffusion across weld zones between the area (b), (d) and (a). The actual weld lines themselves were not clearly visible but were probably marked by patchy and faint grey lines of pearlite and faint and fuzzy white ferritic lines visible in places across the section.

The central area (d) consisted mostly of dark etched irresolvable pearlite and nodular troostite which was mixed with increasing amounts, towards the weld junctions with areas (b), (c) and (a), of ferrite which exhibited an acicular appearance. A distinctive spine of paler etched martensite was visible along the centre of area (d) and this widened out near the tip of the cutting edge which consisted mostly of martensite. The zone where carbon had diffused into areas (b) and (c) also showed a predominantly martensitic appearance near the cuttting edge tip.

The weld junction between the cutting edge, comprising the sandwich of area (b), (d) and (c) and the back part (d) of the blade was also not very clearly marked. No lines of slag were visible across the section although a patchy grey line of pearlite could be seen separating areas (c) and (b) of the cutting edge from the back area (d), but the junction between (d) and (a) had become lost in a generally rather diffuse area of low carbon pearlite.

The martensitic area of the cutting edge steel part (a) near the tip gave a hardness value of 737 HV, but near the centre, where mainly pearlite with some ferrite showed, a hardness value of 236 HV was obtained. The very low carbon iron of the outer edge parts (b) and (c) gave hardness values as follows: part (b), 178 and 175 HV, part (c) 174 and 214 HV.

The structures observed in the sections show that the blade underwent prolonged heating during forging, which caused fairly extensive carbon diffusion across the weld junction between the different parts. The blade was also quenched after its final forging although there is no indication of any subsequent tempering. The heat treatment has rendered the carbon constituent of the central steel part (a), of the cutting edge sandwich, difficult to estimate, at the centre it may be within the region of 0.5 - 0.8% C. Before the assembly of this part of the blade area (d) appears to have been a well prepared homogeneous piece of steel.

S 22, The surviving main part of the blade of a narrow scramasax from the River Thames, at Reading, (Table M), (Museum No. 52.80), its form approximating to the Wheeler type of the 9th - 11th centuries. (Fig. 51). Although the tang end is missing, the remainder of the blade is in comparatively good condition, which is fortunate because at some stage the surface corrosion products appear to have been stripped off by electrolytic reduction. The length is 33.6 cm and the width 2.8 cm. One result of the electrolytic stripping has been to reveal the differential effects of corrosion etching on the surface of the blade, in particular the clear outlining of a narrow pattern-welded strip running along through the centre of the blade on either side. Also still clearly visible were two parallel grooves running along the blade on both sides near the back - a form of decoration very common on Late-Saxon scramasaxes.

A radiograph of the blade showed the superimposed images of the parallel grooves, the partially overlapping images of the pattern-weld strip, visible on either side of the blade, and also showed up a somewhat wavy, but generally straight, wood-grain effect across the rest of the blade.

A narrow section was taken right across the blade at the existing tang end which had clearly been cut (not in antiquity) before, although no record of this survives. Examination of the section before etching revealed a very variable distribution of both large and small 2-phase slag inclusions across the blade. These were mostly irregular in shape and only some showed a tendency to be flattish or ribbon-like. The only distinctly linear distribution of slag inclusions was where they congregated along the weld lines along either side of the central pattern-welded zone (Fig. 55). Much corrosion had clearly taken place, particularly along the larger slag inclusions, which also showed signs of cracking during forging.

Etching with nital further accentuated the uneven variable composition within the different parts of the blade, which could now clearly be seen to consist of three parts, a wide back and a cutting edge between which had been welded a single twisted form of pattern-welded strip. In the pattern-welded strip the main elements showed as alternate light and dark wavy, folded bands. Although much variable size slag was generally visible, a much greater proportion of the slag was concentrated in the darker areas, which contained a great deal, giving a very coarse appearance. Apart from slag, the pale areas comprised fairly pure, very large-grained, ferrite with occasional Neumann bands indicative of cold hammering, visible as straight lines in the ferrite grains. These contrasted (Fig. 55) with the darker areas which were mostly of fine-grained ferrite with some pearlite mainly distributed around the grain boundaries. The carbon content of these areas was very low, mostly less than 0.1%, which really classes it as a low-carbon wrought-iron rather than a steel. The central pattern-welded part (b) of the blade, therefore, consisted of a twisted composite rod containing alternate bands of wrought iron of differing composition.

The composition of the back, part (c), and cutting edge, part (a), was very variable, but in both cases it is predominantly a patchy looking coarse wrought iron. The carbon content was mostly low, below about 0.1%, although in some areas (darkest in Fig. 55) the carbon content, generally lower in part (a), was as much as 0.2 - 0.3% and the varying concentrations of pearlite (differing shades of grey in Fig. 55) are clearly very uneven.

Two hardness readings were taken along the central axis of the low carbon iron cutting edge, part (a), the first nearer the tip of the cutting edge and the second nearer the inner weld. These gave hardnesses of 152 and 143 HV. One of the pale, very large grained ferrite bands of the pattern welded central part (b), gave a hardness of 220 HV. The back piece (c), gave two hardnesses 181 and 184 HV. The relatively much higher value for the pale ferrite band of part (b) probably indicates a higher phosphorus content for this and the other similar bands.

The etching of the section also further accentuated the weld lines between the central pattern-welded part and the back and cutting edge. The reason for this was revealed under higher magnification as the presence of a narrow linear distribution of pearlite amongst the slag (Fig. 55) along both weld lines. This 'grey line' distribution of pearlite must be the result of the absorption of a fairly even, but small amount of carbon, during the forge-welding together of the main elements of the blade. This may have been the result of the use of a fluxing agent which contained carbon in a fairly finely dispersed form. The central pattern-welded strip did not exhibit any sign of a 'grey-line' phenomenon between the light and dark areas, the junctions between which were rather diffuse. The alternate banding of these zones may have been the result of forging out, folding, and forging out again of a single piece of wrought iron of a similarly variable composition to both the back and cutting edge of the blade. A repetition of this forging process (alternatively termed 'piling') two or three times would probably, after a final twisting, have been sufficient to give the final banded effect visible in this case. The very variable content and distribution of slag and pearlite (i.e. carbon) in the back and cutting edge of the blade suggests that these parts in each case consist of a piece of only roughly consolidated bloom of variable carbon content. Fig. 55 shows a diagram of how this blade may have been assembled.

S 9, Scramasax, from the River Thames, (Table M), (Museum of London, A 27086). A fairly short, wide, decorated scramasax or knife with a sharply angled back and missing all but a small part of the tang (Fig. 51). It was found in the River Thames at Hampton and has been classified as a Wheeler Type IV or 'Honey Lane' type of scramasax and, therefore, probably of 10th - 11th century date (Wheeler,[66] 1927, p. 179). The visible decoration consisted of three parallel bands of twisted copper and bronze or brass wires inlaid into the surface near, and parallel to, the back. Near the angle of the back further similar twisted portions of wire have been inlaid so as to join the three parallel bands together. Parallel and next to the three bands of inlaid twisted wire on the site of the cutting edge, traces of a pattern-welded design were visible in the iron corrosion products on the surface of the blade. Its length was 19.0 cm, width 4.5 cm and the maximum thickness at the back 0.55 cm.

Apart from the non-ferrous wires, a radiograph showed up the pattern-welded strip which was visible in the surface corrosion. This pattern was in the form of a sine wave and showed up very sharply, suggesting that the composite piece forming this strip was made from two pieces each initially forged to give a ribbed or corrugated profile along its length and then

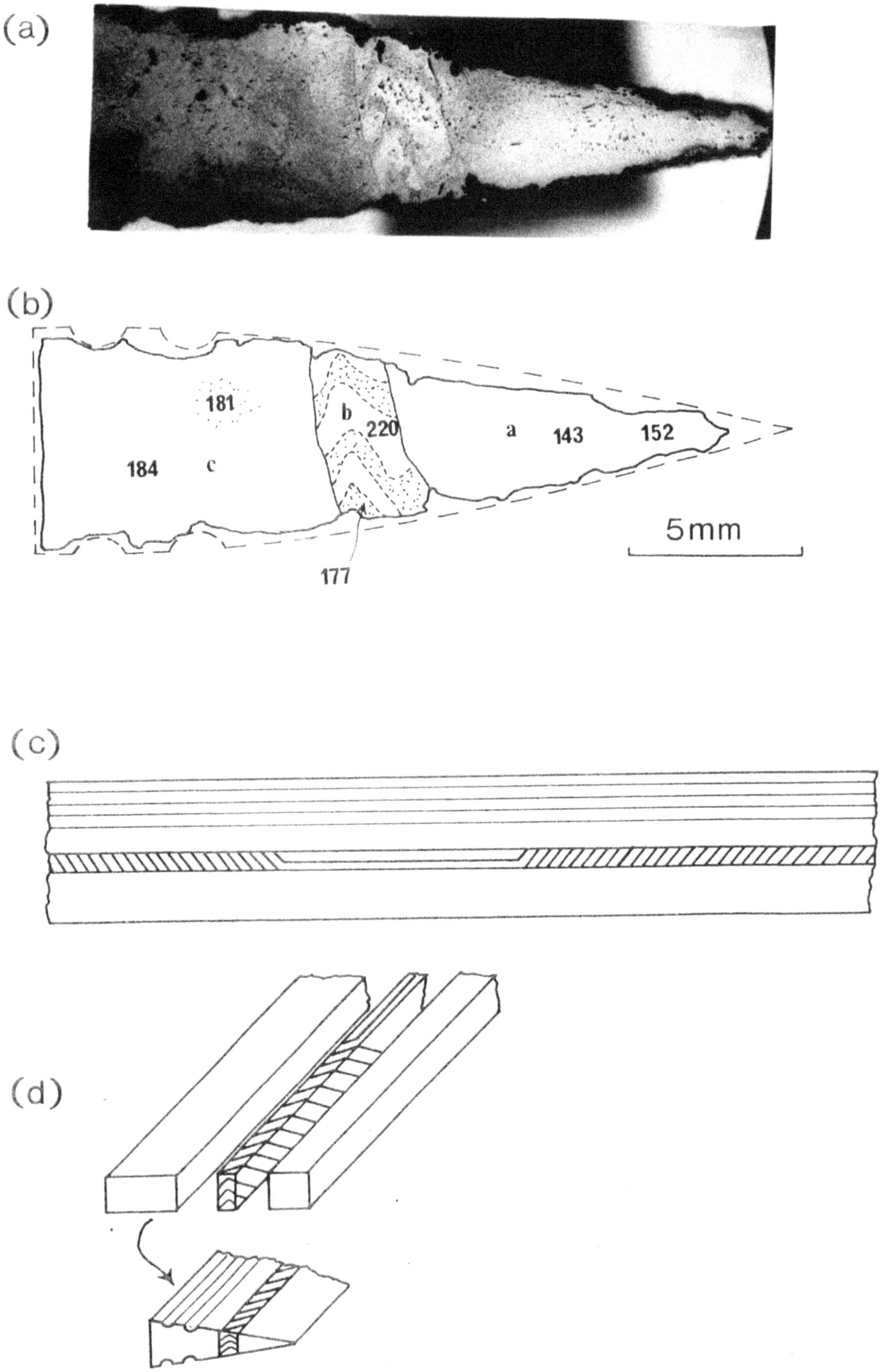


Fig. 55. Section through part of a narrow scramasax No. 22 from the Thames at Reading + surface detail drawn from X-ray and three-dimensional view.

welded together. The back of the blade showed a straight grain effect on X-ray but otherwise appeared to be of one piece, to which the composite piece was welded. The side of the blade towards the cutting edge also showed straight graining although this was restricted to half of this area alongside the composite or pattern-welded piece. This suggests that the cutting edge half of the blade may have been made from two pieces, although this must remain just a possibility indicated by the radiograph.

Sectioning was attempted across the back half of the blade by an existing break and it was hoped that an examination of most parts of the blade would be possible by so doing. Unfortunately excessive corrosion in this area, probably promoted by the presence of the non-ferrous metal inlay work, meant that only a small part of metal was found surviving right at the back of the blade. This piece proved to consist of a very coarse wrought iron with much slag in it. Fig. 56 shows the section after etching with nital. The very large grain structure of the ferrite shows up and the dark patches are a combination of slag inclusions and corrosion surrounding them. This shows up better at higher magnification when the varying size and irregular shape of the slag inclusions can be seen. They can also be seen to consist of two medium-dark grey phases, probably fayalite and a glassy matrix. Also present are stress lines or Neumann bands which indicate that some final cold hammering took place, although not enough to distort the ferrite grains. A series of hardness values ranging between 75 HV and 98 HV were obtained, which are fairly typical of a soft fairly pure wrought iron. The metal used here for the back of the blade appears to be a poor quality piece of wrought iron, with much slag, little more than a roughly consolidated piece of bloom.

S 34, Scramasax, found in the River Thames at the old ford at Kempsford, Gloucestershire. (Swindon Museum, No. 790308). (Table M).

The fairly straight cutting edge and the sharply angled back allow this blade to be grouped as one of the variants of the Wheeler Scramasax Type IV or Honey Lane type which have been dated approximately to the 10th/11th centuries. (Fig. 51). The length was 24.6 cm and the width 2.0 cm. The blade surface had been fairly heavily etched by corrosion although the metallic core had survived well. Different parts of the surface had etched or corroded preferentially so as to reveal a chevron pattern formed by two twisted composite rods welded side by side (these two bands ran down the central part of the blade on either side). The direction in which the chevron pattern pointed reversed on each side so that it was impossible to deduce whether or not two similar pairs of twisted composite strips had been welded back to back.

Two offset V-shaped sections were cut from the central area of the blade. When viewed unetched, both sections showed distinct although fairly fine, broken lines of small slag spots and ribbons which ran across the section and marked the positions of welds between the different parts of the blade. It appeared that the central part of the blade consisted of just two twisted composite rods welded side-by-side. The back area showed only a few small slag inclusions while the central, composite parts of the blade contained rather varying amounts of inclusions which were uneven in both size and distribution. The area of the cutting edge contained much 2 and 3 phase slag inclusions and varying from small spots to large irregular inclusions. These were grouped mainly towards the centre of this area with

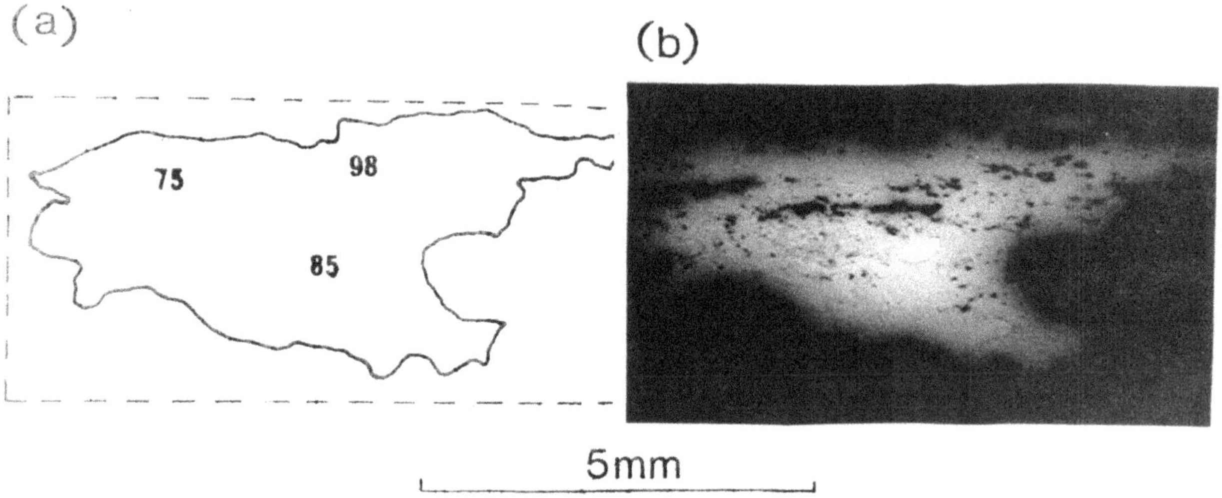


Fig. 56. Section through a narrow scramasax No. 9 from the Thames at Hampton, Middlesex.

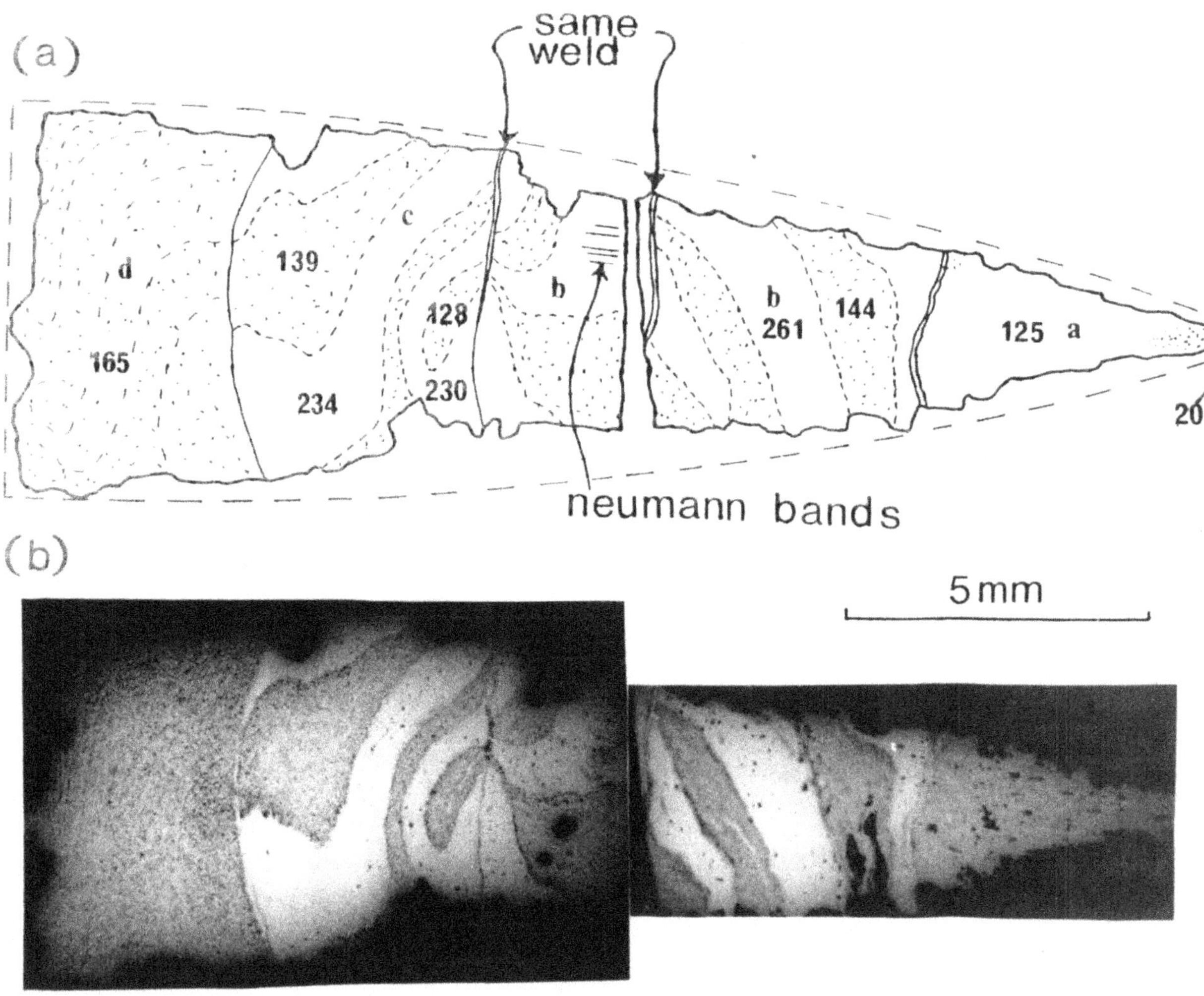


Fig. 57. Section through a scramasax No. 34 from the Thames at Kempsford.

the concentration of slag being much lower along the margins towards the surface and the tip of the cutting edge.

When etched with nital the structure of the two sections showed up quite clearly and their appearance was quite striking as can be seen in the macrograph of Fig. 57b. The explanatory diagram in Fig. 57a shows the structure visible on the macrograph in simplified form.

The blade could now quite clearly be seen to be of four parts, a cutting edge piece (a), two composite pieces (b) and (c), and (c) and (d) were marked by dark grey lines of pearlite, with a very acicular appearance, superimposed upon the broken lines of slag spots which was all that had indicated the weld position before etching. Faint traces of a similar line of pearlite were just visible amongst the slag spots along the weld junction between parts (a) and (b).

The cutting edge part (a) mostly showed up as a pale area of ferrite of a variable medium to large grain size, with much slag in the central area. A narrow diffuse pale grey margin of pearlite lay along and just inside the weld line but the carbon content of this margin was clearly very low - probably less than 0.1%. A grey area of very acicular pearlite occupied the area of the tip of the cutting edge. It exhibited an almost martensite-like appearance and would appear to represent a quenched structure in an area whose carbon content was too low to give a martensitic structure under the conditions in which the quenching took place. The carbon content is difficult to estimate but is probably unlikely to be more than about 0.1 - 0.2%. The presence of the margin of pearlite near the weld, as well as that of the tip, may indicate that this part of the blade was partially carburised before the final assembly of the blade. At the weld some decarburisation appears to have taken place, hence the pale zone between the grey margin and the weld. If surface carburisation did take place then the corresponding pearlitic areas that might have been expected to be visible on the upper and lower surface areas of the cutting edge, must have to be removed either by decarburisation - less likely since the pearlitic area at the tip survived - or by surface grinding after the final forging of the blade. There was no evidence that the tip might have been welded as a separate piece into the cutting edge. The acicular appearance indicates a Widmanstätten distribution of pearlite in ferrite.

The composite parts (b) and (c) both showed up as a distorted sandwich structure, in each case made up of three white and three grey bands.

The pearlite of the grey bands was of a medium grain size and had a very acicular appearance similar to that of the thick grey lines of pearlite marking the weld junction between (b), (c) and (d). These grey pearlitic zones contained only a little slag. Allowing for the presence of the pearlite the carbon content of these bands was probably no more than 0.1 - 0.2%.

The white zones consisted mainly of very large grained ferrite with quite a lot of slag - mostly as small blobs and partially flattened ribbons but with few larger inclusions. Neumann banding was also visible in the ferrite grains which would indicate some cold working.

The junction between pearlitic and ferritic areas showed up very sharply and although there were almost no tell-tale lines of slag spots these must indicate weld lines. The sharp distinctions show no sign of decarburisation or carbon diffusion across the weld junction, but by contrast the pearlite actually appears denser along most parts of these boundaries. These denser margins of pearlite may have been present as a similar phenomenon as the grey lines of pearlite observed along the welds between areas (b), (c) and (d) but have been less obvious because of the masking effect of the adjoining grey pearlite bands.

The back part (d) of the blade showed as a fairly even low slag distribution of acicular grey pearlite and ferrite probably representing a carbon content of about 0.2%, slightly higher than in the grey bands of (b) and (c). Near the surface of the back area the pearlite and ferrite had partially segregated into narrow bands, giving a kind of wood grain effect on the section. This may have been the result of a small area, locally rich in phosphorus causing the segregation, the banding being created by the forging out of this strip or bar of metal before the final assembly of the blade.

The tip of the cutting edge (a) gave a hardness of 204 HV, whereas the inner, iron, part gave a hardness of 125 HV. On each of the bands of pale, very large grain ferrite and grey ferrite + pearlite in the pattern-welded piece (b) and two each of these bands in pattern-welded piece (c) were given hardness tests with the following results: (b) very large grain ferrite 261 HV, (b) ferrite and pearlite 144 HV; (c) very large grain ferrite 230 and 234 HV; (c) ferrite and pearlite 128 and 139 HV. The much higher values for the very large grain ferrite bands probably indicates a high phosphorus content for these pale bands. The back part (d) of the blade gave a hardness of 165 HV.

The blade appears to have been of four parts, a back piece (d) of fairly homogeneous carbon steel, two twisted composite or pattern-welded rods consisting of alternate bands of high phosphorus (pale) and low carbon iron (grey), and a cutting edge piece of partially carburised wrought iron. The cutting edge must have been partly ground down at the sides removing further evidence of carburisation which might have been expected.

S 4, from Dorset (?). (Dorset County Museum). (Table M). A scramasax with an angled back and a gently curved cutting edge (Fig. 51). It appeared to be fairly lightly and evenly corroded with a good metallic core, however, the surface corrosion layer was quite thick and masked what lay beneath. The condition was fairly typical of finds from riverbeds or other waterlogged environments, although the find site is not known. Its length is 31.6 cm with a maximum width of 3.3. cm and in shape it is similar to the Wheeler Type III/IV scramasax found at Walthamstow dated to the 9th - 10th centuries and so a similar date may be likely here. This specimen has been discussed elsewhere.[66]

A radiograph showed the blade to be of fairly complex construction and to be pattern welded. A double band each side showing a clear diagonal grain pattern ran down the blade from the tang parallel to the upper part of the back and meeting the surface of the lower part of the back near the angle. The absence of any criss-crossing in this pattern showed that two pairs of similar twisted composite rods had been welded behind one another. On the radiograph the double diagonal pattern was outlined on either side by a narrow strip which showed up a straight grained effect. The remaining areas, the narrow margin by the upper part of the blade and the wider area by the cutting edge showed little or no signs of graining. At least 8 separate parts to the blade were indicated by the radiograph.

Two wedge shaped sections were cut from the blade, one from the back and one from the cutting edge, the cuts being staggered so as to give an overall composite transverse section without cutting the blade in half or losing too much strength. The section from the back was taken 10 cm down from the hilt or shoulder and the section from the cutting edge about 3 cm further down the blade. Unetched a very low slag content showed although many corrosion pits were visible.

After etching with nital, a very clear structure showed (Fig. 58) and the blade could be seen to have been made up of 8 main components, (a) - (h) in Fig. 58 a. The edge part (a) gave a rather banded appearance with darker bands consisting mainly of dense, mostly unresolvable pearlite, with some ferrite interleaved with paler bands, predominantly of martensite, with some pearlite. There were no obvious signs of welds between the bands but a skilfully welded piled structure would seem most probable for the piece, probably originally of high carbon steel and wrought iron parts welded, forged out and subjected to fairly prolonged heating although not for long enough to give a homogeneous medium carbon steel (Fig. 58 c and d). The carbon content was difficult to estimate but probably varies between about 0.4 - 0.6%.

The pale etched 'plain' pieces (b) and (g) had been welded to either side of the pattern welded zone and they each consisted of very large grained ferrite.

The pattern welded core consisted of four twisted composite rods (c) and (f), welded side by side and back to back as shown in Fig. 58. The very large, roughly diamond shaped, corrosion pit visible in Fig. 58 occupies the position of a very large 3 phase slag inclusion (of wüstite, fayalite in a glassy matrix), some of which still survived and is just visible in Fig. 58. This very large inclusion was trapped in the centre of the pattern welded part of the blade when the four composite components were welded together. The composite pieces (c) and (f) consisted of alternate pale bands of very large grain ferrite and darker grey bands of small grain ferrite plus a little pearlite at most, giving a carbon content of about 0.1 - 0.2% i.e. the whole of the pattern welded part of the blade, as well as the adjacent strips appeared to consist mainly of wrought iron with some fairly low carbon iron in the grey bands of the pattern welded pieces.

The back part (h) consisted of a piece of fairly high carbon steel, probably in the region of 0.5 - 0.7%. Fig. 58b shows its rather patchy appearance varying between paler areas of martensite and darker areas of fine grain virtually irresolvable pearlite with some ferrite.

The edge part (a) was given four hardness tests, three from the martensitic areas and the other from one of the dense pearlitic areas. The first three gave values of 775, 555 and 647 HV and the fourth a hardness of 318 HV. The two pale etched, very large grain parts (b) and (g) gave hardnesses of 230 and 210 HV. The darker, grey lower carbon iron of pattern welded parts (d), (e) and (f) gave hardnesses of 184, 185 and 152 HV respectively. One of the martensitic areas of the back part (h) gave a hardness of 647 HV and one of the fine grain pearlitic areas gave 311 HV. The comparatively high values for all the very large grain ferrite areas in part (b) and (g) probably indicates a high phosphorus content.

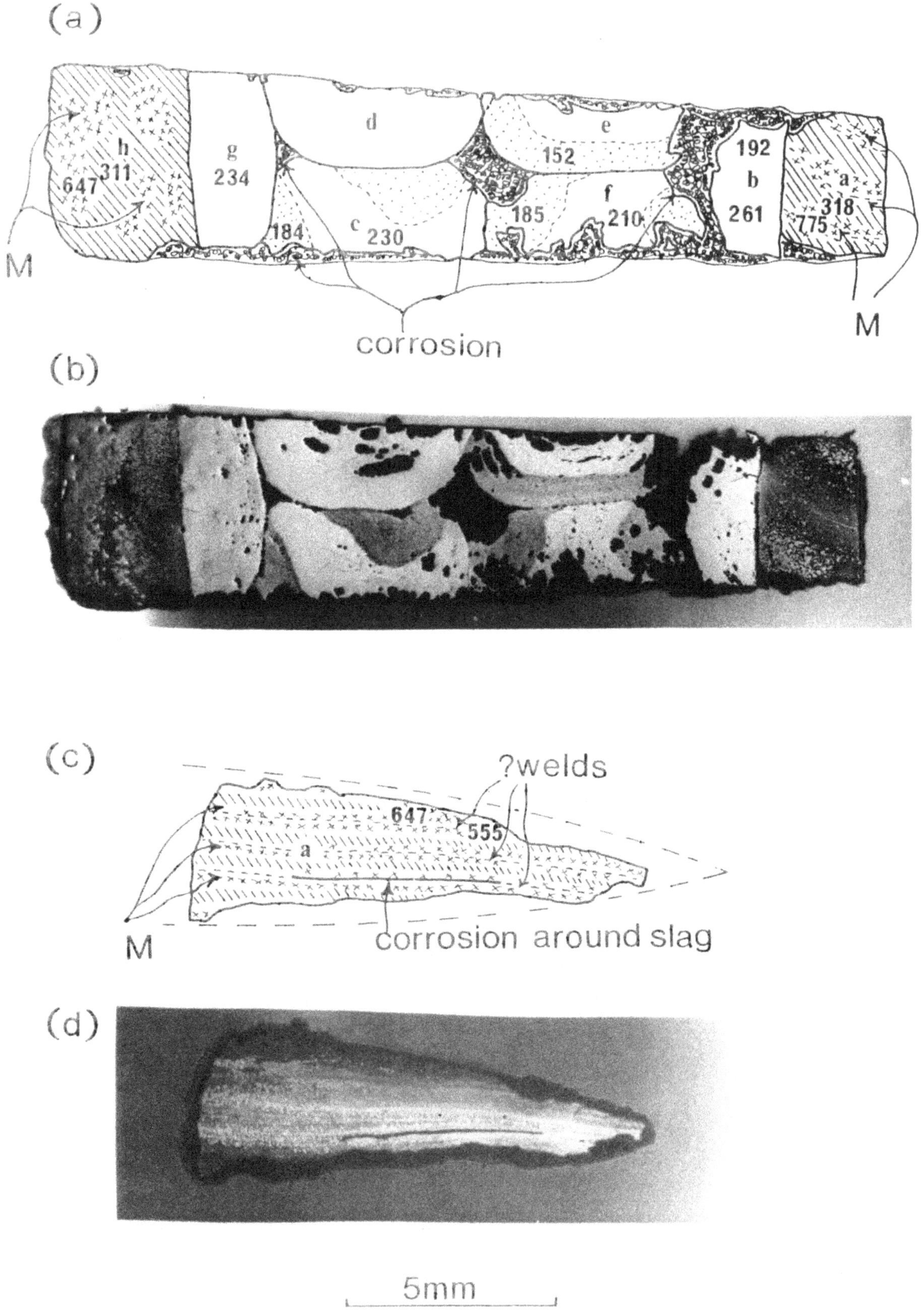


Fig. 58. Section through scramasax No. 4 from Dorset

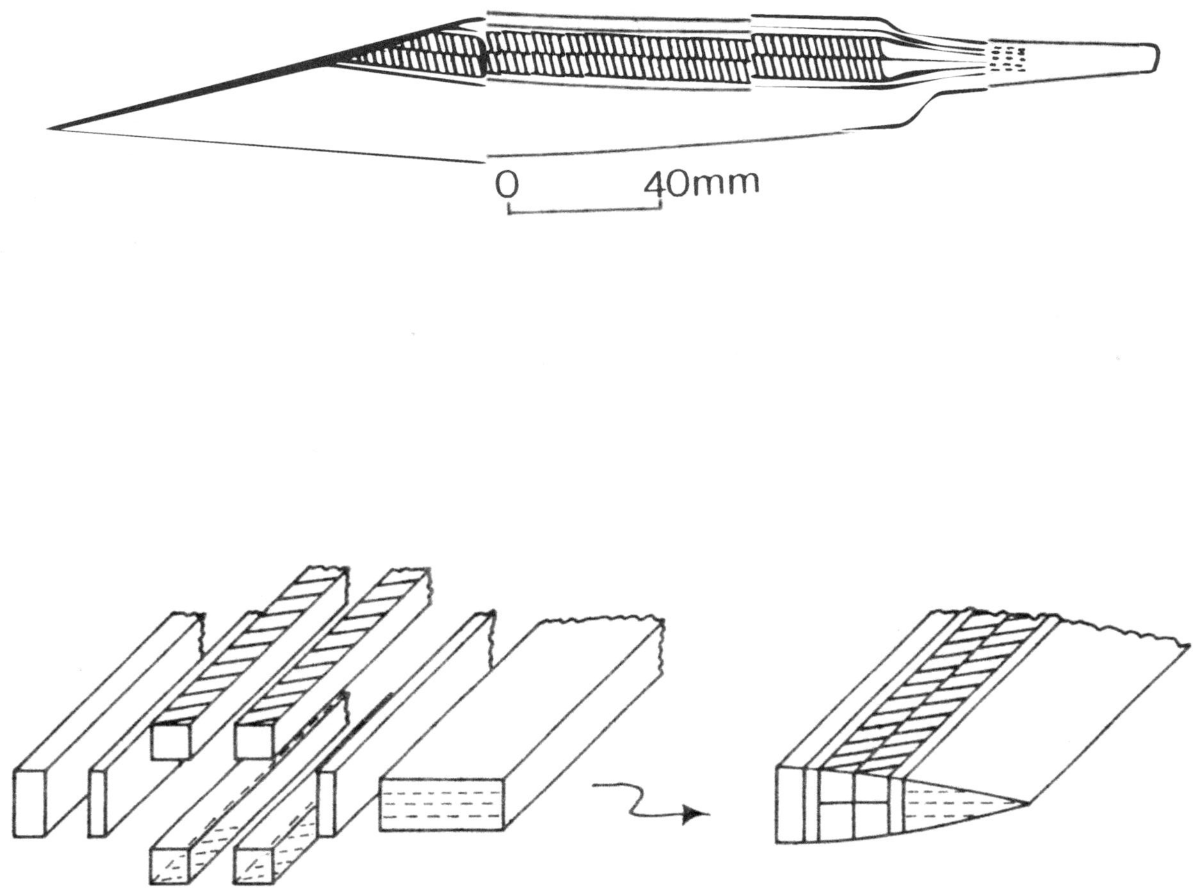


(e)

Fig. 58 (continued) Surface view drawn from X-ray and three-dimensional reconstruction.

After final assembly, welding together and forging out, the blade was quenched, although not efficiently enough to give a fully martensitic microstructure to the steel parts (a) and (h). The structure does, however, appear to be quite even, which may be indicative of a deliberate slower or 'slack' quench.

(n) Daggers and Kris

Short, piercing weapons, need to be even less metallographically sophisticated than long slashing weapons like scramasaxes and swords and we are not surprised to find the Roman example is made of wrought iron. Unfortunately, the sample was very limited in this group, and the other examples, two Malaysian kris, are rather exotic with their wavy pattern-welded blades. Clearly the pattern was meant to be seen, but like the Anglo-Saxon sword, we wonder if the pattern fulfilled any physical purpose, apart from being a by-product of the forging process.

The Roman site at Brancaster produced a dagger blade. A segment was removed from one edge. It was entirely ferritic with variable grain size. It contained some very large, lightly worked, slag inclusions and some slag stringers. There was some carbide films (cementite) in some of the grain boundaries. The carbon content would be about 0.05% C which agrees with the hardness of 119 HV. Clearly, a fairly pure iron.

A very different form of dagger is the Malayan kris. One of the two examined probably dates from the 17th century. This has been made to give a pattern-welded structure (Fig. 59, TI).

A section from the tip measured 2.5 x 11.5 mm (Fig. 60a). It consisted of a central band of ferrite + pearlite 0.5 mm thick containing about 0.5% C with a Widmanstätten distribution. The hardness was 197 HV1 which is reasonable for this carbon content and would not indicate high phosphorus. The pearlite was rather more spheroidal than lamellar and fairly coarse.

The central band was bordered on either side by a piled structure of ferrite and slag (Fig. 60a and b). The ferrite had a hardness of 156 HV1. The piling near the centre is finer than on the outside and the ferrite bands were found to alternate, with continuous lines of high-nickel iron. These are certainly harder than the ferrite matrix, i.e. greater than 300 HV (50g) compared with 210 HV (50g).

An electron probe scan from the high carbon core across some of the laminations gave the results shown in Fig. 61, which should be compared with the photomicrograph. Both show bands which tend to decrease in frequency towards the outer surface. The bands are 5 - 15 μm thick and the ferritic areas between are 20 - 60 μm wide. The nickel content of the bands is of the order of 2 - 3%.

These bands are very like the white lines in the Goltho knife and the 'V' zones in Chilton's wrought iron. In both cases these are most probably due to surface enrichment during oxidation prior to folding and piling. In the case of the kris they are certainly harder than the ferritic matrix ( 300 HV) and would appear to contain some martensite. Since the core has

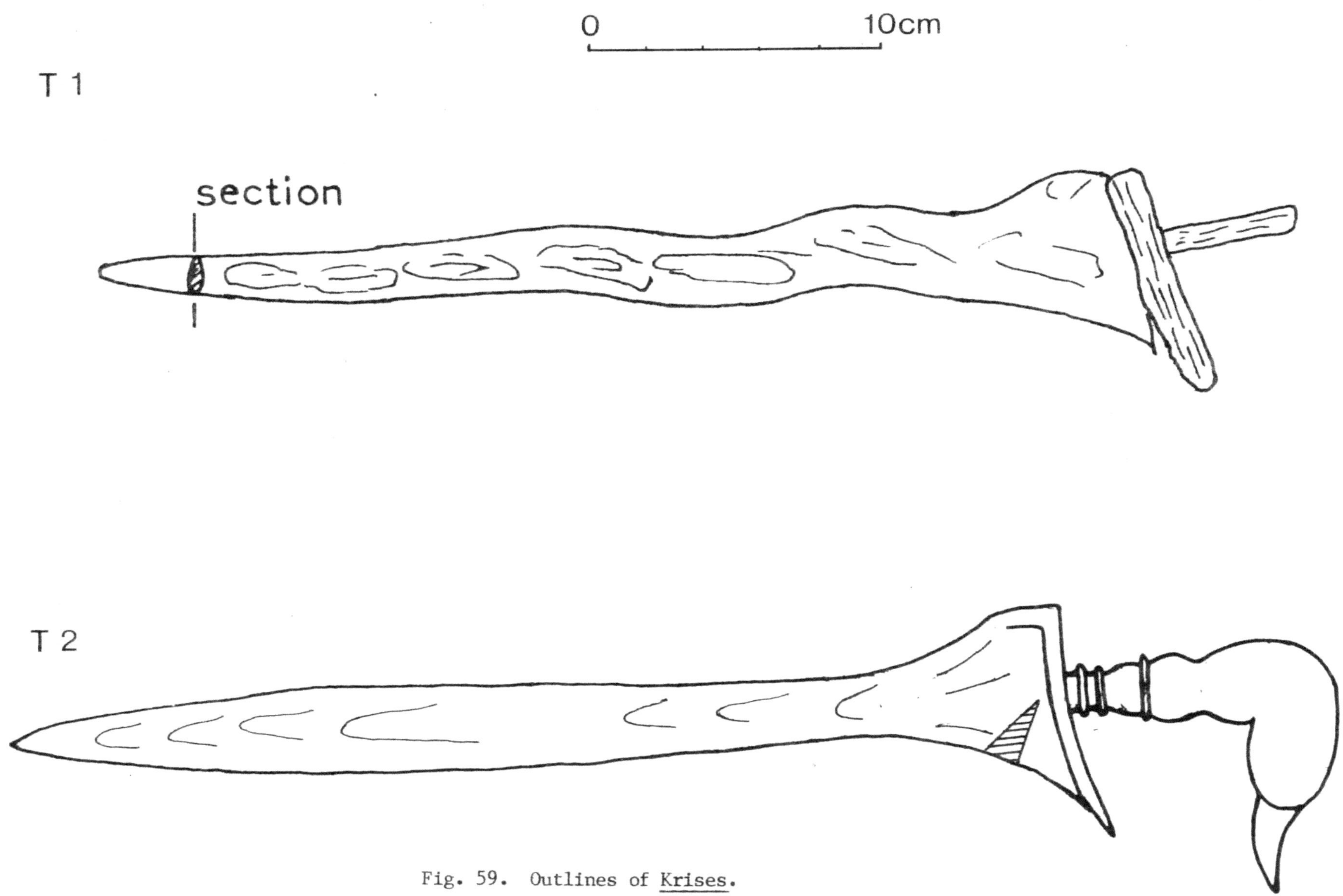


Fig. 59. Outlines of Krises.

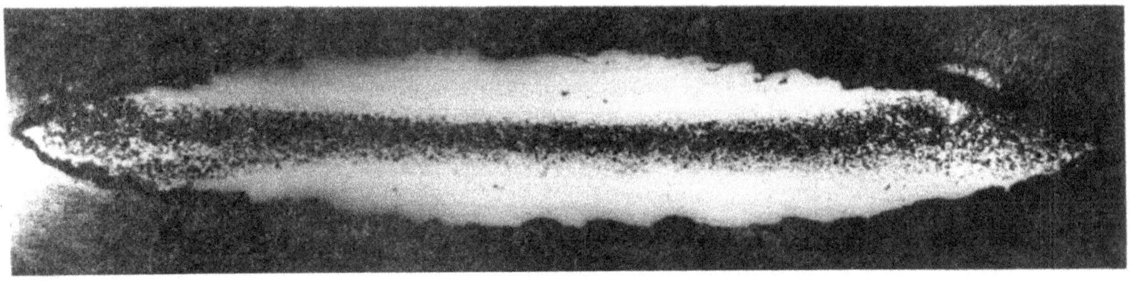

(a) 1mm

(b) 0.1mm

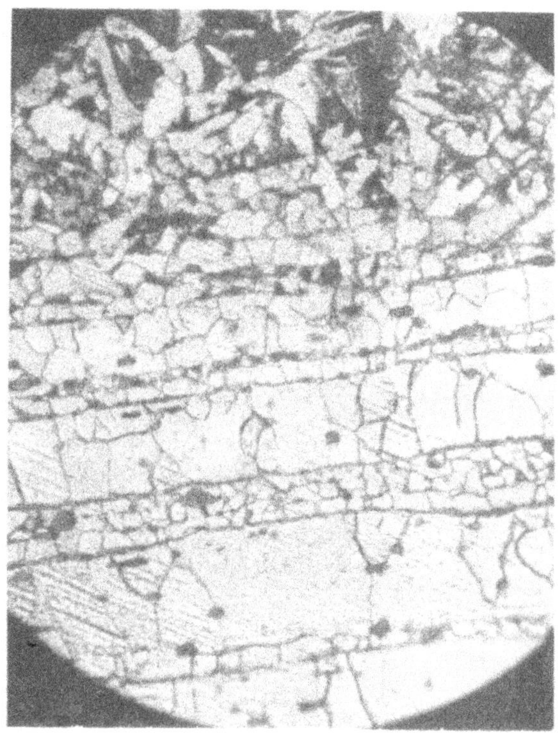

(c) 0.1 mm

Fig. 60. Photomicrographs of structure of Malayan Kris; T.l. (a) Whole section (b) Near core, X 100 (c) same, X 400.

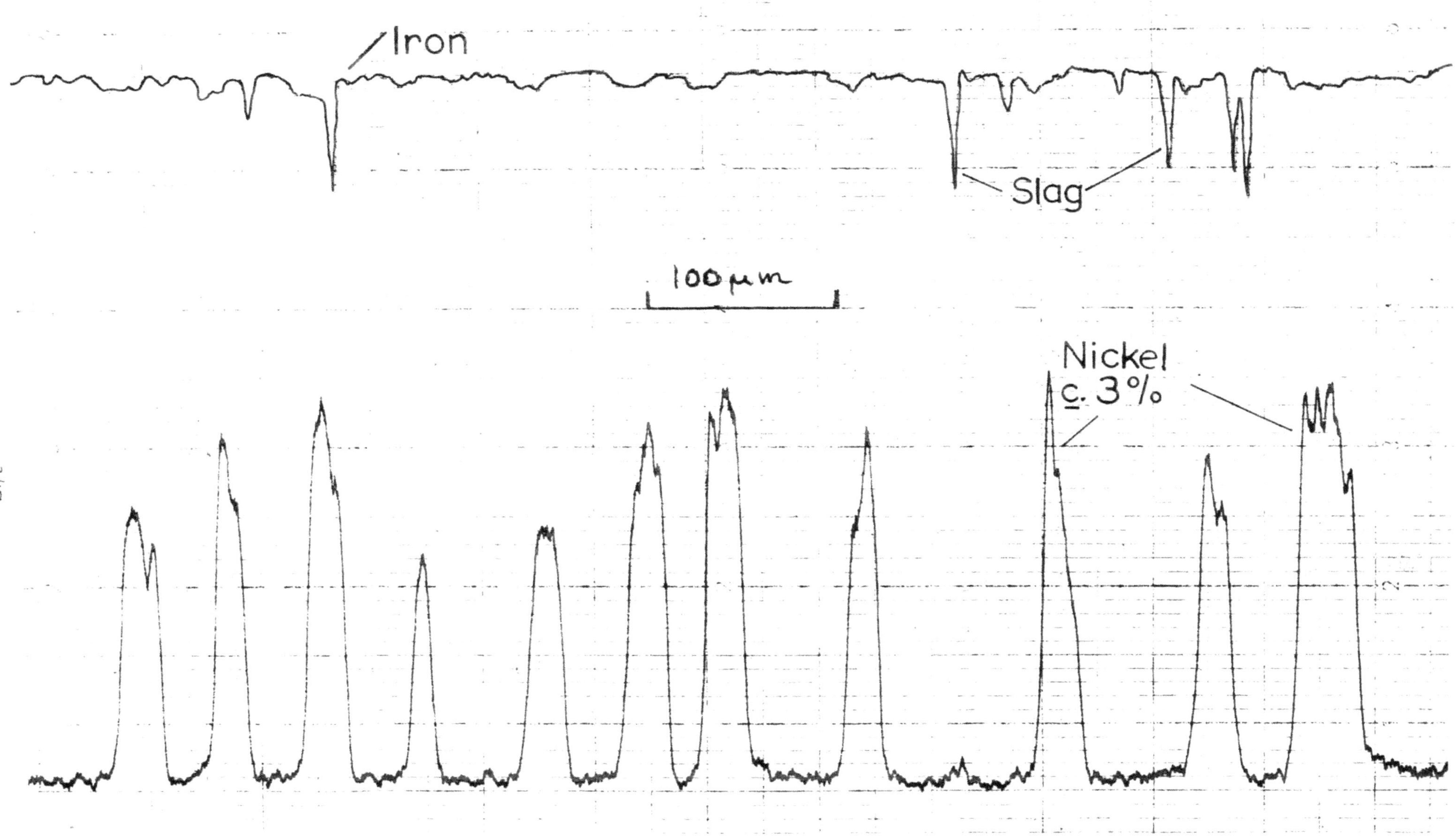


Fig. 61. Electron probe scan across blade, T.1, showing variation in nickel content.

not been quenched, it is possible that the bands are air-hardened steel, which would be expected if the composition was in the range 3 - 5% Ni and 0. % C.

A second kris (S57, T2 in Fig. 59b) thought to be 19th or 20th century was loaned for examination by Dr. Ian Glover of the Institute of Archaeology, University of London. It is a typical example of a pattern-welded SE Asian blade 46.2 cm long and not unlike the previous example. But the lower part of the blade was rather pitted by corrosion, while the upper part showed the typical 'relief map' structure.

A narrow wedge-shaped section was removed from the flared shoulder of the hilt part of the blade and when etched the structure shown in Fig. 62 became clearly visible. It was clear that the section consisted of four parts welded together (Fig. 62).

Part (a) formed the cutting edge as well as the core of the blade and as in the previous case consisted of a piece of homogeneous high carbon steel, this was well-spheroidised pearlite, with ferrite, indicating a carbon content of about 0.6 - 0.7% and having a hardness of 189 HV. The central piece (b) consisted of alternate pale bands of ferrite interleaved with ferrite and pearlite. In this area the carbon content was low, 0.1 - 0.2%. This part suggested a piled structure of alternate bands of iron and low carbon steel repeated several times (Fig. 62).

The surface showed a similar banded structure and it is possible that this banded structure is accentuated by being made of laminations of high and low nickel iron, formed either by oxidation enrichment during forging, or by high nickel steel. No further work was done to decide which process, if any, was involved.

The cutting edge was formed by grinding away one side of the assembled pieces after they were welded together. The results of this oblique grinding on either side was that the core part (a), was exposed to form the cutting edge and whatever final polishing and etching that was done to the blade, caused the characteristic 'contour map' pattern to be exposed.

During and after the final forging and before this final grinding and finishing, the blade must have been subjected to prolonged heating, followed by slow cooling, which left the steel core of the blade in a relatively soft spheroidised state. This may have been part of a deliberate toughening process.

### (p) Swords

The swords examined in this section belong mostly to the Anglo-Saxon period i.e. mostly from the 6th to the 11th century. The reason for the comparative frequency of the earlier swords is that most come from inhumation burials in large cemeteries at a time when it was the custom to bury the dead with their weapons and badges of office. Many of the later swords of this period are the result of (probable) ritual deposition in rivers - there are many from the Thames - and marshes. Here the lack of oxygen in the bottom muds have led to reduced oxidation and good preservation. This practice is found during the later Roman period and became common in this country in the Late Saxon period.

(a)

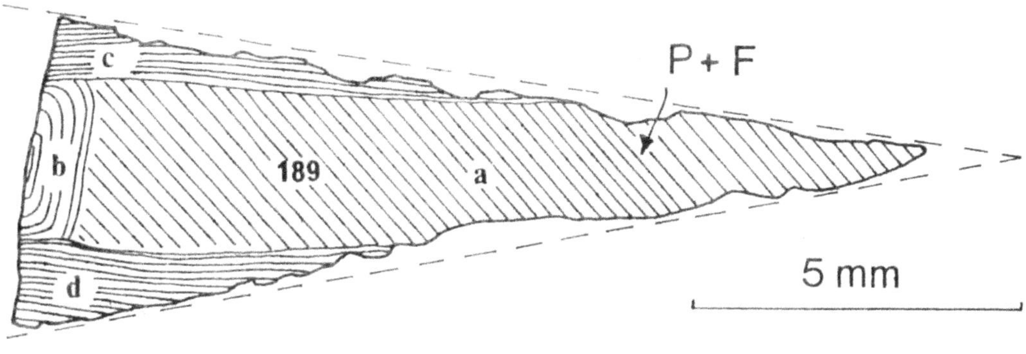


(b)

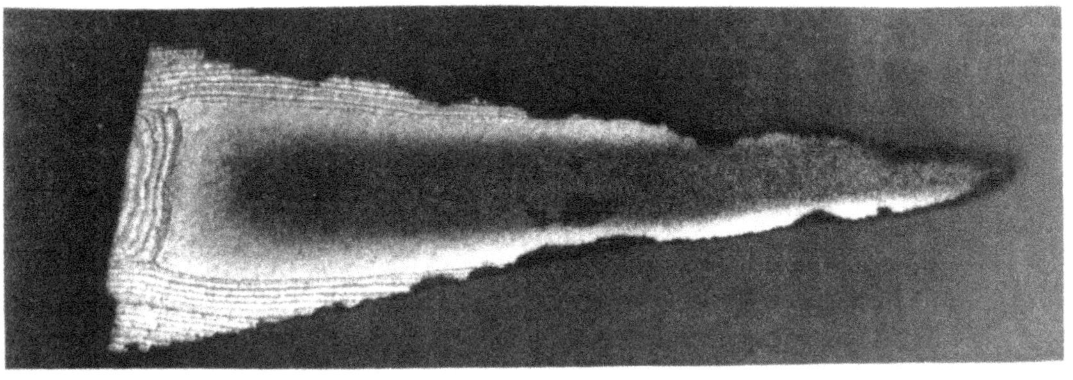

(c)

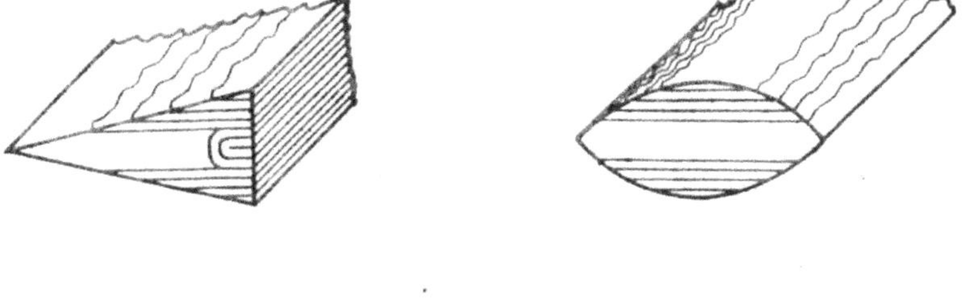

Fig. 62. Section through Kris No. 57. (T2).

Today, after finding, it is the normal practice to radiograph most weapons and this should reveal the use of pattern welding. As this method of construction is revealed before sectioning by corrosion along the smithed welds, a new sword such as a replica will not show it. Also a blade that has rusted within its scabbard may not show pattern welding very clearly, either because corrosion has been reduced by the presence of the scabbard, or because the non-ferrous metal of the scabbard obscures the pattern on the blade.

In this section all the hardness readings were taken with a load of 100g.

S 32. Sword, Provenance Uncertain; (Reading Museum, 131:61). The surviving blade fragment of a short sword which was broken, just below the hilt (Fig. 63) Table N. Its very distinctive leaf-shaped blade is very similar in shape and size (its length being 38.6 cm and its width 3.7 cm) to leaf-shaped sword blades of the late Bronze Age. The condition of the blade was very characteristic of river finds of iron that have been immersed and at least partially buried in sediments for a long time. Parts of the blade survived with only a little surface pitting whilst other parts, particularly in the centre, had been corroded almost right through and the corrosion appeared to have spread slowly outwards from focal points, attacking different parts within the blade selectively, so as to reveal what appeared to be a distinctly layered structure to the blade. There appears to be no particular reason to doubt that this blade is anything other than an early Iron Age copy of a late Bronze Age form of sword. Transitional Bronze Age/Iron Age objects, forged from bloomery metal so as to look like traditional forms, previously only made from cast bronze, are rare, but have been found from time to time.

A V-shaped transverse section extending just over half way across the width was cut from the central, wider, part of the blade (Fig. 64). When viewed unetched, the areas of the section nearer the surface of the blade gave a striated appearance, caused by many broken lines of slag ribbons and spots, which ran parallel to the surface on either side. Fairly extensive corrosion had taken place around some of the larger flattish inclusions. The slag was mostly two phase, light and dark (wüstite and fayalite) in appearance. The central area of the section was fairly slag-free with only a few small flattened slag inclusions present.

When etched with nital, three fairly clear zones were visible, but as can be seen in the macrograph of Fig. 64 the divisions between these zones were rather blurred. The diagram of Fig. 64 shows a simplified view of this structure.

Most of the upper half of the section visible in the macrograph was occupied by zone (a) which consisted of very large grained ferrite, with much two phase slag (see above) distributed as parallel broken lines along the length of the section. This banded or striated appearance is suggestive that the zone or piece (a) was piled as well as extensively forged before the final assembly of the blade. A little pearlite occurs in places along the slag lines and also near the junction with area (b) where it is probably a result of carbon diffusion from (b) to (a) during the final forging of the blade.

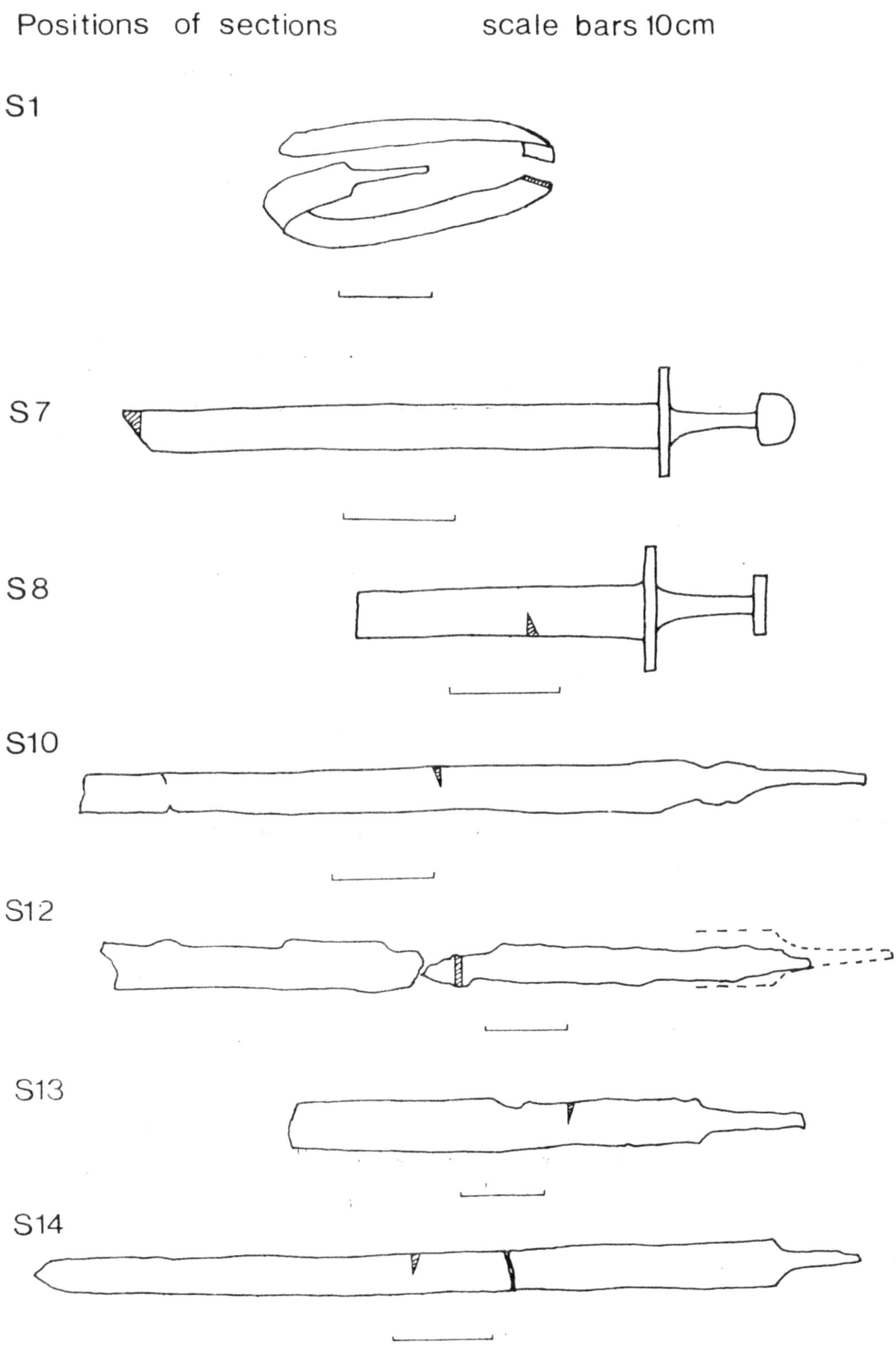


Fig. 63. Outlines of swords showing positions of sections.

S15

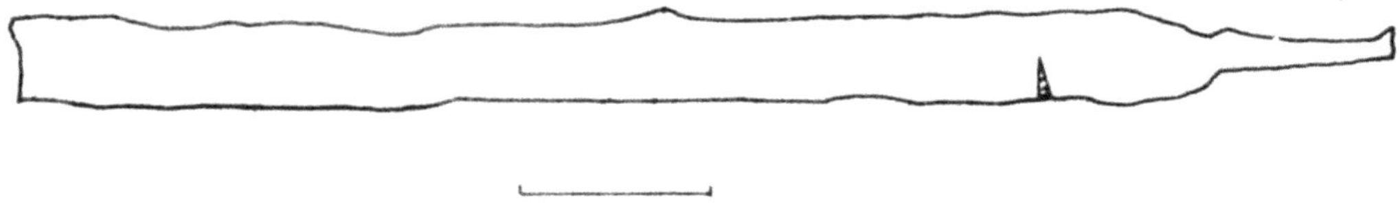

S16

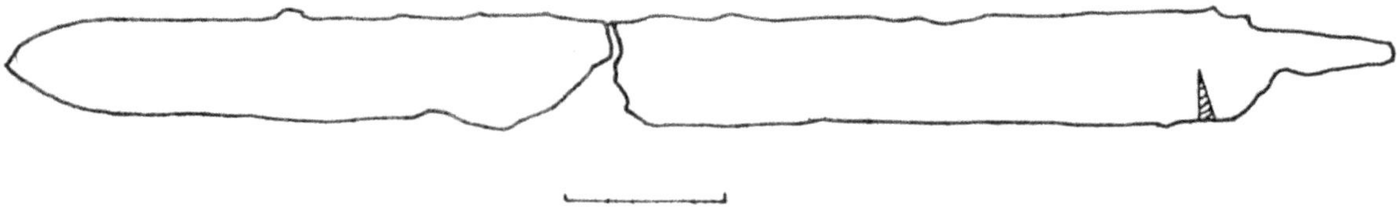

S17

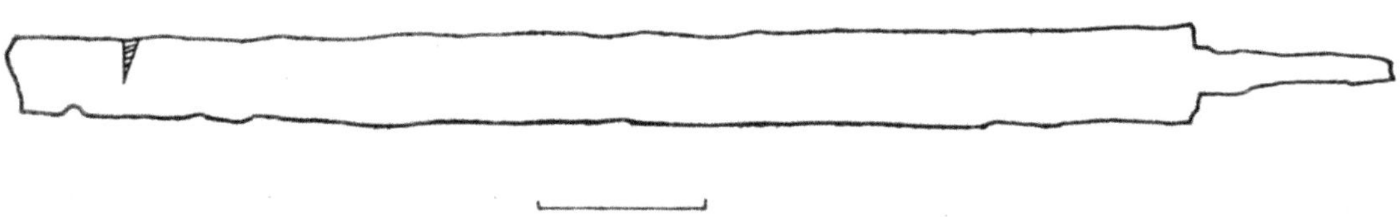

S18

S19

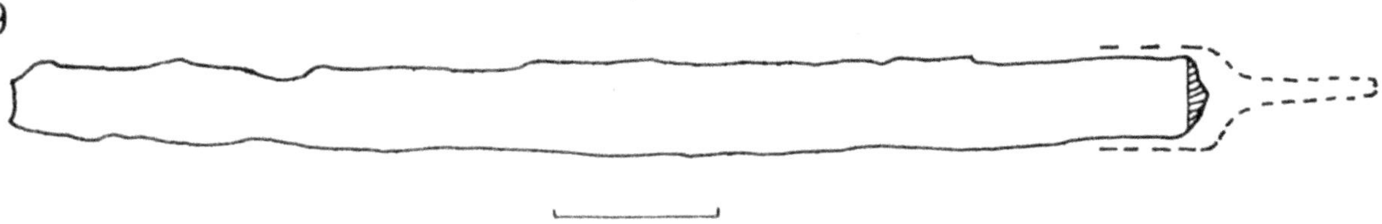

S20

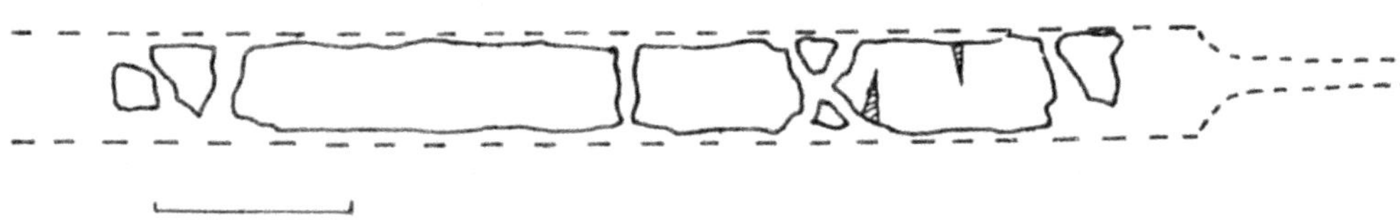

S23

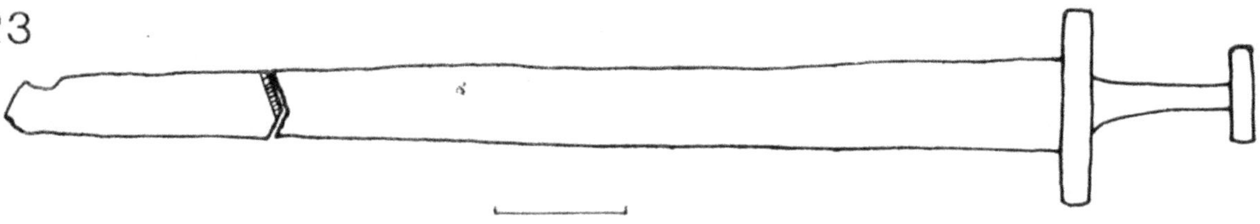

S24

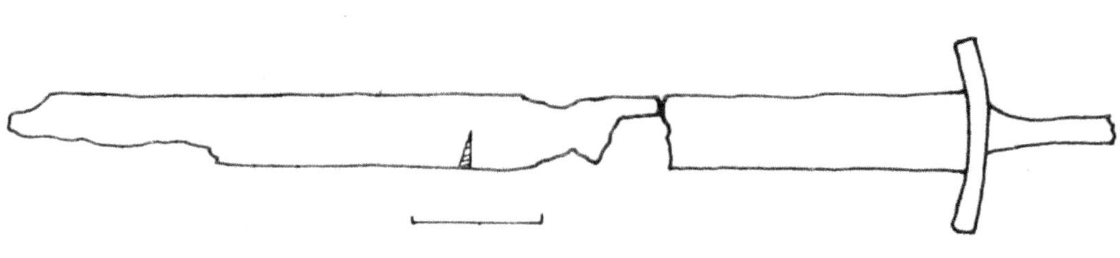

Fig. 63(2).

S25

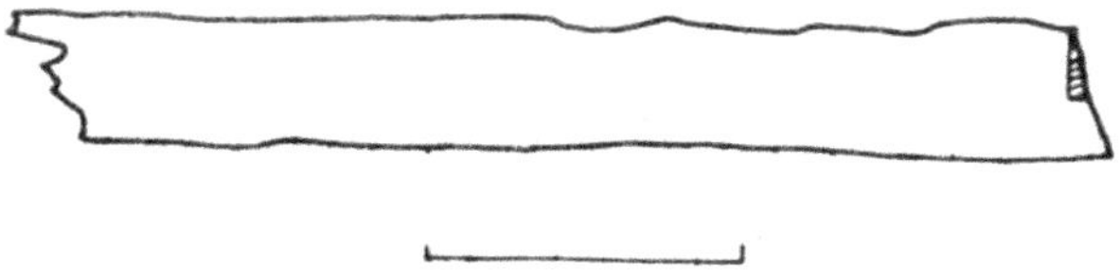

S26

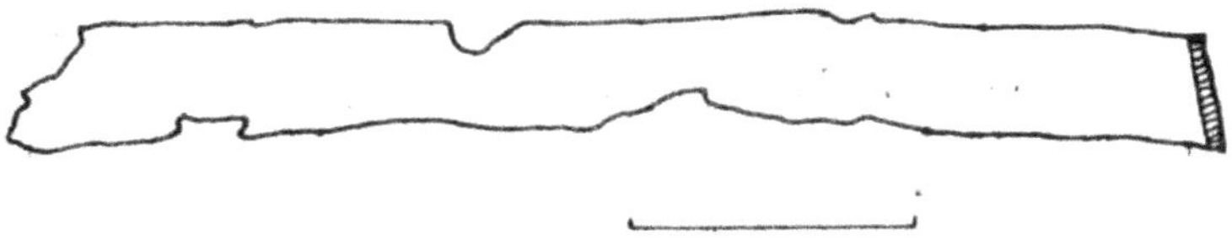

S27

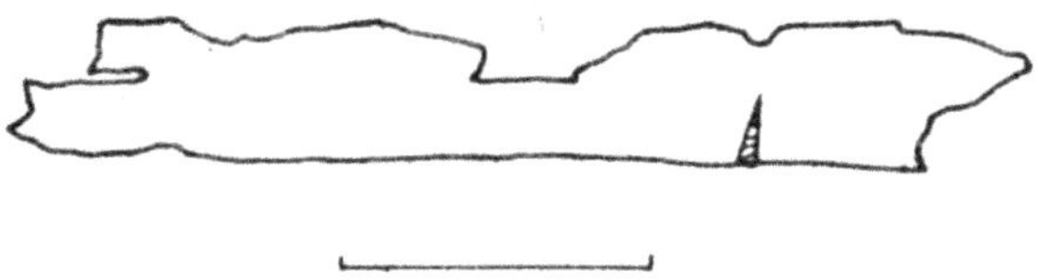

S28

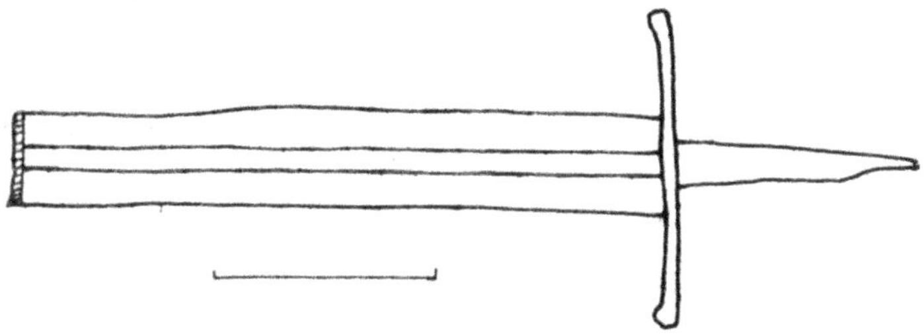

S29

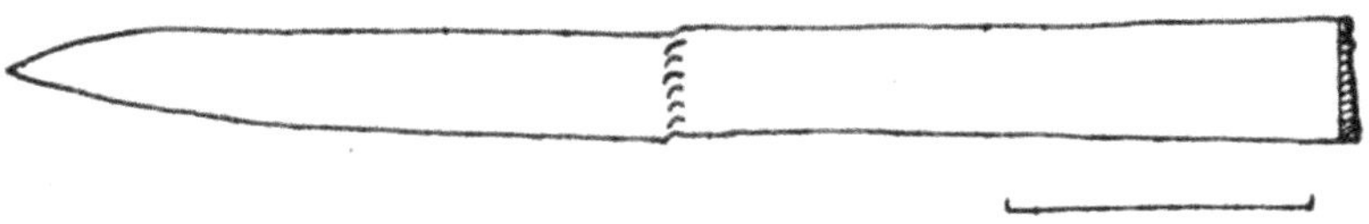

S30

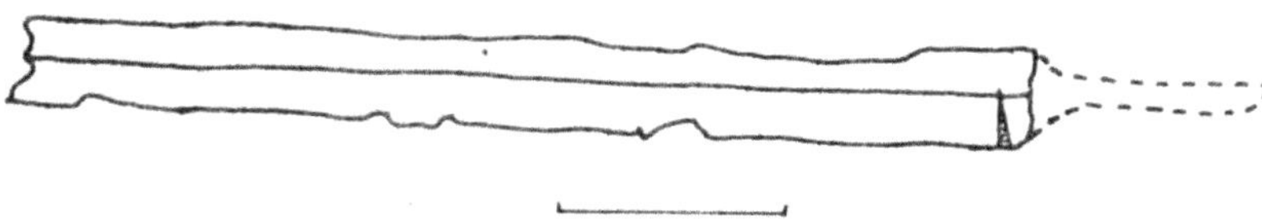

S31

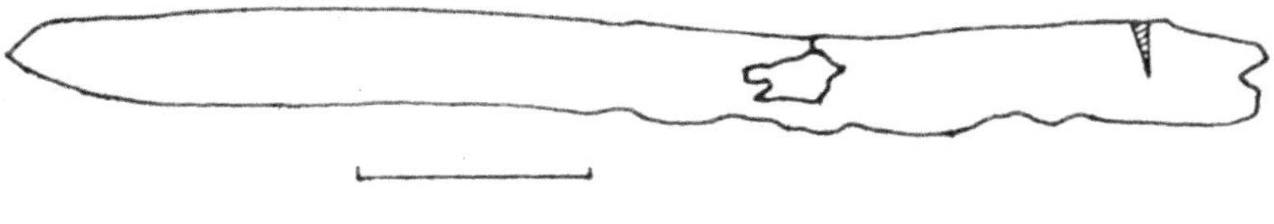

S32

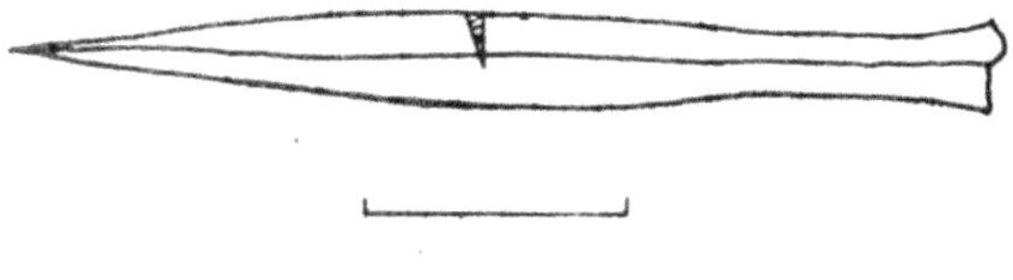

Fig. 63(3).

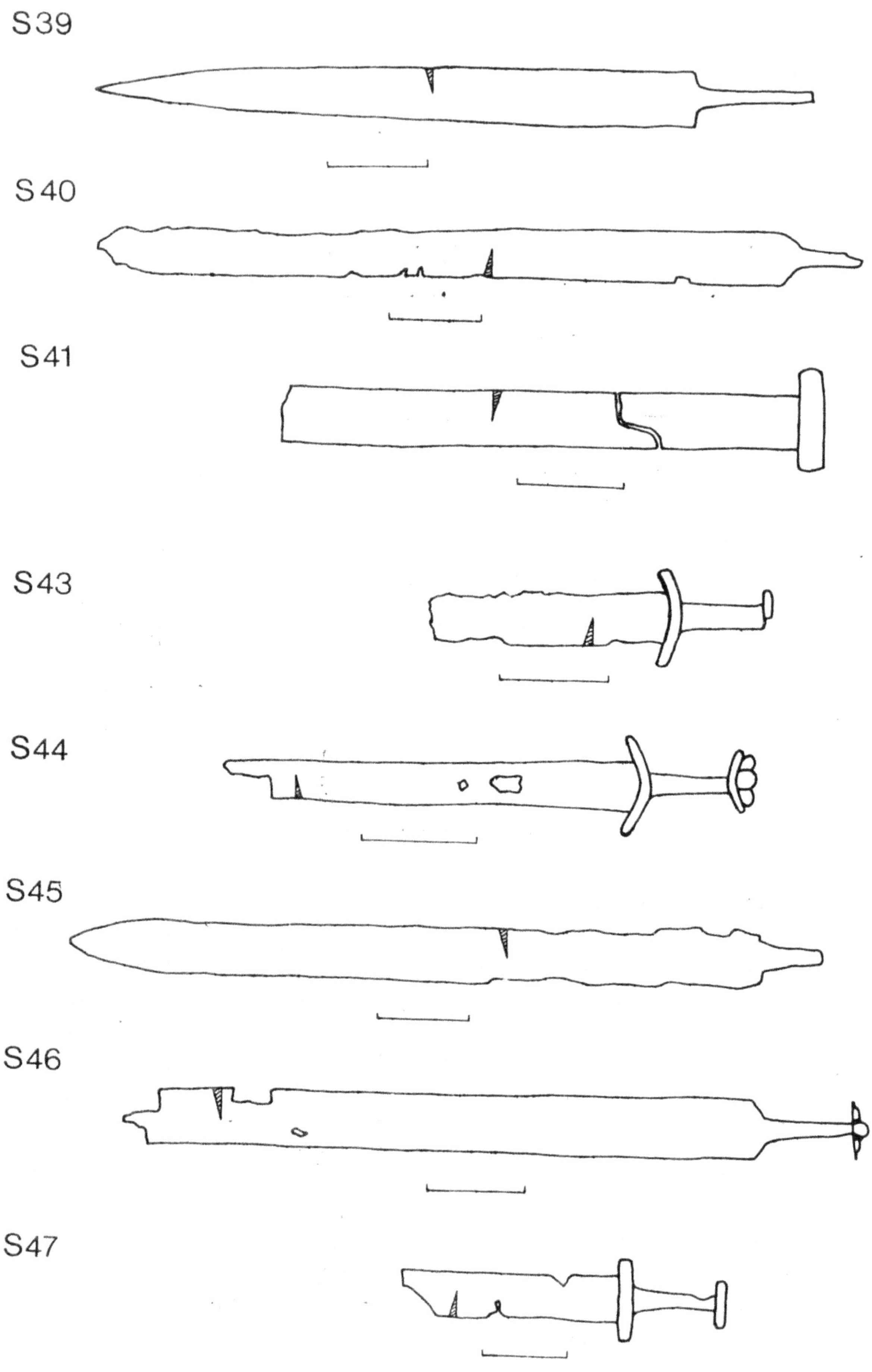


Fig. 63(4).

S48

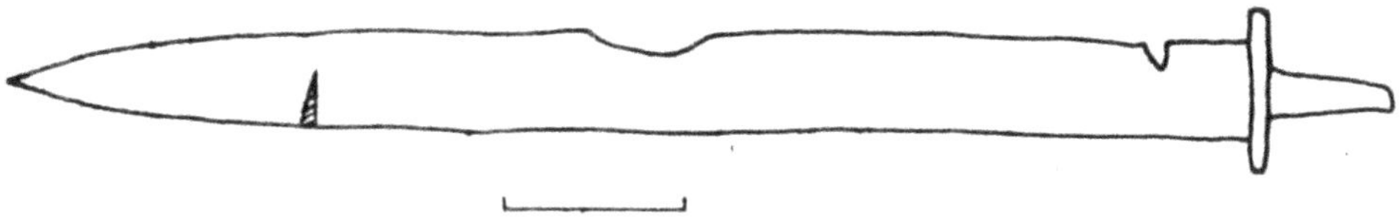

S49

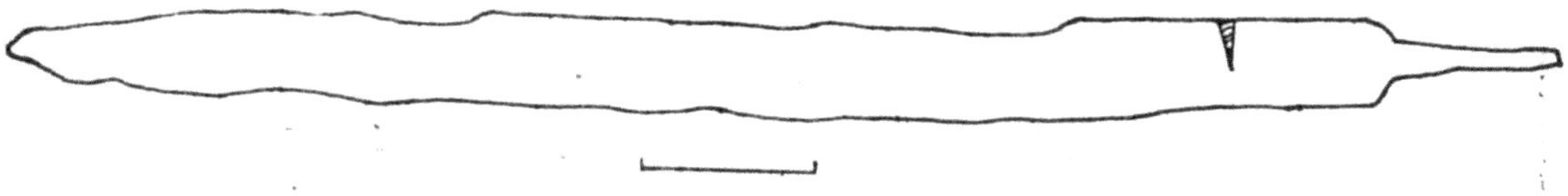

S51

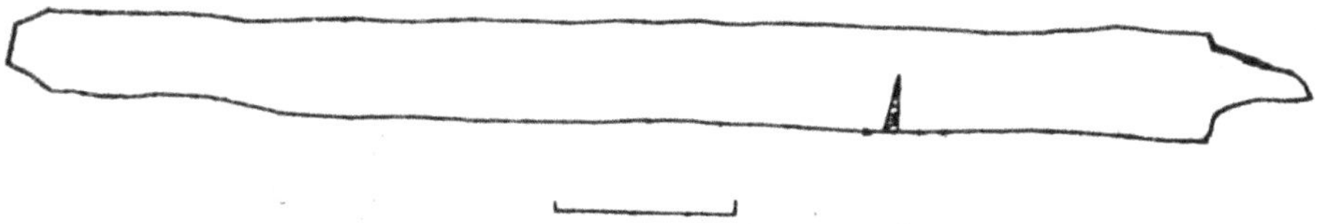

S52

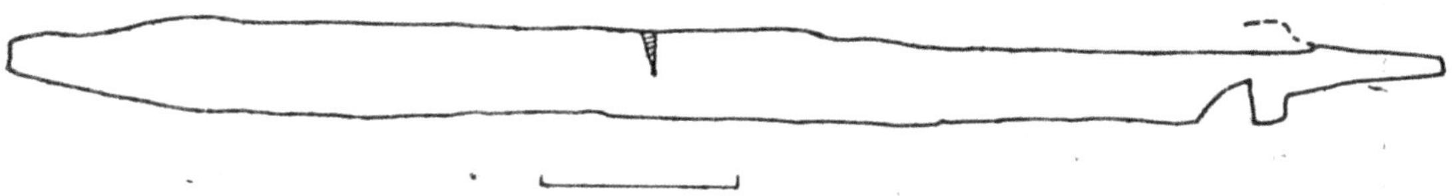

S53

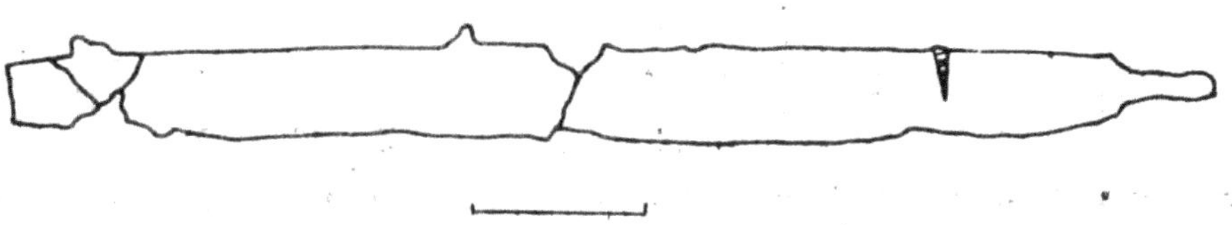

S54

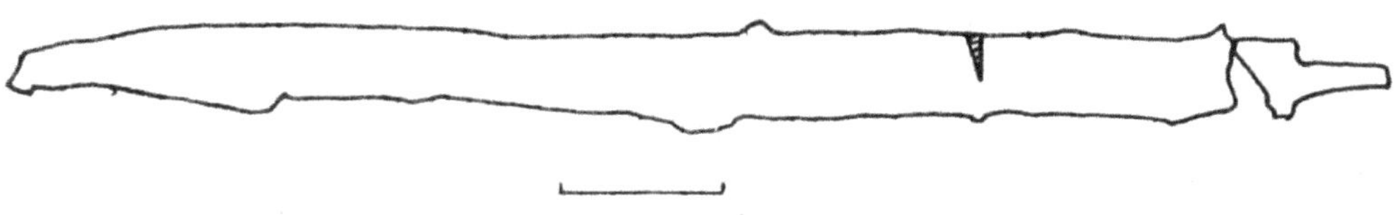

S55

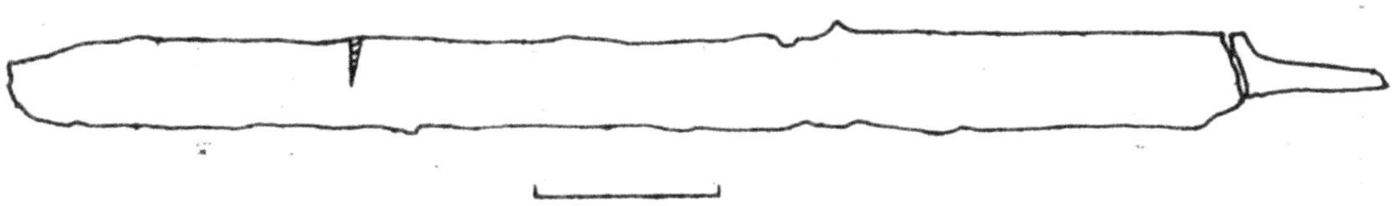

S56

Fig. 63(5).

Table N. Summary of Swords.

| No. | Site | Site Ref.No. | Date | Structure/Hardness HV Edge | Back/Core | Remarks |
|---|---|---|---|---|---|---|
| S 32 | Unknown (Reading Museum). | 131–61 | EIA | F+P/112–129 | | |
| S 29 | Thames at Little Wittenham. | Reading 141.60 (TCB) | LIA | F+P/214–329 | | |
| S 27 | " " " | 106.61 (1) | LIA | F+P/178–F/120 | | |
| S 39 | Whittlesey, Cambridge. | Peterbro' 28;65 | R–B | F+P+/185 | F+P/149 | Type 1 |
| S 13 | Kent. | Maidstone, Tomlinson Collection | 5–7th cent. | F+P/171–230 | | |
| S 20 | Lenham, Kent. | Maidstone 3.1977 | 6th cent. | P/298 | F+P/179 | PW |
| S 45 | Mitcham, Surrey. | Mus. London. 56.106/13 | " | F+P/200–221 | F+P/215 | PW |
| S 10 | " " | 56.106/14 | " | F+P/203 | F/96–123 | PW |
| S 55 | Sarre or Bifrons, Kent. | Maidstone KAS836 | 6–7th cent | M+T/581 | F/181 | PW |
| S 54 | " " | Maidstone KA834 | " | F(P)164–197 | F/172 | PW |
| S 53 | Sarre, Kent. | Maidstone | " | F/144 | F+P/159–217 | PW |
| S 52 | " " | Maidstone KAS823 | " | F/187 | F+P/159–162 | PW |
| S 51 | " " | Maidstone KAS820 | " | F+P/156–189 | | – |
| S 12 | Wickhambreux, Kent(?) | Maidstone KAS | " | F+P/206 | F/177–202 | PW |
| S 49 | Bifrons, Kent (?) | Maidstone Tomlinson Coll. KAS620 954 | " | F/158 | F/118–178 | PW |
| S 1 | Loveden Hill, Lincs. | Lincoln; 6.66 | " | F+P/165 | F/119–202 | PW |
| S 40 | Chesterton, Cambr. | Peterbro'. 16/1979/1 | 5–7th cent. | P+T/272 | F/230–243 | PW |

Table N. (continued).

| No. | Site | Site Ref.No. | Date | Structure/Hardness HV Edge | Back/Core | Remarks |
|---|---|---|---|---|---|---|
| S 14 | Eastrey, Kent. | Maidstone, A-S.121 | 6-7th cent. | F/234 | F/230-243 | PW |
| S 18 | Aylesford, Kent. | Maidstone | " | F+P/212 | P+F/187-212 | PW |
| S 19 | Wickhambreux, Kent (?). | Maidstone | " | F+P/241 | F/138-185 | PW |
| S 16 | Sarre, Kent. | Maidstone KAS818 | " | F(P)/170 | F+P/163-172 | PW |
| S 17 | " " | Maidstone KAS830 | " | M+T(564 | F+P/236 | PW |
| S 56 | Bifrons, Kent. | Maidstone | 6-7th cent. | F/152 | F/164-214 | PW |
| S 15 | Holborough, Kent. | " | 7th cent. | F/114 | F/129-217 | PW |
| S 31 | Caversham, Reading, (SU73867511) | Reading 51.80 | 7-9th cent. | M+T/775 | F/149-217 | PW |
| S 46 | Brentford, Middx. | Mus. Lond. 0.2112 | " | TM/647 | M/788-F/212 | PW |
| S 41 | Thames, Vauxhall. | Mus. Lond. A.13591 | 8-10th cent. | TM/490 | F/305 | PW |
| S 23 | Thames | Reading 112.66/1 | 9-10th cent. | TM+P/762 | P/428 | Inlaid |
| S 7 | Unknown (Thames?) | Mus. Lond. C2260 | " | M+T/477 | M+T+P/268 | Homog. |
| S 8 | " " | Mus. Lond. C2258 | " | F+SC/172 | F/81 | Steel/Fe |
| S 48 | Rochester, Kent | Maidstone | " | M/564 | F/163 | PW |
| S 47 | Thames, London Waterloo Br. | Mus. Lond. A3670 | 10th cent. | P(F)/239 | F+P/144 | Steel |
| S 24 | Cleeve, Goring on Thames | Reading 256:636(TCB) | 9-11th. cent | M+T/647 | F/221 | PW |

PW = Pattern-Welded.

Table N. Summary of Swords.

| No. | Site | Site Ref.No. | Date | Structure/Hardness HV Edge | Back/Core | Remarks |
|---|---|---|---|---|---|---|
| S 43 | Thames, nr. London. | Mus. Lond. A 17923 | 10-11th cent. | P/256 | P+F/179 | PW |
| S 25 | Reading, Oxford Rd. | Reading 16:26 | " | P(F)/272 | P+F/170 | Inlaid |
| S 44 | Thames at Brentford | Mus. Lond. A. 24419 | 11th cent. | P/325 | F/157 | PW |
| S 26 | Thames at Wallingford Bridge | Reading 1608:64 | 12-14th. cent. | P+F/195 | F/133 | Iron + Steel |
| S 28 | Unknown | Reading 56:80(HL) | 13-14th. | TM/496 | TM/393 | Steel |
| S 30 | Sunbury Weir, Middx. | Reading | Late Med. | P+F/236-171 | | Steel |

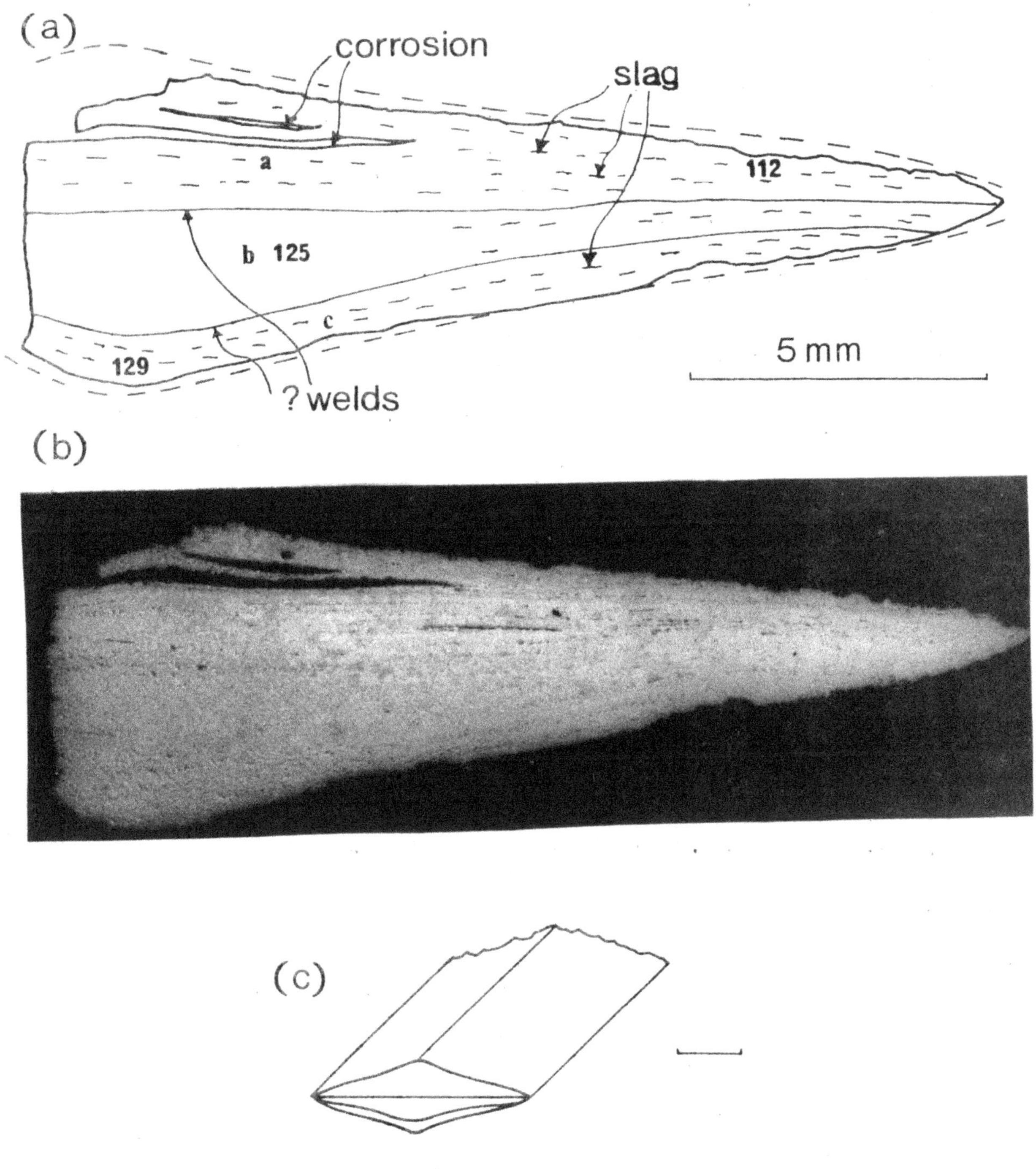


Fig. 64. Section through sword blade S 32 from Reading Museum.

The upper zone (a) was very similar to (c) although the parallel lines in (c) were a little less marked in appearance, although the difference was minimal. These parts consisted mainly of very large grained ferrite with lines of slag with a little pearlite present near the weld boundary with (b), again probably the result of carbon diffusion from (b) to (a) during the final forging of the blade. Somewhat more pearlite was present along some of the slag lines within area (c) than was the case with (c) but the carbon content was still very low, hardly more than 0.05% so that zones (a) and (c) represent strips of well forged, probably piled wrought iron of a high slag content that formed the outer parts of the blade.

The central area (b) was rather different from (a) and (c). It was fairly slag free and consisted of a fairly even distribution of pearlite, showing a low carbon content of about 0.1% and there did not appear to be any banding in this area. The pearlite occurred in a fairly large grain structure and showed a fairly marked tendency towards a non-equiaxial Widmanstätten type of distribution. This can be seen in the macrograph of Fig. 64 which also shows the marked difference in appearance between (a), (c) and (b). This contrast almost certainly means that area (b) was represented by a separate and differently prepared core piece of very low carbon steel that was sandwiched between the other pieces (a) and (c), the whole assembly then forged welded together. In each case the innermost line of slag spots of both (a) and (c) is most likely to mark the weld position. The three main parts of the blade identified as (a), (b) and (c) in the section gave hardnesses of 112, 125 and 129 HV respectively.

The more markedly ferritic areas inside the boundaries of (b) with (a), particularly near the cutting edge where very little pearlite remained, show that considerable decarburisation must have taken place before or during the final assembly of the three separate parts of the blade. Some carbon diffusion appears to have taken place across the weld boundaries between (a), (c) and (b) after they were welded together, probably during the final forging of the blade. This suggests that the final forging was fairly prolonged which would tend to be confirmed by the large grain size of the pearlite part (b). This large grain size is the probable result of grain growth within the austenitic temperature range which must have occurred at about 875°C or above, i.e. at or above the upper critical point for a 0.1% carbon steel. The very large grain size of the ferrite with parts (a) and (c) probably represents austenite grain growth that took place before the final assembly of the blade, therefore the temperature of the final forging is unlikely to have risen much above 925°C - approximately the temperature that the ferrite of these areas would re-crystallise as austenite. This means that the probable final forging temperature for this blade was between about 875° and 925°C.

S 29. Sword from the Thames at Little Wittenham, Berks (now Oxford). (Reading Museum, 141.60 (T.C.B.)). The lower surviving half of a sword blade (Fig. 63) found in the river Thames at Little Wittenham. It was bent sharply, at right angles, roughly in the middle and although the surface appeared to be fairly heavily pitted in places, the corrosion effects were mostly fairly even and the metallic core of the blade appeared to be quite well preserved. The blade had a convex profile but no other distinguishing characteristics which might indicate its date survived. The sharp bend in the blade was probably the result of more recent contact with a heavy object (?dredger) rather than having been bent nearer the time of deposition in the river. Date: ? late Iron Age (or later Medieval).

A transverse section was cut from just inside the broken end right across the blade. When viewed unetched, a rather variable slag content could be seen. It was fairly low and evenly spread as spots and flattened ribbons over the centre of the blade, whereas towards the cutting edges the proportion of slag increased somewhat and tended to be aligned in flow patterns, suggesting folds within the metal of the cutting edges. Some much larger, partially corroded slag inclusions suggested weld lines between the cutting edge and core zones of the blade section.

When etched the structure shown in the macrograph of Fig. 65 was revealed. A composite assembly consisting of two edge pieces and a core itself probably made up from 3 different parts. A simplified interpretative diagram of the macrograph is shown in Fig. 65c.

The cutting edge zones (a) and (b) show up fairly well in the macrograph outlined partly by slag inclusions and partly by the grey pearlitic areas, which in both cases appeared to be folded around a paler central area, which consisted mainly of ferrite with a small proportion of pearlite near the centre, with a carbon content of less than 0.1%. The proportion of pearlite gradually increased on either side to give a maximum carbon content of c. 0.3% in the darker grey areas. The pearlite of these edge zones was of varied grain size - much of it large and in a distinctive Widmanstätten spiky or columnar distribution.

The main core part (c) of the section consisted almost entirely of ferrite, although some carbon had diffused inwards from the edge parts (a) and (b). It also appeared that the core area (c) itself consisted of three parts, with the smaller areas (d) and (e) being fairly clearly visible on the macrograph. (d) and (e) appeared to have originally consisted of ferrite, although there had clearly been extensive carbon diffusion inwards from across the junctions with the edge parts (a) and (b).

The weld junctions between the different parts of the blade section were mostly fairly clear, as can be seen from the macrograph (Fig. 65). These junctions were mainly marked by lines of slag spots and ribbons and also by more massive slag inclusions which shows the weld junction between parts (a), (c) and (d). Also visible in the macrograph is a white line running down the centre of edge zone (a) approximately half way to the tip, which appears to double back on itself and run back the other way to give the appearance of two white lines running parallel to one another. Several broken lines of slag spots and ribbons ran parallel to the white line down one side of zone (a) then turned away from the tip and ran back the other side to give a folded appearance to this part of the section. The other edge zone (b) bore a marked similarity to (a) although the weld line between (b) and (c) was not clear.

The parallel slag lines and white line clearly visible in cutting edge piece (a) and, to a lesser extent, in (b), may be the result of forging and piling a strip of wrought iron, which was finally folded along its length. Both (a) and (b) appeared to have a steel jacket in each case which shows up as a grey pearlitic zone, gradually becoming paler towards the centre, the carbon contents varying from about 0.3% at the outside, to about 0.1% at the centre. This is probably fairly clear evidence that these cutting edge pieces were carburised after forging and (possible) piling and folding.

The weld junction between (a) and (d) is partly marked by a white line clearly visible on the macrograph (Fig. 65). The white line visible in

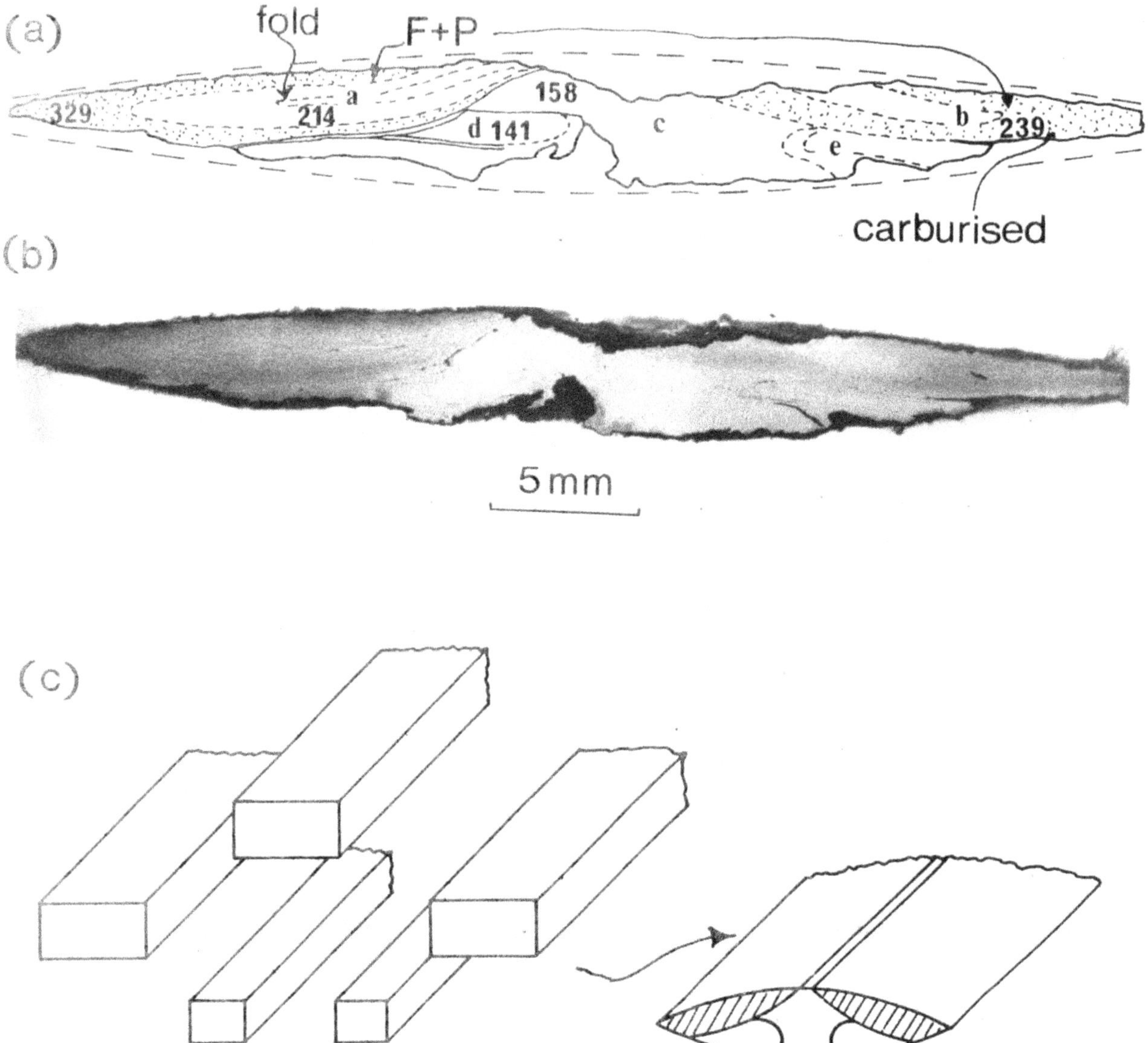


Fig. 65. Section through sword blade S 29 from the Thames at Little Wittenham.

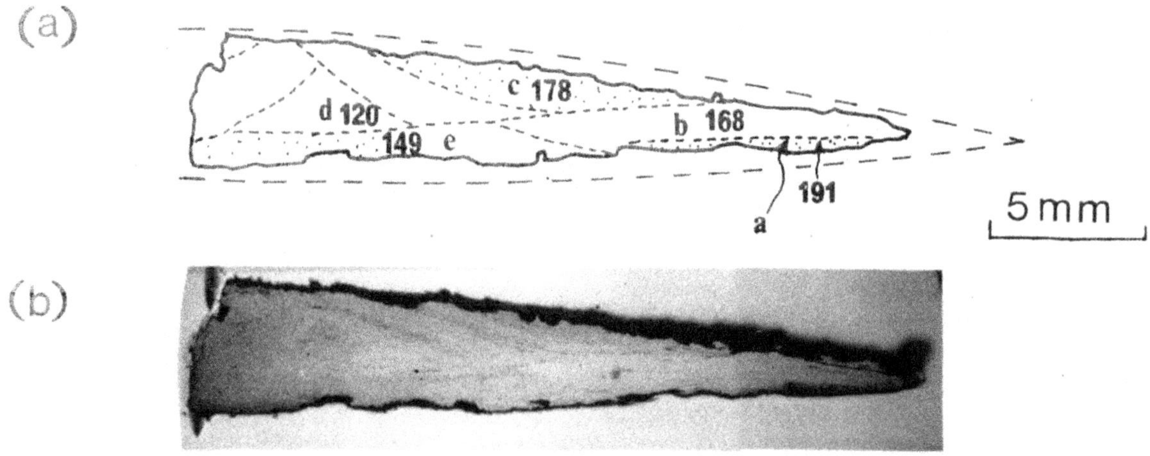


Fig. 66. Section through sword blade S 27 from Little Wittenham.

part (d) and slag lines visible in (d) and (e) may indicate that a piling technique was used in the preparation of these pieces.

The main central core area (c) did not contain readily discernible slag lines - it had a generally low slag content and appears, therefore, to have been a piece of well-prepared wrought iron.

The hardness of the cutting edge part (a), was 329 HV, near the tip, where the carbon content was higher, and 214 HV near the centre where it was lower. The iron of the central core pieces (c) and (d) gave hardnesses of 158 and 141 HV respectively, whereas the higher carbon part of the other cutting edge, (b), had a hardness of 239 HV.

S 27. Sword from River Thames at Little Wittenham, Berks (now Oxford). (Reading Museum, 106.61/1). The surviving fragment (Fig. 63) from the central part of another sword blade found in the River Thames at Little Wittenham, its condition being fairly typical of river finds. The ends were both corroded right away so that it was impossible to say whether the breaks were the result of corrosion or that the blade was already broken beforehand. The metal survival towards the middle of the fragment was quite good although here, as was especially noticeable towards the ends, heavy differential corrosion etching had produced a very pronounced wavy wood-grain effect. Not much in the way of corrosion products could be seen which may indicate that this fragment has been electrolytically cleaned since its discovery, a process which would have highlighted the corrosion etching effects. The blade had a convex profile but a lack of distinguishing characteristics make even an approximate date difficult to assign. It could be later Iron Age (or later Medieval).

A V-shaped transverse section extending approximately half way across the blade was taken from where the metal appeared to survive best. Unetched, a very variable slag content was visible, mostly as 2 phase flattened ribbons and spots.

When etched it became clear (Fig. 66) that a number of different components were present in the blade. These showed up as different areas of metal with contrasting carbon and slag contents. The different areas were irregular in size and shape and whilst within each area the slag ribbons were mostly parallel to one another, the direction of these varied from area to area. Each area was either outlined by a discontinuous line of slag spots and ribbons, or was clear from the change in orientation of the slag inclusions.

An interpretative view of the structure visible in the macrograph, Fig. 66b, is given in the diagram, Fig. 66a. The grey areas represent pearlitic zones of variable carbon content, mostly 0.1-0.3%. The pearlite was generally fairly fine grained and partially lamellar and partially amorphous in form, although some patches in area (a) exhibited a large grained Widmanstätten formation (Fig. 66), probably a result of localised overheating during forging. The predominant partially lamellar form of the pearlite is probably indicative of fairly slow air cooling of the blade after the final forging.

The pale zones consisted mainly of ferrite into which some carbon had diffused from adjoining pearlitic areas, the diffusion being rather uneven. Much of area (b) showed a large grain structure which included an even distribution of a short rod- or needle-like formation (Fig. 66), probably of iron nitride or carbide. The variability of the slag content between pearlitic areas, as well as between pearlitic and ferritic areas, can be seen in the macrograph in Fig. 66 which shows the junction between various parts in the centre of the section. Hardness measurements were made and the results are as follows: (a) ferrite + pearlite, 0.2–0.3% C - 191 HV; (b) ferrite + pearlite - 168 HV; (c) ferrite + pearlite, 0.2–0.3% C - 178 HV; (d) ferrite - 120 HV; (e) ferrite + pearlite, 0.1–0.2% C - 149 HV.

It appears that this blade was made by welding together a series of wrought iron and low carbon steel strips, probably about 18 in all, which were interleaved and overlapped and forged out along the length of the blade. A three dimensional reconstruction of this is shown in Fig. 66. The wavy wood grain effect has resulted from differential corrosion along the slag of weld lines between the components, and along the slag ribbons within each component piece.

S 39. Sword from Funtham's Lane, Whittlesey (T.L.238968); (Peterborough Museum, 28:65). This sword (Fig. 63) was found in 1965 in a claypit owned by Central Brick Company. It lay 2.5-3m below modern ground level, under material containing 2nd - 4th century A.D. pottery. The sword is complete and is very well preserved, probably as a result of having been buried in waterlogged conditions low down near the clay-land subsoil. It is 0.72 m long with an unusually thin blade with a flat profile and maximum thickness 3 mm. A well-preserved bone grip survived in place on the tang and traces of a possible fabric scabbard lining were still adhering to the blade which measures 47 mm at its widest.

A late 1st, or early 2nd century date for the sword may be indicated by the material found above it and it is probably of the spatha type[95], its length comparing more closely with the two spathae from Canterbury (Webster et al,[68] 1982, p. 185), rather than the shorter gladius. The sword may have belonged to a Roman soldier, possibly attached to the fort at Longthorpe which lay about 5 miles to the West of the findspot.

A V-shaped section was cut from the blade approximately two-thirds of the way up from the tip. The sectioning was done at right angles to the cutting edge and extended to just over half way across the blade.

When the section was examined before etching a fairly small amount of slag was visible in the form of small spots and parallel flattened ribbon-like formations. None of these could easily be interpreted as a weld line.

After etching a relatively carbon rich (dark) zone (b) of pearlite and ferrite could be seen running along the centre of the section (Fig. 67). The carbon content of this zone varied between approximately 0.3% nearer the cutting edge and 0.2% nearer the centre of the blade. The carbon content of the paler zone (a) and (c) on either side varied between almost pure ferrite (the areas lacking in black spots in Fig. 67e) and about 0.2%.

Discontinuous lines of slag ribbons were visible (Fig. 67d, f) on either side of the darker central zone. These appear to mark the position of welds between the outer and central portions of the blade, which have otherwise become very blurred by the extensive diffusion of carbon across the welds from the higher to the lower carbon zones. The paler outer zones contained more parallel slag ribbons than the darker, carbon rich central

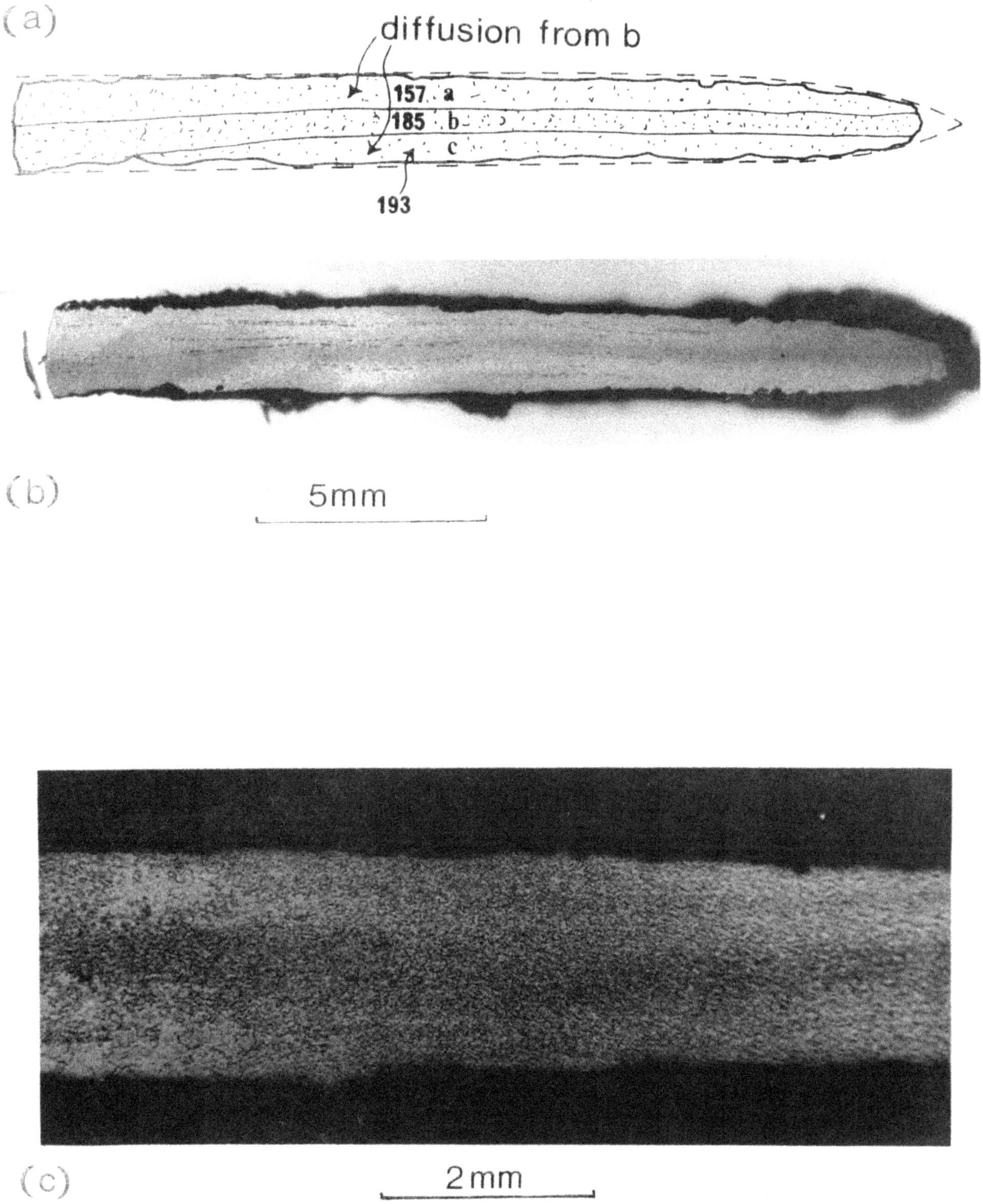


Fig. 67. Examination of Roman sword S 39 from Funtham's Lane, Whittlesey and micrographs: (a) section; (b) Macro X 7.5; (c) Macro X 35; (d) Micrograph of central strip near cutting edge showing ferrite + pearlite X 450; (e) Micrograph of central strip near centre of blade X 225; (f) Micrograph showing full section near centre X 60.

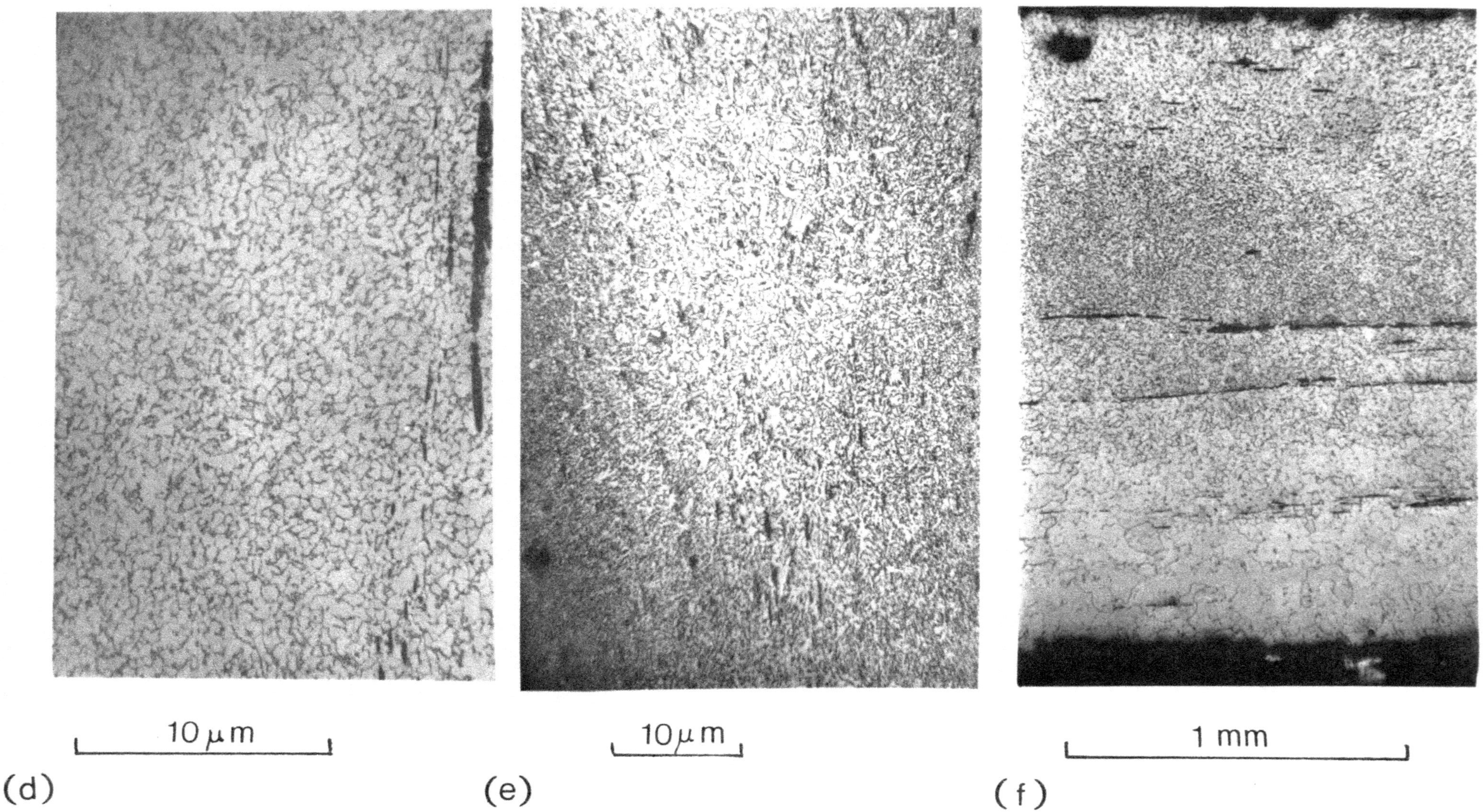


Fig. 67 (continued).

area. The outer zones also exhibited some darker banding within them where there was no indication of further weld lines. The three main zones (a), (b) and (c) gave hardnesses of 157, 185 and 193 HV, at three points across the centre of the section. (a) consisted of ferrite with some pearlite and (b) and (c) of ferrite with rather more pearlite.

It appears probable that the blade was constructed of three main parts; a central fairly low carbon steel strip welded between two very low carbon wrought iron strips to form a sandwich. The central steel strip may have begun as a well-forged strip of wrought iron, carburised for long enough (several hours) to give a fairly even carbon distribution throughout, or alternatively, it may have been formed from a steel portion of a bloom, well forged to give an even carbon and low slag content. It is not clear from the observed section which was the case. The somewhat streaky and banded nature of the outer portions are probably the result of the fairly extensive forging and piling of the wrought iron of these strips, which may have originally contained localised patches of low-carbon pearlite.

Without further treatment this blade would have been typical of a Type 1 blade construction. In this case however, the subsequent diffusion of carbon from the central, to the outer parts of the blade, has been so great that the original structure has almost disappeared, to give the appearance of a generally more homogeneous low carbon steel which is presumably what the smith was trying to achieve. After the sandwich type of assembly and subsequent forging, the blade must have been held for a fairly long time (probably several hours) at about 750°C in the part of the forge where reducing conditions predominated. This is indicated by the lack of surface decarburisation from the blade. The carbon of both the central and outer zones of the blade is distributed as fairly even fine-grained pearlite whose structure is mainly irresolvable (optically), although a somewhat spheroidal or granular form is just visible in some areas - mainly those of a lower carbon content.

S 13. Sword, from Kent: (Maidstone Museum, Tomlinson bequest). The upper, approximately two thirds, part of a sword blade (Fig. 63) of which the provenance is uncertain. Other finds from the Tomlinson bequest in a similar condition reputedly[69] come from the Anglo-Saxon cemetery at Bifrons and these include the unidentifiable remains of a sword (K.A.S. 620, 1954.20) and a weaving batten (K.A.S. 620, 1954.19). The surviving blade fragment was 61.8 cm long by 5.8 cm wide with a very approximate thickness of 0.5 cm. It was in a poor state, the surface having already corroded and flaked off and what was left was deeply pitted although it did appear to have some metallic core surviving. The lack of provenance is unfortunate as there were no particular aspects to the blade to suggest any particular date range. The overall form was of a wide blade with a shallow convex or flattened profile and as such could belong to any date range between the late Iron Age to the late Saxon period although probably not outside these limits. An Early Saxon, 6-7th century A.D., date may be most likely in view of the possible provenance.

A V-shaped section extending approximately halfway across the width was cut from part of the blade 16.0 cm from the upper end. Before etching a very varied slag content could be seen across the section. This differed from quite low in some parts to quite high in others and the variations occurred roughly in horizontal bands across the section. The slag occurred as small spots or well flattened inclusions, both one phase - dark

(? fayalite) and two phase - light and dark (? wüstite and fayalite or a glass).

When etched with nital the structure shown in Fig. 68 showed up. It consisted of a number of horizontal laminations, possibly six altogether, and a separate, welded on, cutting edge. This cutting edge piece, (a) in Fig. 68, appeared to have been a strip welded on to one side of the already partly assembled blade to form one of the cutting edges and it occupied about half the thickness of the blade at this point. It seems probable that this arrangment was mirrored on the opposite side of the blade and the reconstruction diagram of Fig. 68c has been drawn to show this. A large curved ribbon-like slag inclusion was trapped in the weld between piece (a) and the rest of the blade. Part (a), itself, varied gradually between medium grain size ferrite mixed with a little pearlite near the inside, and large grained ferrite near the tip of the cutting edge. The ferrite was equiaxed in each case. At most the carbon content was well under 0.1% so that part (a) appears to have been a piece of wrought iron containing a little carbon, possibly an accidental inclusion although it may have represented an attempt to produce a low carbon steel which subsequently became partially decarburised.

The remainder of the blade section consisted of a possible total of six horizontal bands, (b) - (g) in Fig. 68. These were partially discernable as areas of differing, quite low - quite high, slag content and were partly divided by broken lines of slag inclusions which coincided with faint grey lines of pearlite, and in places these were highlighted by slight traces of white lines which did not show up well against a pale ferritic background. These lines appear to mark weld positions between the horizontal parts or laminations of which the main part of the blade was composed. The slag content of bands (c) and (g) was low, but was quite high in (b), (d), (e) and (f) and the mainly well flattened inclusions were aligned along the main (horizontal) axis of the section. Much Neumann banding was also visible across this main part of the section indicative of some cold hammering after the final hot forging of the blade.

Hardness readings were taken across the section and covered bands (c) to (g). The results were as follows: (c), 171 and 185 HV; (d), 215 and 230 HV; (e), 204 and 215 HV; (f), 181 and 183 HV. These values suggest a fairly high phosphorus content throughout the blade.

The main part of the blade appeared to be a piled structure of wrought iron laminations. The extra piece welded along one edge (and probably the other as well) may have been a largely unsuccessful attempt to provide a harder and tougher cutting edge.

<u>S 20. Sword, from Lenham, Kent.</u> (Maidstone Museum, 3.1977). The very fragmentary and barely recognisable remains of a sword from an Anglo-Saxon grave found by chance during building work. Parts of two graves were uncovered and recorded[70] and the sword and a number of associated finds retrieved. These were almost completely mineralised and now only survive in a very fragmentary state (Fig. 63). An iron shield boss would appear to indicate a date probably within the 6th century A.D. and the graves probably belonged to a more extensive cemetery of this date. The main blade fragment to survive intact, although with little or no metal left inside it, was joined (by corrosion) to a knife or small scramasax possibly from the adjacent grave. The broken sections showed the blade to have had a

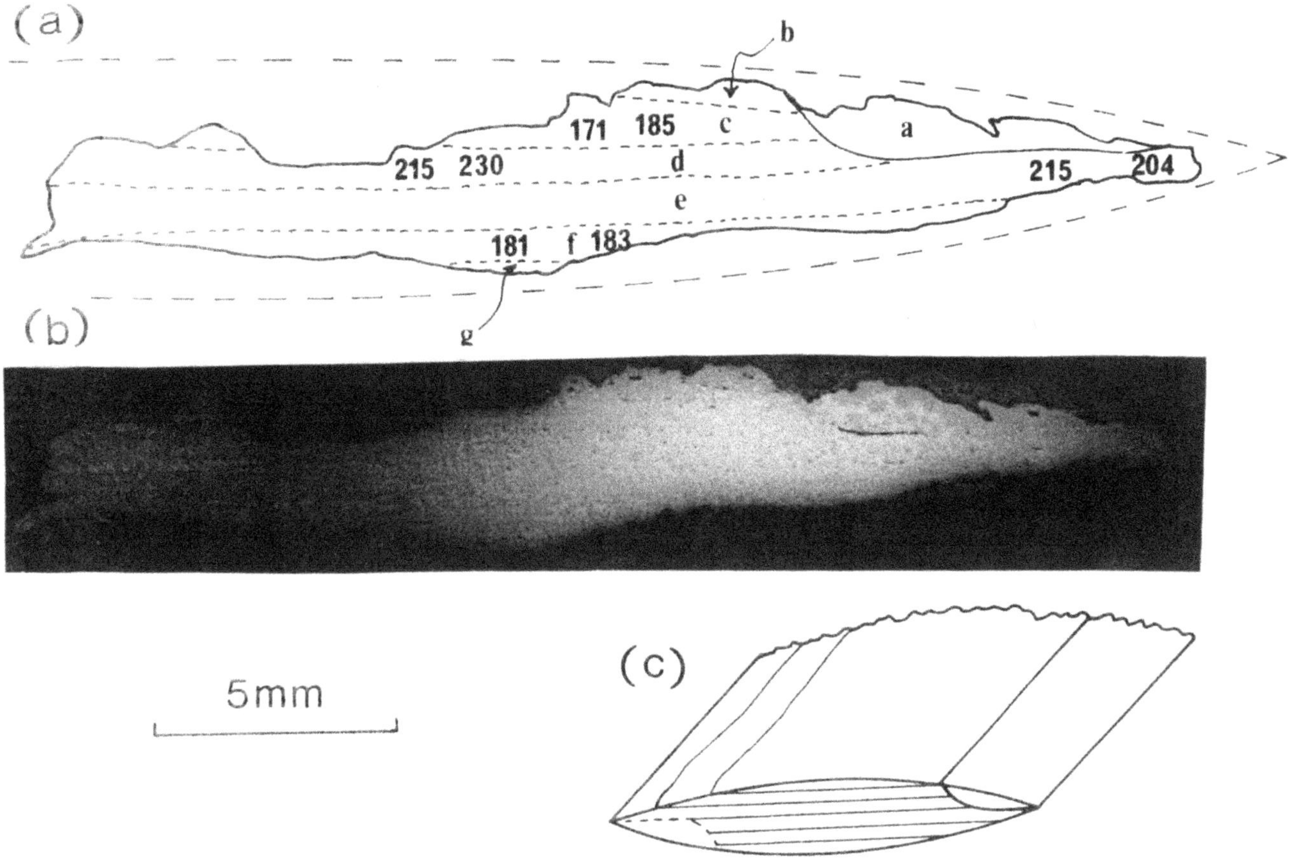


Fig. 68. Section of sword S 13 from Kent and three-dimensional view.]

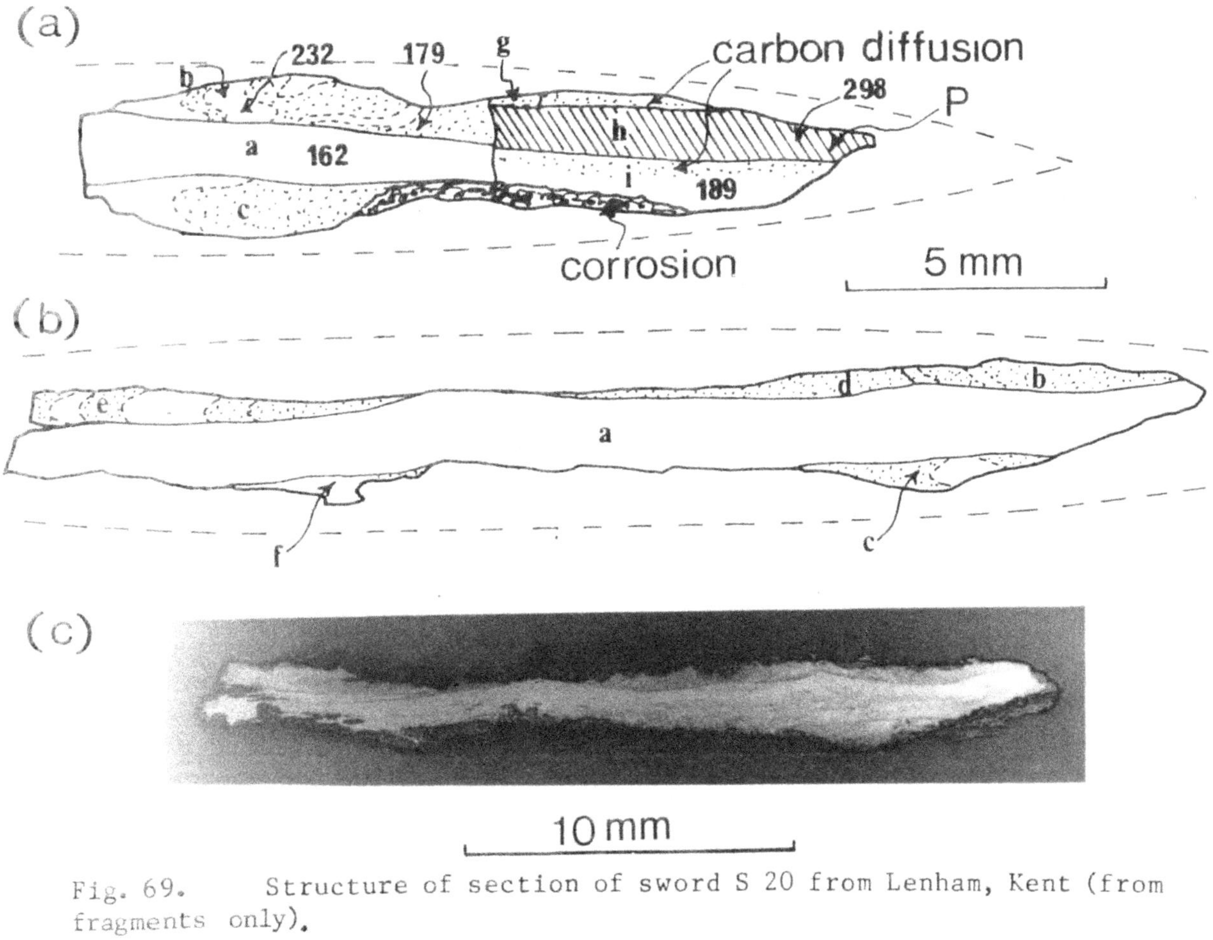


Fig. 69. Structure of section of sword S 20 from Lenham, Kent (from fragments only).

flattened convex profile. From the photographs taken when the sword was found it would appear to have been about 80 cm in length and its width was approximately 5.4 cm. Its original thickness can only be estimated from the corroded section but was probably about 0.5 - 0.6 cm. The outer parts of the sword blade fragments were coated with a thick orange-brown sandy incrustation which probably included the remains of a scabbard.

Several radiographs showed slight indications of a pattern-welded core to the blade, which appeared to consist of three adjacent bands.

One of the incrusted surface fragments was, fortunately, found to have attached to it a small piece of blade which still contained some metal. This was removed and radiographed separately with good results. The radiograph showed that the central core part did indeed consist of three adjacent pattern-welded bands and also that these were superimposed on a further set of three adjacent bands. The chevron pattern created by the twisted elements showed up quite clearly in places and it also appeared that these probably alternated, at least in part, with untwisted or straight 'grained' portions. Unfortunately from such a small fragment it was impossible to say whether or not this formed part of a design which alternated between twisted and straight grained elements along the whole length of the blade. The cutting edge part did not exhibit any observable structure on X-ray. Fig. 69c shows a reconstruction of the pattern indicated by the radiograph of the small blade fragment.

Two transverse metallographic sections were taken from the small surviving metallic blade fragment so that the combined results from them would provide some information about both the pattern-welded core and the cutting edge. When viewed unetched the 'core' section showed an uneven and varied slag content which appeared to be mostly present as one phase (dark grey - ? fayalite) fairly small flattened spots and ribbons mostly segregated along broken lines giving a somewhat banded appearance in places. Slag was also present near the edge of the section, as 2 phase spots, ribbons and also as larger irregular nodular inclusions. The section that included part of the cutting edge showed less slag inclusions - mainly fairly small spots and ribbons confined to the central part of the wider end of the section.

Etching with nital (Fig. 69) showed the section of the core area to be composed of a central piece or strip (Fig. 69, (b)) on either side of which the three adjacent composite rods, seen in radiograph, had been welded and only quite small areas (Fig. 69b, (b)-(e)) of these rods survived.

The central core piece (a) consisted mostly of very large grained ferrite with some streaks of slag spots, ribbons and flattish patches with some pearlite - mostly very little at the grain boundaries. Much of part (a) was covered with short rod-like nitride or carbide needles (Fig. 69) which tended to be concentrated in the areas of largest ferrite grain size. Much Neumann banding was also present in a patch near the centre of (a) (Fig. 69a and b), indicating that some final cold hammering took place.

From what remained of the pattern-welded pieces, (b)-(e) it could be seen that they were similar, with alternating pale and dark bands with a quite sharply folded appearance indicating the twisted nature of these rods. The pale bands consisted mainly of very large grained ferrite which in a few places contained a lot of slag spots and ribbons although overall the slag

(d)

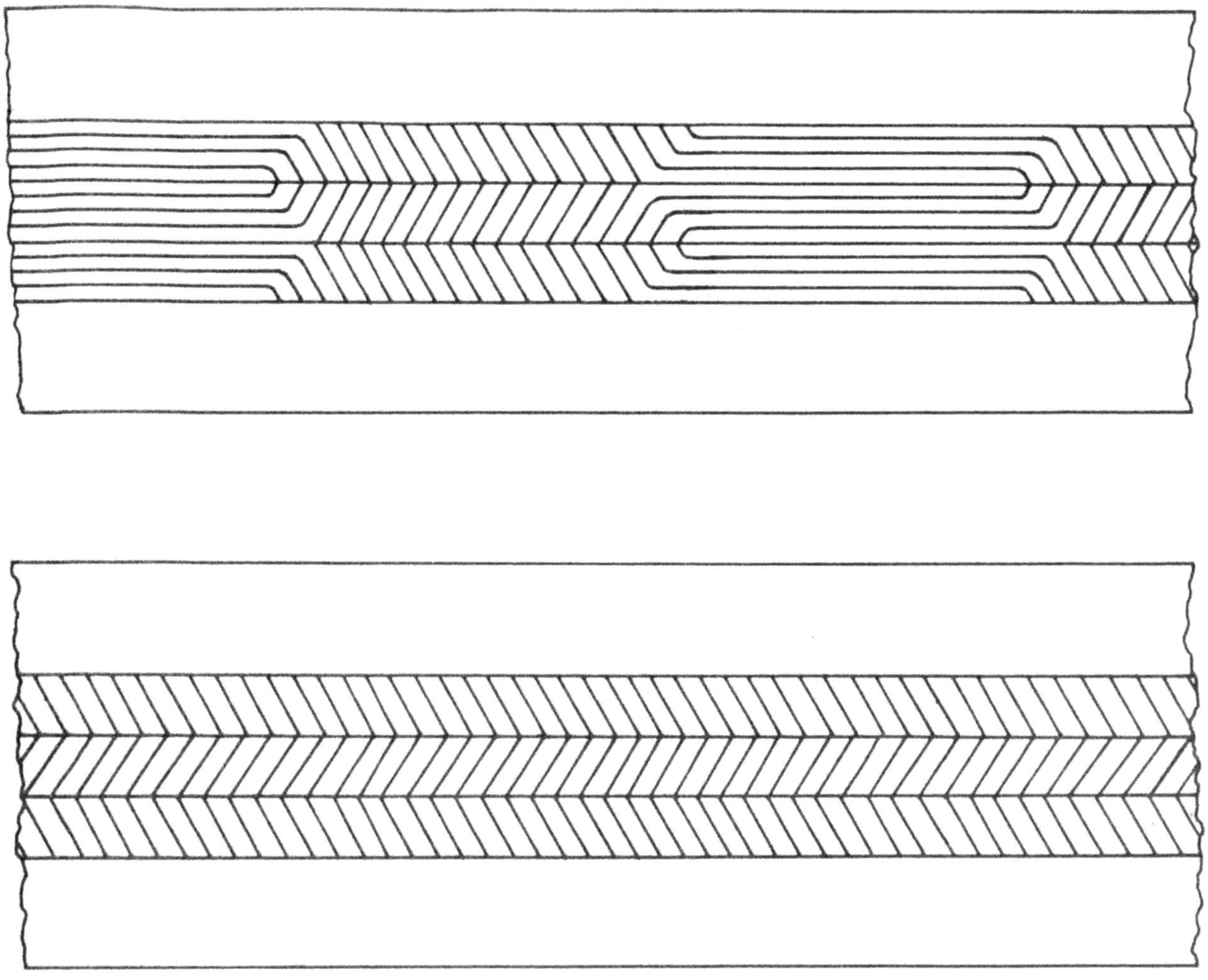

(e)

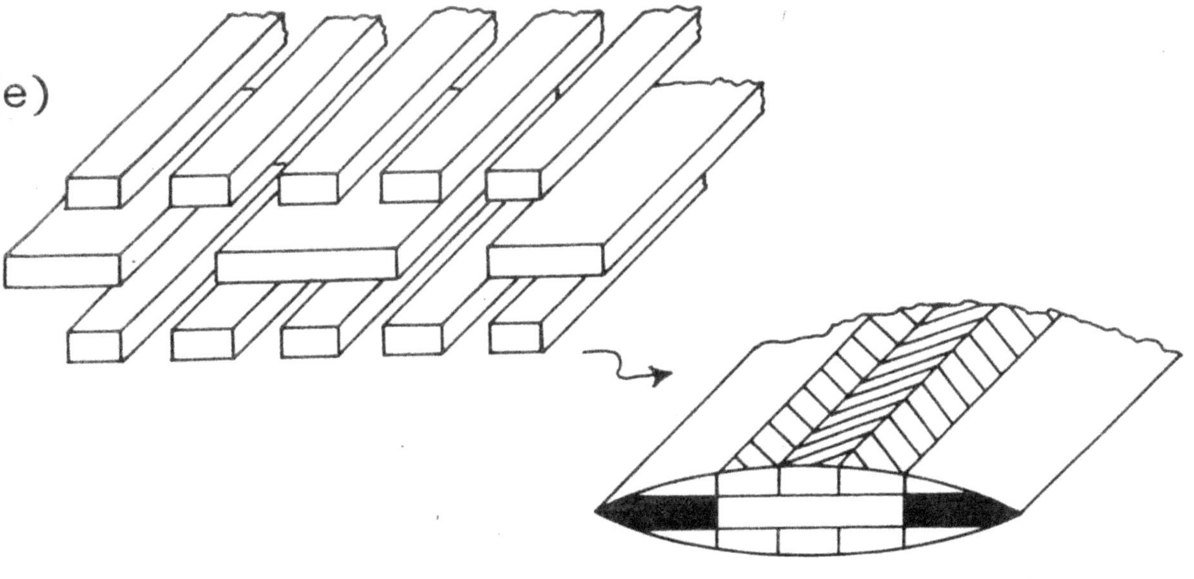

Fig. 69 (continued) surface detail drawn from X-ray and three-dimensional view.

content was quite low. The slag also followed the lines of the folds which tended to emphasise this appearance. Nitride or carbide needles were again visible mainly in the areas with the largest ferrite grains.

The darker or grey bands consisted of fine grain pearlite and ferrite with very little slag. The pearlite was difficult to resolve optically but a lamellar structure was visible in places; the carbon content for these grey bands was approximately 0.1%, i.e. low carbon iron, not greatly different to the wrought iron of the rest of the section.

The welds between the pattern-welded pieces and the central core piece (a) were of good quality with only a few flattened slag inclusions showing along them. They were shown up mainly by thin but fairly clear grey lines of pearlite.

On the second section (Fig. 69a) the central core strip (f) and pattern-welded rod parts, (g) and (h) compared closely with those seen on the first section (and described above). The weld junction between the core and the cutting edge was not easy to identify as it was not clearly marked either by slag inclusions or by the thin pearlite lines that marked weld positions in the core zone of the blade sections.

The cutting edge contained little slag and appeared to be composed of a sandwich of three horizontal bands, g, h, and i, although there were no obvious signs (slag, pearlite or ferritic lines) to mark any welds between. The very dark central band (h) consisted mostly of large grained lamellar pearlite with a carbon content of about 0.6–0.7% i.e. a fairly high carbon steel. Some of the largest grains exhibit a Widmanstätten structure. The paler grey outer bands (g) and (i) consisted mostly of a fairly even medium-fine grained mixture of ferrite and pearlite, partially lamellar and partially unresolved. The change between the outer bands (g) and (i) and the central band (h) was fairly abrupt and even on either side. The cutting edge therefore appears to have been a sandwich of 2 pieces of low carbon iron (similar to the pearlitic bands of the composite rods) with a piece of high carbon steel between. The skill in welding these pieces was clearly very high.

Five hardness readings were taken on section (c) which included part of the cutting edge and part of the core of the blade. The central steel band (d) of the cutting edge gave a value of 298 HV and the other lower carbon part (f) a value of 189 HV. The central iron core had a hardness of 162 HV. One of the darker low-carbon iron bands in the pattern-welded part (b) gave a value of 179 HV whereas an adjacent very large grain ferrite band has a hardness of 232 HV. This probably indicates a high phosphorus content for this and for the other pale coarse grained ferrite bands observed in the pattern-welded parts.

S 45. A Sword from Mitcham, Surrey. (Museum of London, 56. 106/13). The blade was complete except for part of the tang (Fig. 63). It came from the Anglo-Saxon cemetery found here and is dated to the 6th century by comparison with Wheeler (Wheeler, 1927, p. 175). The blade was well-corroded and was 84.9 cm. long. Its width was 6.0 cm max. and its thickness about 0.5 cm max.

Radiography showed that it was pattern-welded with a double chevron pattern running down the central area (Fig. 70, c). This was composed of 4

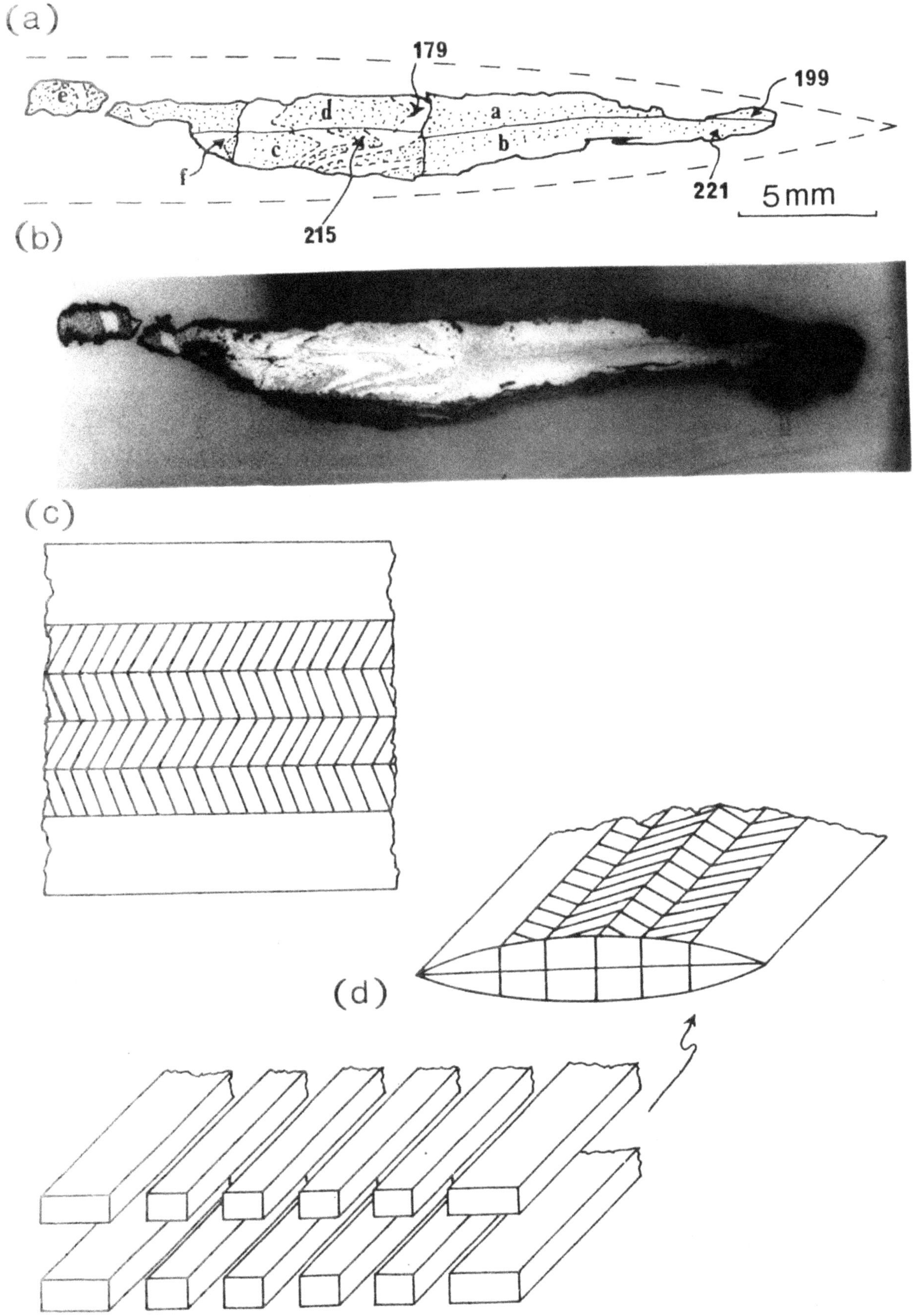


Fig. 70. Section through sword S 45 from Mitcham, Surrey X 5. + surface detail drawn from X-ray and three-dimensional views.

twisted rods welded side by side.

A section was taken to about half way across the blade, 31 cm down from one shoulder. (Fig. 70). Slag stringers of varying density were visible. After etching it was clear that 6 separate parts had been welded together; two halves of the cutting edge and 4 core parts, each composite. The two halves of the cutting edge were mirror-images of each other along either side of a central weld. This weld was bordered on both sides with a line of lamellar pearlite containing no more than 0.1% C. Away from the weld the structure was coarse-grained ferrite.

The composite parts consisted, as usual, of bands of ferrite alternating with bands of ferrite + pearlite. There was no central core-piece. Neumann banding was again seen in both the ferrite of the core parts and the cutting edge suggesting that the blade was subjected to cold work after hot forging. The hardness of the edges was 199 - 221 HV and the ferrite of the composite rods, 179 HV, while that of the darker areas was 215 HV.

Assuming symmetry on either side of the blade, it would appear to have consisted of 8 central, composite, twisted, core components welded together back-to-back in two sets of four. To this were welded the cutting edges which themselves consisted of two halves. The blade was essentially of wrought iron, although some degree of carburisation had been given which would have had little effect on the properties of the weapon.

S 10. Another Sword from Mitcham, Surrey. (Museum of London, 56: 106/14). This sword had lost its tip, its pommel and the end of the tang and parts of its cutting edge. It was found in an Anglo-Saxon cemetery and is similar to another sword, C. 2444, given a 6th cent. date by Wheeler.[66] The blade length was 79.5 cm and the width 5.8 cm. (Fig. 63).

A wedge-shaped section extending a little more than half the width was cut from the blade 28 cm down from one shoulder. The slag content was fairly low but two different formations of slag inclusions could be seen. It was clear that two sets of triple composite pieces had been welded onto a separate core piece to give a pattern-welded surface. Etching revealed the structure shown in Fig. 71d. The cutting edge consisted of two halves (a) and (b) with an additional small part (c) forming the tip. All the parts were mostly ferrite although the tip has a small amount of lamellar pearlite which tended to a Widmanstätten structure, this probably had about 0.4-0.5% C. Diffusion had taken place across the welds. (Fig. 71a).

The structure of the pattern-welded pieces were very similar consisting of medium to coarse-grained ferrite. Slag stringers showed some pieces to be twisted and other straight.

The central core piece (j) consisted of a single piece of iron, coarse grained ferrite with Neumann bands. In the weld lines were narrow darker lines of pearlite which were interspersed with two-phase slag inclusions.

The steel part of the cutting edge had a hardness of 203 HV while the hardness of the ferrite was 115-128 HV. The iron core (j) had a hardness of 96-123 HV and the pattern-welded strips, 104 to 123 HV.

This blade has been made wholly of wrought iron except for a small strip of steel at the tip of the cutting edge (Fig. 71e).

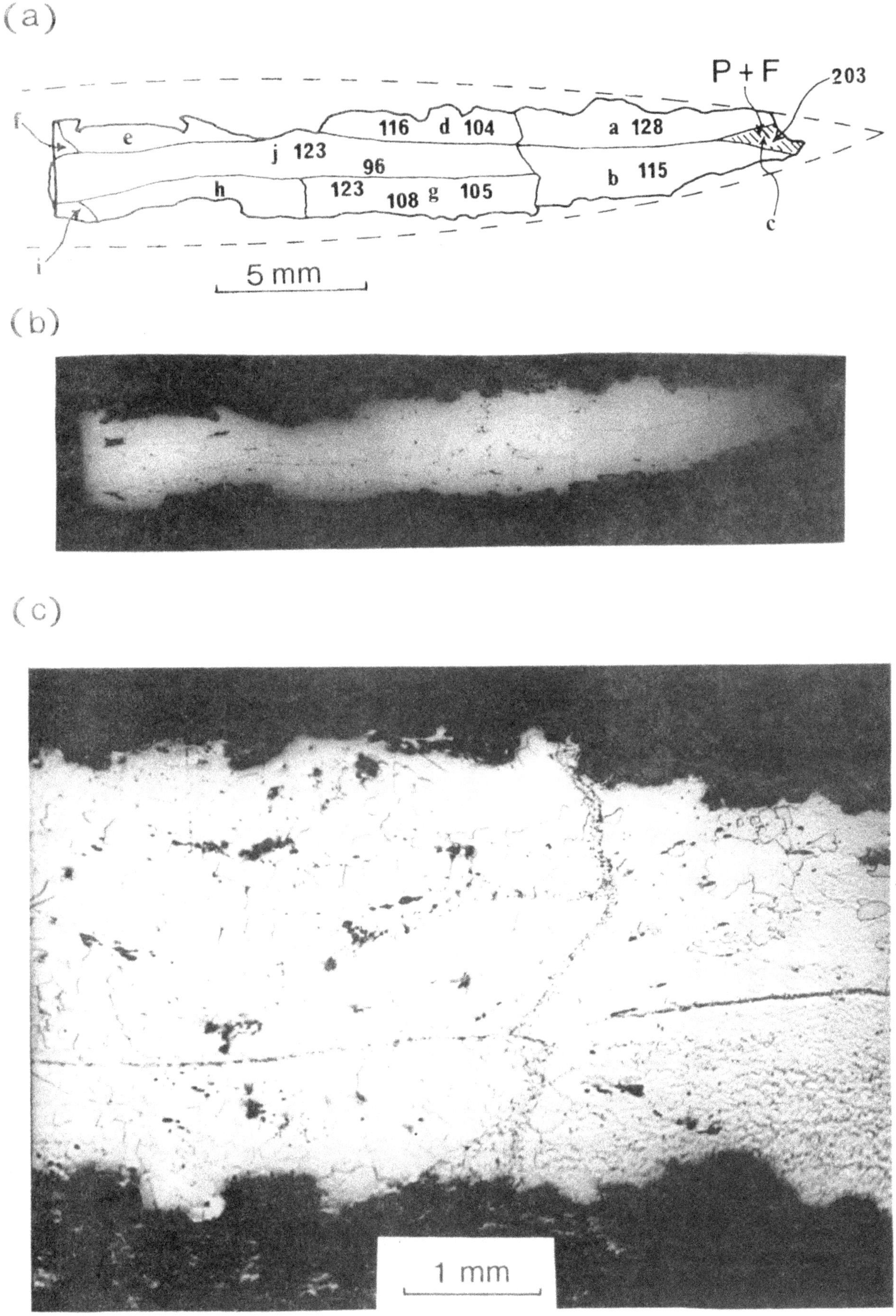


Fig. 71. Section through sword S 10 from Mitcham, Surrey (b) X 5, (C) X 25.

(d)

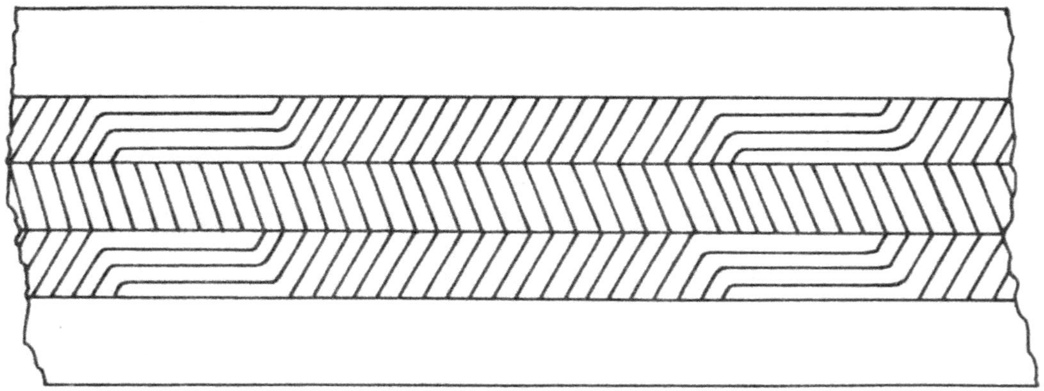

(e)

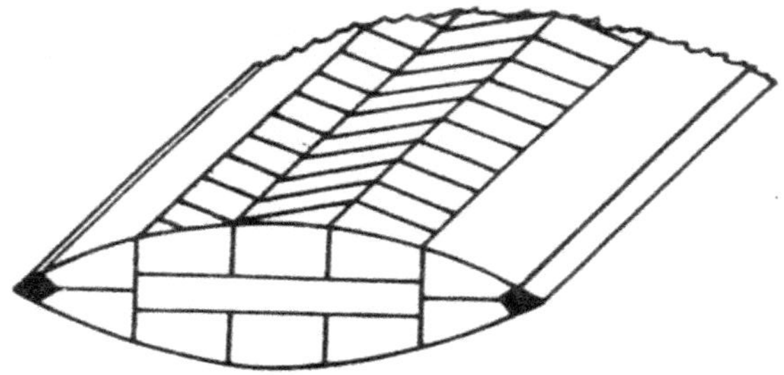

Fig. 71 (continued) surface detail drawn from X-ray and three-dimensional view.

S 55. Sword from Sarre or Bifrons, Kent. (Maidstone, Museum, KAS 836). A nearly complete sword blade with the end of the tang, the tip and parts of the edge missing. It was in poor condition, heavily mineralised but included some of the scabbard. Records are uncertain but appearances would favour Sarre as the find-spot. It would date from the 6th. cent. A.D. The length was 82.5 cm and the width at most 5.6 cm. (Fig. 63).

Radiography showed it to be a typical pattern-welded blade with at least one set of three twisted composite rods welded side-by-side with adjacent twists in the same direction.

A section was cut 20 cm from the tip-end of the blade. The slag content was low but big inclusions marked the weld between the edge and the rest of the blade. After etching it was clear that the composite rods had been welded to a central core piece and then the single piece cutting edge welded to the pattern-welded core composite. The structure of the cutting edge varied with patches of martensite and troostite in a matrix of ferrite and pearlite (Fig. 72). The proportion of martensite decreased away from the cutting edge. The central core piece consisted of ferrite and pearlite with variable grain size and had a carbon content of about 0.1%. The composite pieces had the usual banded appearance.

Hardness readings were taken from the edge piece; one in the martensitic area near the tip of the edge gave a value of 581 HV; the second in fine pearlite had a value of 234 HV, and the third in the ferrite + pearlite zone near the weld had a value of 211 HV. The hardness of the low carbon core was 181 HV. The banded composites gave 206 HV in the ferritic areas and 164 HV in the darker areas, which suggested that the paler, ferritic areas had a higher phosphorus content.

This blade had been quenched, probably in water after final forging, resulting in a hard tip to the cutting edge. The rest of the blade was largely unaffected by the quenching.

S 54. A Sword from Sarre or Bifrons, Kent. (Maidstone Museum, KAS 834). In poor condition with two main fragments, but missing most of the tang, tip and cutting edge. What remained included the remains of the scabbard. The blade was 78.3 cm long and about 6 cm wide. It came from either the cemetery at Sarre or that at Bifrons and can therefore be dated approximately to the 6th century A.D.[71-72] (Fig. 63).

Radiography showed it to be pattern-welded with a pair of twisted composite rods welded side by side, with the twists in opposite directions to give a chevron effect (Fig. 73c).

A section was cut across the blade 16.5 cm down from the break nearest the hilt. The slag content was low. Four main parts were visible in the half section. The cutting edge was a single piece with medium to large grained ferrite with a little pearlite indicating a carbon content below 0.1%. (Fig. 73). Slag inclusions showed that it had been folded and forged before being welded to the core part of the blade. It had a central core piece of ferrite and lamellar pearlite with a carbon content of about 0.1% which tended to decrease at the welds. Twisted composites formed the surface but these were badly corroded.

Three hardness readings were taken along the central axis of the edge of

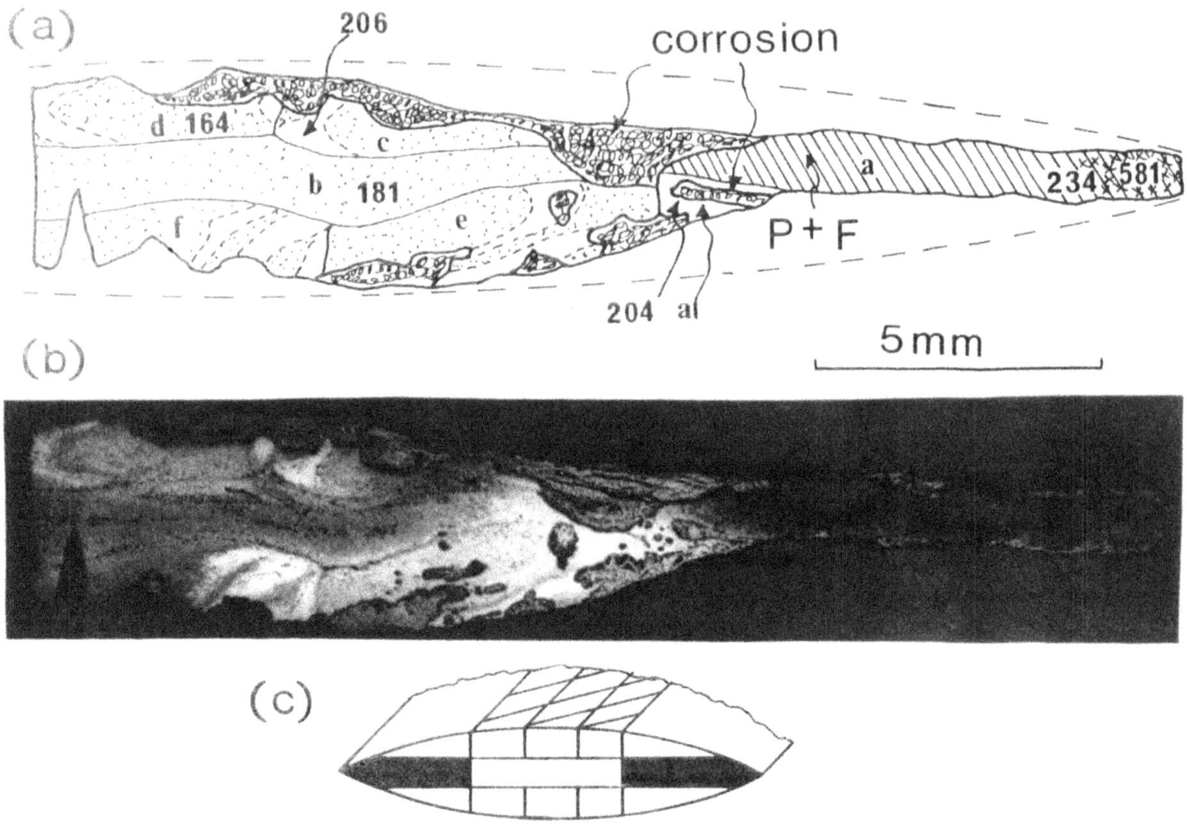


Fig. 72. Section through sword S 55 from Sarre or Bifrons, Kent + three-dimensional view.

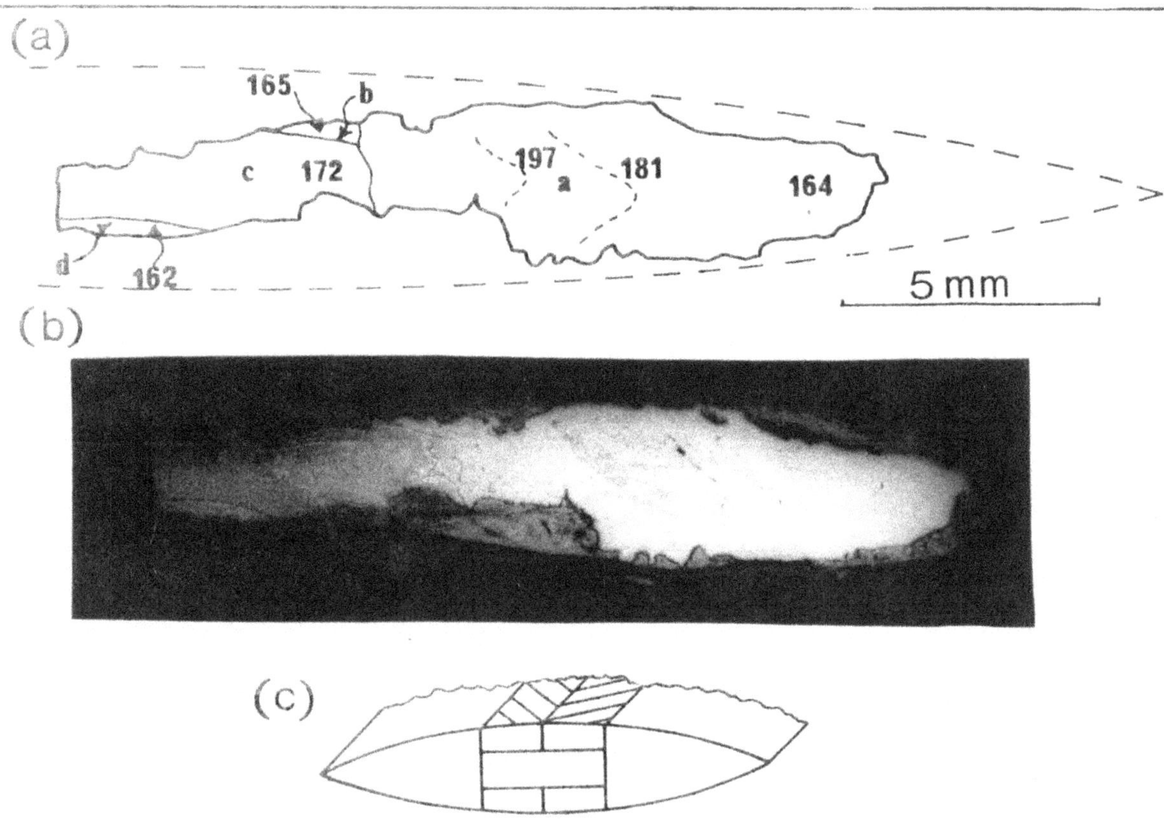


Fig. 73. Section through sword S 54 from Sarre or Bifrons, Kent + three-dimensional view.

the blade. From the tip of the edge these were as follows: 164, 181 and 197 HV. The ferrite core piece had a hardness of 172 HV and the pattern-welded composites had hardnesses of about 164 HV. The blade appears to have been air-cooled after its final forging.

S 53. A Sword from Sarre, Kent. (Maidstone Museum KAS 825). This sword is now in five pieces with the end of the tang, tip and some parts of the blade missing. It was excavated in 1864 from Grave 250 of the Anglo-Saxon cemetery at Sarre (Brent,[72a] 1865, p. 317). The blade was highly mineralised and bore the remains of the scabbard. In profile it had a flattish convex shape; it was 73.5 cm long and about 5.4 cm wide. It dates approximately from the 6th century A.D. (Fig. 63).

Radiography showed that it had a fairly narrow central pattern-welded zone, similar to that of S 54 except that here we have a clear chevron pattern (Fig. 74c). Both sides had areas of twisted rods running in opposite directions.

The metallographic section was cut from one side of the blade 14 cm down from the hilt (Fig. 74). Slag inclusions indicated the main weld lines and it appeared that the blade consisted merely of pattern-welded composites and edges with no central core-piece. The two edge parts consisted of coarse grained ferrite with nitride or carbide needles. The composites showed pale and darker bands alternating, the pale bands being wholly ferrite while the darker band consisted of ferrite with some lamellar pearlite indicating a carbon content of about 0.1%. The edge parts had hardnesses of 143 and 144 HV while the composites gave hardnesses of 210 and 217 HV in the ferritic parts and 153 and 159 HV in the ferrite + pearlite zones, indicating the usual high phosphorus in the former.

S 52. A Sword from Sarre, Kent. (Maidstone Museum, KAS 823). An almost complete blade in 5 fragments but with parts of the edge missing near the hilt. It was excavated in 1864 from Grave 113 in the cemetery at Sarre, Kent. (Brent [72a] 1865, 175). It was in poor condition and covered with a thick incrustation which included the remains of the scabbard. The length was 76.6 cm and the width 6 cm and it dated approximately to the 6th century A.D. (Fig. 63).

It was pattern-welded with 4 twisted composites welded back-to-back to another set without a separate core piece (Fig. 75). The section was cut halfway across the width of the blade 42 cm down from the hilt. The slag content was low. The half section showed eight components. A weld-line divides the two halves of the cutting edge which is mostly coarse grained ferrite but has a narrow pearlite zone near the weld. Nitride or carbide needles were visible in some places. The composites had the usual banded structure with darker bands of ferrite + pearlite and lighter ones of medium to large grained ferrite. The hardnesses of the edge parts was in the range 149 to 187 HV. The darker bands of the composites were 159 and 162 HV while the lighter bands were 174 and 189 HV.

Assuming symmetry across the blade, the cutting edges consisted of two halves welded together while the main part of the blade consisted of two quadruple sets of twisted composites welded back-to-back.

S 51. A Sword from Sarre, Kent. (Maidstone Museum, KAS 820). A blade

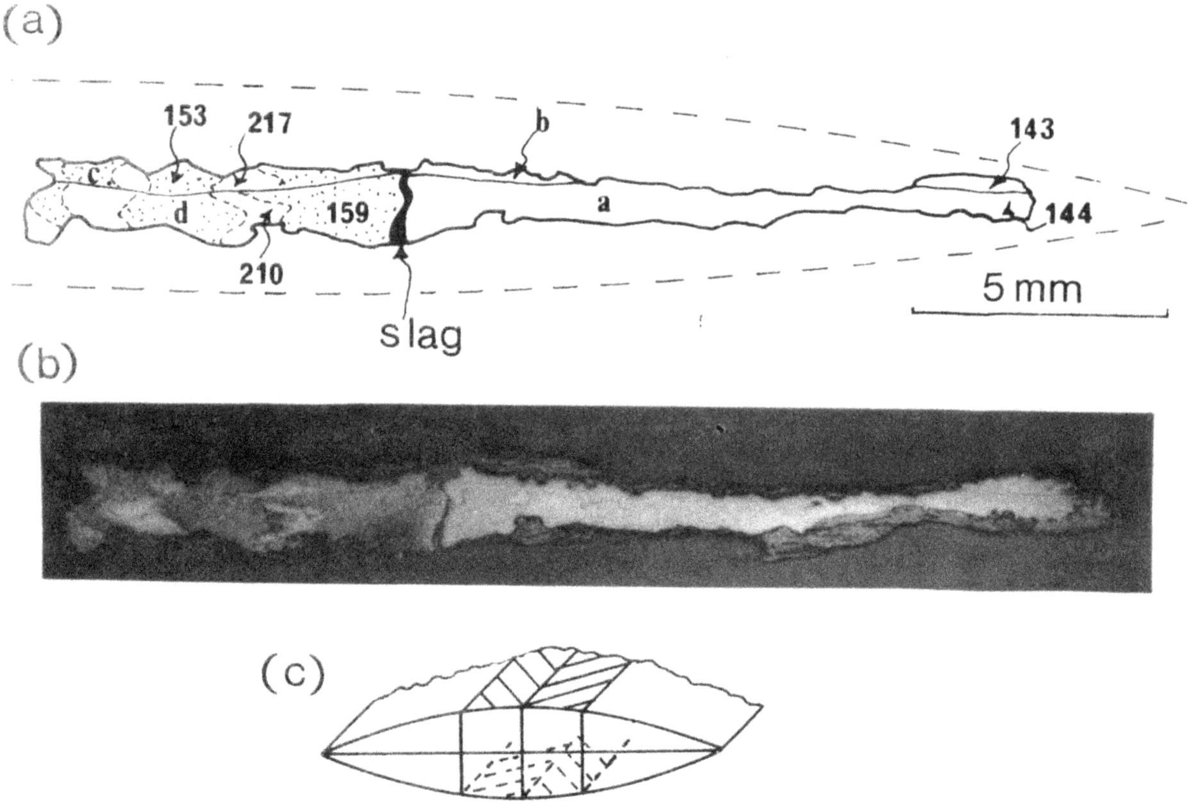


Fig. 74. Section through sword S 53 from Sarre, Kent
+ three-dimensional view.

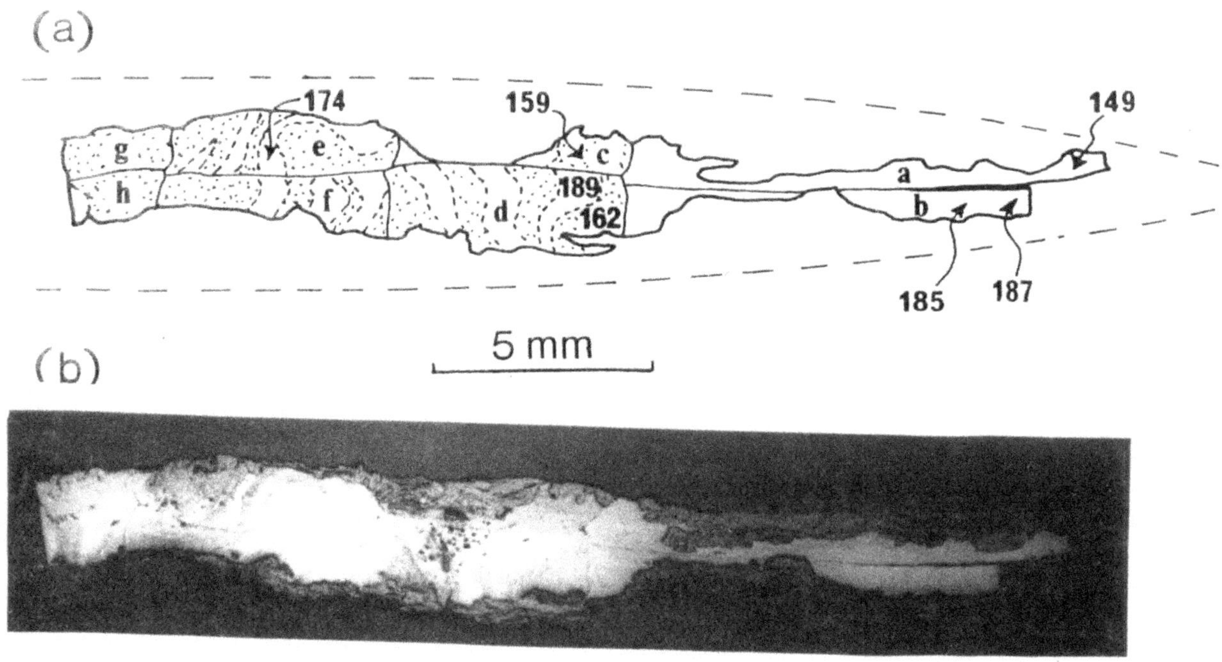


Fig. 75. Section through sword S 52 from Sarre, Kent.

(c)

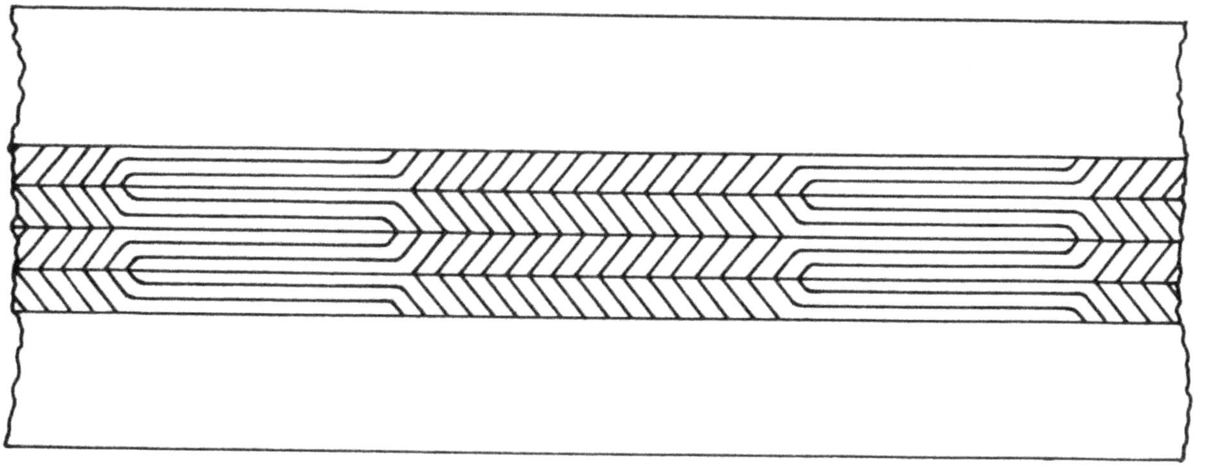

(d)

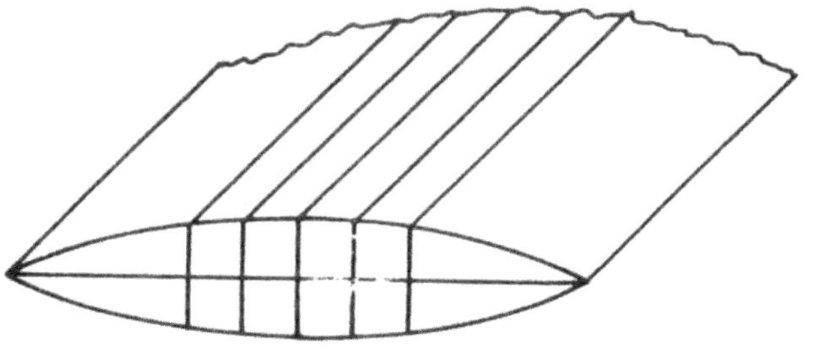

Fig. 75 (continued) surface detail drawn from X-ray and three-dimensional view.

missing the tip with the tang in two pieces. It was excavated in 1863 from Grave 64 of the cemetery at Sarre and is dated to the 6th century A.D. (Brent, 1864, p. 168). The tip appears to have been broken off when excavated and the sword may have been broken before burial. It was very corroded and the remains include those of a scabbard. The length was 75 cm and the width 5.8 cm. (Fig. 63).

A wedge-shaped section from the blade was cut 22 cm down from the hilt. It clearly had a laminated or banded structure unlike that of a pattern welded blade (Fig. 76). The paler bands consisted of equiaxed ferrite and pearlite, lamellar in places, and the carbon content was between 0.1 and 0.2%. The darker bands had more pearlite and probably contained about 0.3% C. Hardness readings across the section showed great uniformity and were mostly in the range 156-174 with occasional values reaching 189 HV. The structure indicates that the blade was made from 5 separate parts welded together and forged out to give a triple-decker sandwich. The various parts had been very skillfully welded together leaving little trace of the welds, apart from slight variations in carbon content.

S 12. Sword, from Wickhambreux, Kent. (Maidstone Museum). The very corroded remains of a sword blade consisting of two main fragments and flakes, some of which were adhering to the main piece. The sword was found in 1886 at a gravel quarry at Wickhambreux in Kent. It was found with another sword (see S 19 below) and appears to have come from a small portion of a cemetery which included a bronze bowl and a blue glass vessel (Dowker,[73] 1887). (Fig. 63). The two main fragments only roughly join together but together indicated a total length of about 90+ cm. No recognisable part of the tang or tip was present. They were in a very poor state with much of the edge missing and with the barely recognisable mineralised and flaky remains of a scabbard forming the outer parts of the flakes on either side. The blade appeared to have been about 6.0 cm or a little less wide and the flaky state of the remains was such that only a very rough estimate of thickness of about 0.5 cm could be made. Date, approximately 6-7th century.

Radiographs of the sword fragments showed it to be a composite weapon with a pattern-welded central zone of an unusual type similar to that seen on the radiograph of the sword from Aylesford (see S 18 below). This central zone was unchanged down the surviving parts of the blade and showed up as a central and two outer twisted bands with a distinct criss-cross effect showing up along these. There was quite a large overlap showing in the criss-crossed aspect of these twisted bands suggestive that two twisted bands might in each case be involved across the thickness of the blade. Interleaved between the twisted bands were two bands which showed a distinct straight grain patterned effect. Along the outside part of the blade on either side, where they survived at all, the cutting edges also showed traces of a straight grain effect on the radiograph, although this was not as pronounced as that seen in the straight grained bands of the core.

A transverse section extending across the blade was taken from near one of the broken ends (Fig. 77). When examined unetched a very varied slag content could be seen. Most noticeable were very large roughly triangular or three-pointed star shaped inclusions which occurred at intervals approximately along the main axis of the section. These slag inclusions most probably became trapped in each case at the weld junction between three separate parts of the blade. These large slag inclusions were distinctly

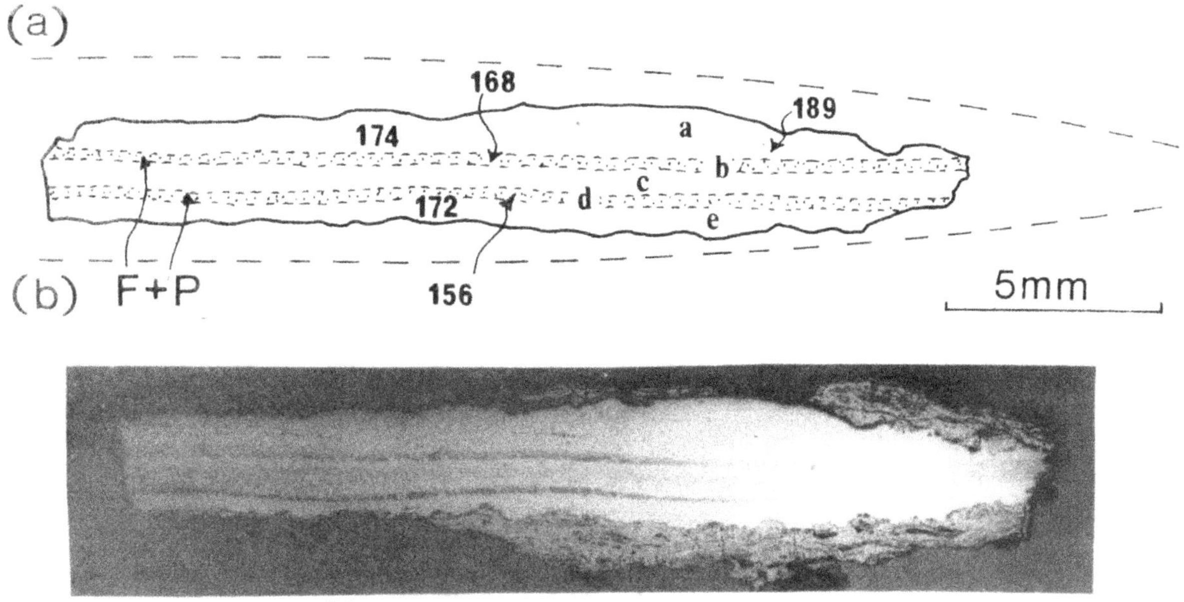


Fig. 76. Section through sword S 51 from Sarre, Kent.

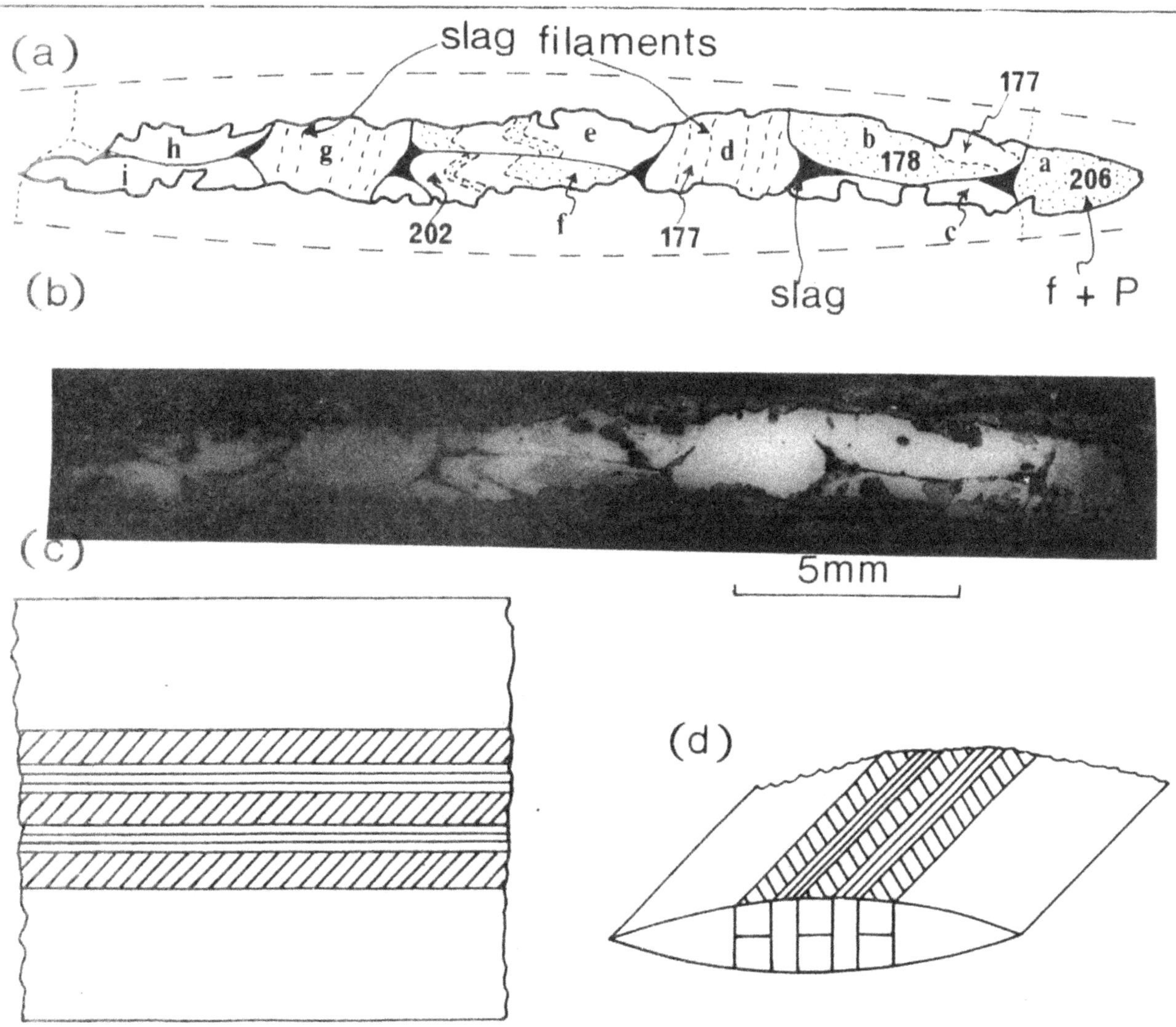


Fig. 77. Section through sword S 12 from Wickhambreux, Kent.

two phase - mostly a dark background or matrix (? a glassy constituent) which surround paler areas (? of wûstite ). The slag content across the section was otherwise mostly quite low although in localised patches it was higher and consisted of small spots or ribbon-like inclusions which appeared mostly dark but against the background a paler phase showed in some cases.

When etched with nital the structure visible in the photomacrograph of Fig. 77 showed up. Part of one cutting edge only survived here, (a) in Fig. 77a, as well as eight composite parts of the core of the blade, (b)-(i).

The cutting edge, (a), appeared to be one piece of fairly homogeneous low carbon iron or mild steel. This consisted of an even and equiaxed distribution of fine grained ferrite and pearlite which was partially lamellar, partially rather granular or spheroidised, and partly unresolved. The slag content here was low. Some decarburisation appears to have occurred along the weld between the cutting edge (a) and the core piece (b) and (c). This presumably took place before or during welding as there was not much sign of carbon diffusion across this weld.

The core part of the blade section consisted of three pairs of twisted composite pieces, (b)/(c), (e)/(f) and (h), (i), welded back-to-back and each pair was divided from the next by a single untwisted piece, (d) and (g) in Fig. 77. The banded nature of the composite pieces was just discernable after etching although it was not clear at all. There appeared to be approximately eight bands or parts to each composite piece. The twisted parts showed up partly as alternate pale and somewhat darker bands which gave a folded appearance (particularly in part (e)), and partly by the flow patterns of slag inclusions which gave a similar appearance. The slag content was mostly low and was only intermittently higher where it occurred in small spots and ribbon-like inclusions. These inclusions appeared mostly as two phase; medium grey (? fayalite) with a darker grey matrix or background (? a glassy constituent).

The pale bands of the twisted composite pieces consisted of very large grained ferrite with a few Neumann bands visible indicating that some final cold hammering took place. The other bands while sometimes noticeably darker were often very little different in shade to the pale bands. Where darker they consisted of medium-large grained ferrite with some pearlite, similar in form to that seen in the cutting edge area (a), but the carbon content here was much less, with a maximum of approximately 0.1% or less and this decreased where the bands became paler where only a little carbon was visible at the grain boundaries. These bands were, therefore, little more than wrought iron even where the carbon content was at its highest.

The only banding visible in the ares (d) and (g), which exhibited quite clear straight graining on the radiograph, were occasional and very intermittent vertical (on Fig. 77a) lines of small slag inclusions which showed the piled nature of these composite pieces and that the plane of this forging process was at right angles to the eventual main axis of the blade section. Parts (d) and (g) both consisted solely of fairly even medium-large grained ferrite with small amounts of carbide visible at the grain boundaries, sometimes visible as poorly formed lamellar pearlite. This can best be described as piled wrought iron with some, possibly accidental carbon content.

Five hardness readings were taken across the blade section and the results were as follows: part (a) near the tip of the cutting edge, carbon content about 0.3%, hardness 206 HV; part (b), a band of medium grain-size ferrite with some grain boundary carbide, hardness 178 HV; part (b), band of very coarse grained ferrite, hardness 177 HV; part (d), medium grain-size ferrite, hardness 177 HV; part (f), band of very coarse-grained ferrite, hardness 202 HV.

S 49. A Sword, probably from Bifrons, Kent. (Maidstone Museum, KAS 620. 1954). A very corroded, but more or less complete sword blade with the end of the tang missing. It was one of a group of objects which formed the Tomlinson bequest, and may have come from the cemetery at Bifrons. The length of the blade was 89 cm and the width 5.3 cm but too corroded to determine the true thickness (Fig. 63). Date, approximately 6th century.

Radiography indicated pattern welding with straight and twisted patterns (Fig. 78c). It would appear that the central part of the blade consisted of two quadruple sets of composite rods welded back-to-back.

A section was cut half-way across the blade, 13 cm down from the hilt (Fig. 78). It had a low slag content. The surviving parts of four composite rods were visible but there was no separate core piece. The composites consisted of coarse-grained ferrite with small amounts of grain-boundary carbide and darker bands of ferrite + pearlite with a carbon content of 0.1% or less. In some of the bands, needles of carbide or nitride were visible.

The edge part had hardnesses of 143 and 158 HV near the tip and the mainly ferritic composite bands had a hardness of 118 HV and 159 HV while that of the paler coarse-grained ferrite bands was as high as 178 HV.

The centre of the blade consisted in all of eight pieces and the individual pieces consisted either of wrought iron or of a mixture of wrought iron and low carbon steel.

S 1. A Sword from Lovedon Hill, Lincoln. (Lincoln City and County Museum, 6.66). The blade was more or less complete but in two pieces and bent double in two places. The surface was heavily corroded although the underlying metal survival was good. The total length of the blade was 87.6 cm and the maximum width 4.8 cm. The sword was found in Grave 31 of the cemetery at Loveden Hill during excavations in 1956[74]. The cemetery has been dated to the 6-7th century A.D. (Fig. 63).

Radiography showed that the central part was pattern-welded. It appeared to be a mixture of chevron and straight-grained effect (Fig. 79c).

A cross-section was cut from one side of a break in the blade. Unetched, the slag lines along the main welds showed up clearly. When etched it was clear that the pattern-welded section consisted of two sets of five composite rods welded back-to-back without a separate core-piece (Fig. 79d). Each of the cutting edges consisted of a single piece. Each was of medium to coarse-grained ferrite with some pearlite.

The pattern-welded part (f) had completely corroded away. The pattern-welded areas (c) and (d) were very similar to one another with pale bands of

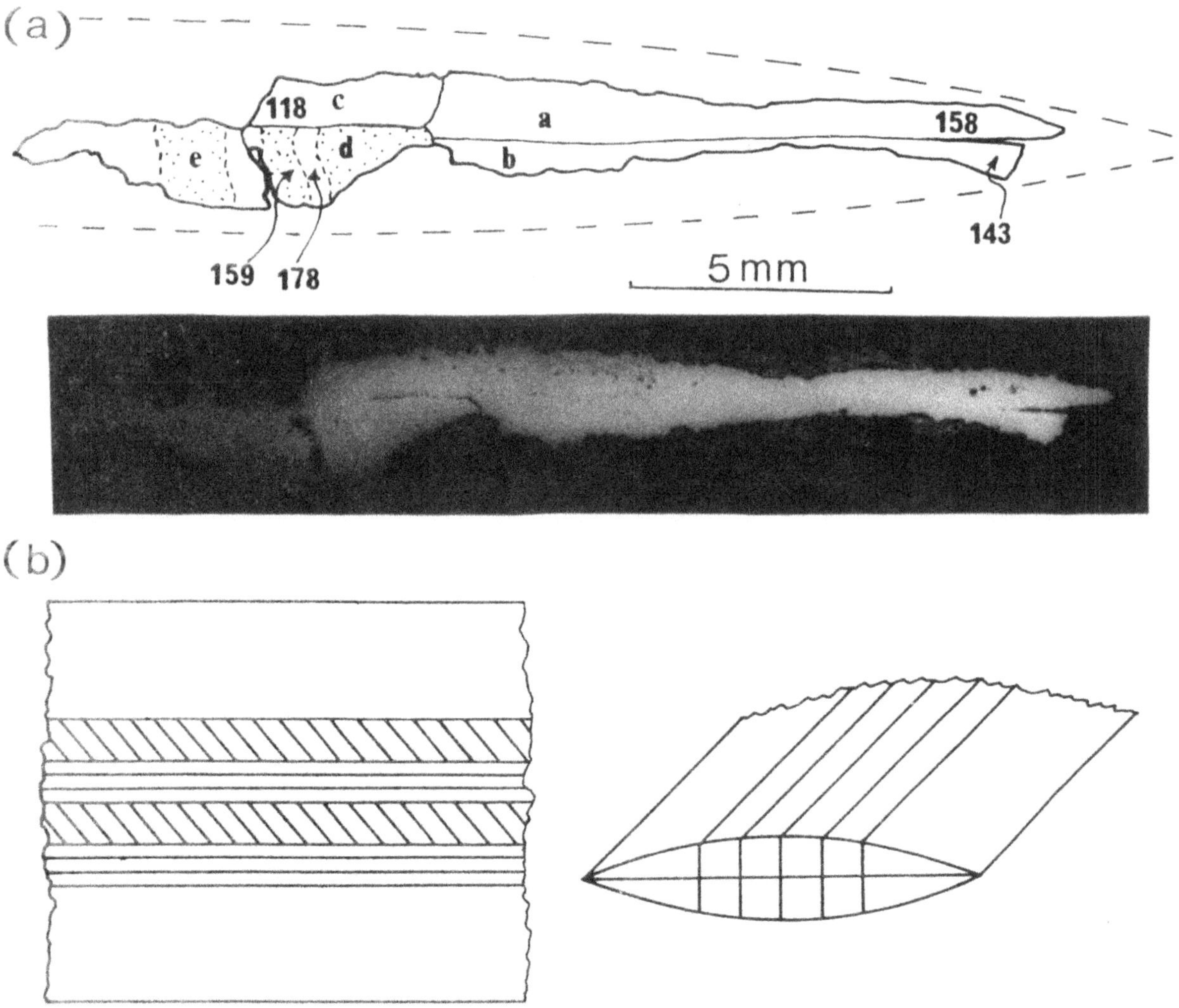


Fig. 78. Section through sword S 49, probably from Bifrons, Kent.
+ surface detail drawn from X-ray and three-dimensional view.

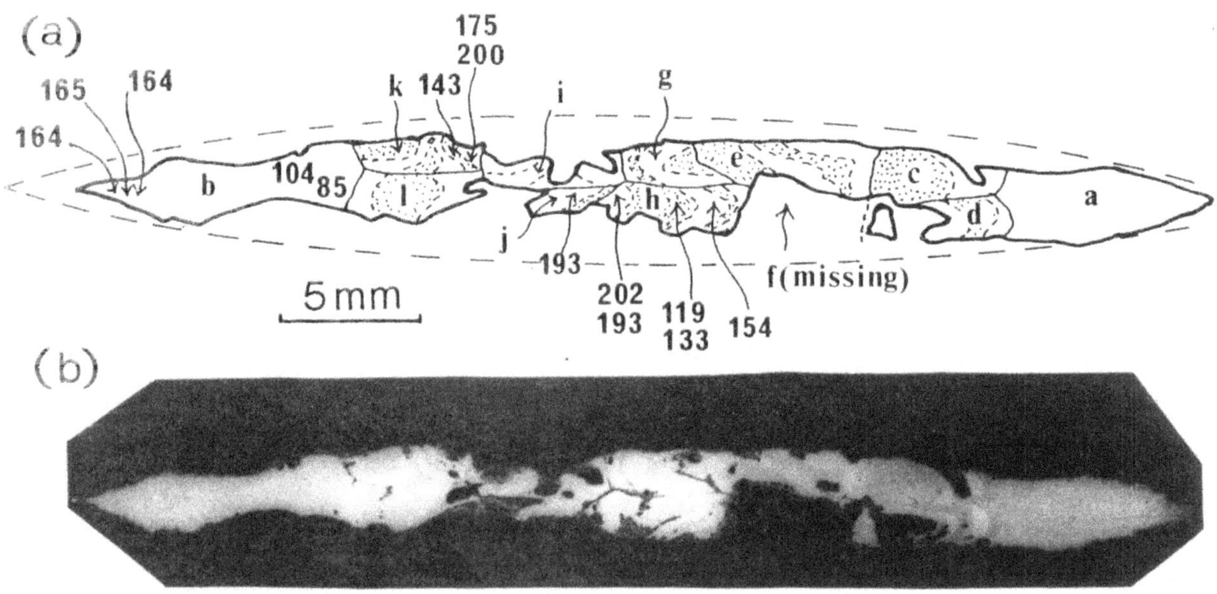


Fig. 79. Section of sword S 1 from Loveden Hill, Lincoln.

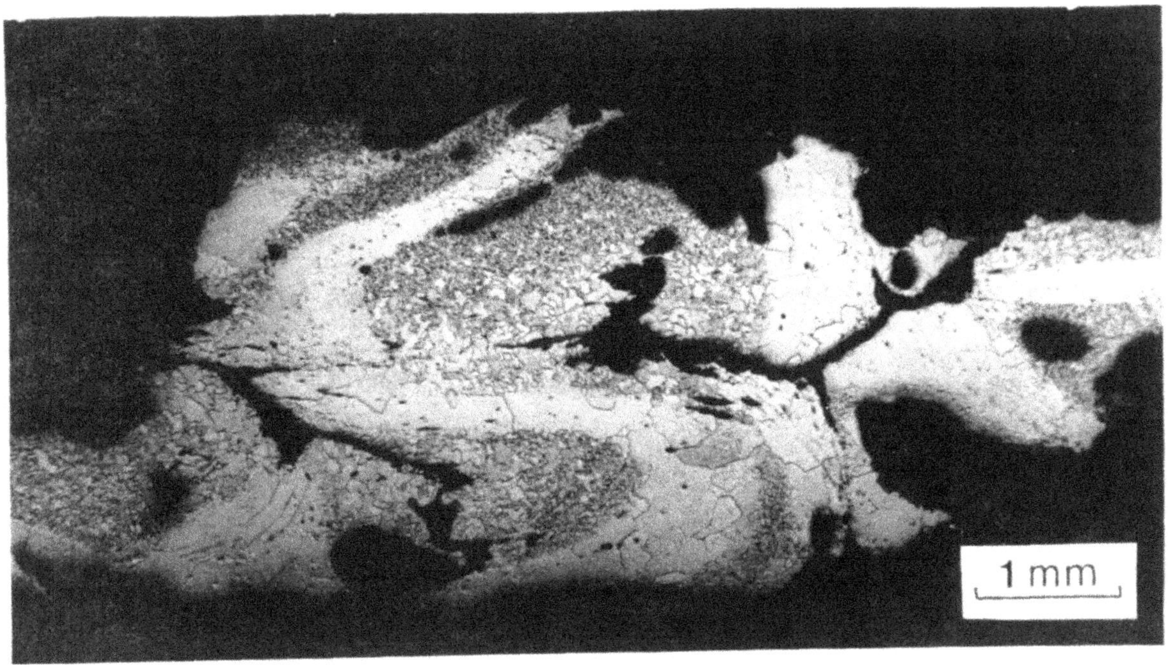


(c)

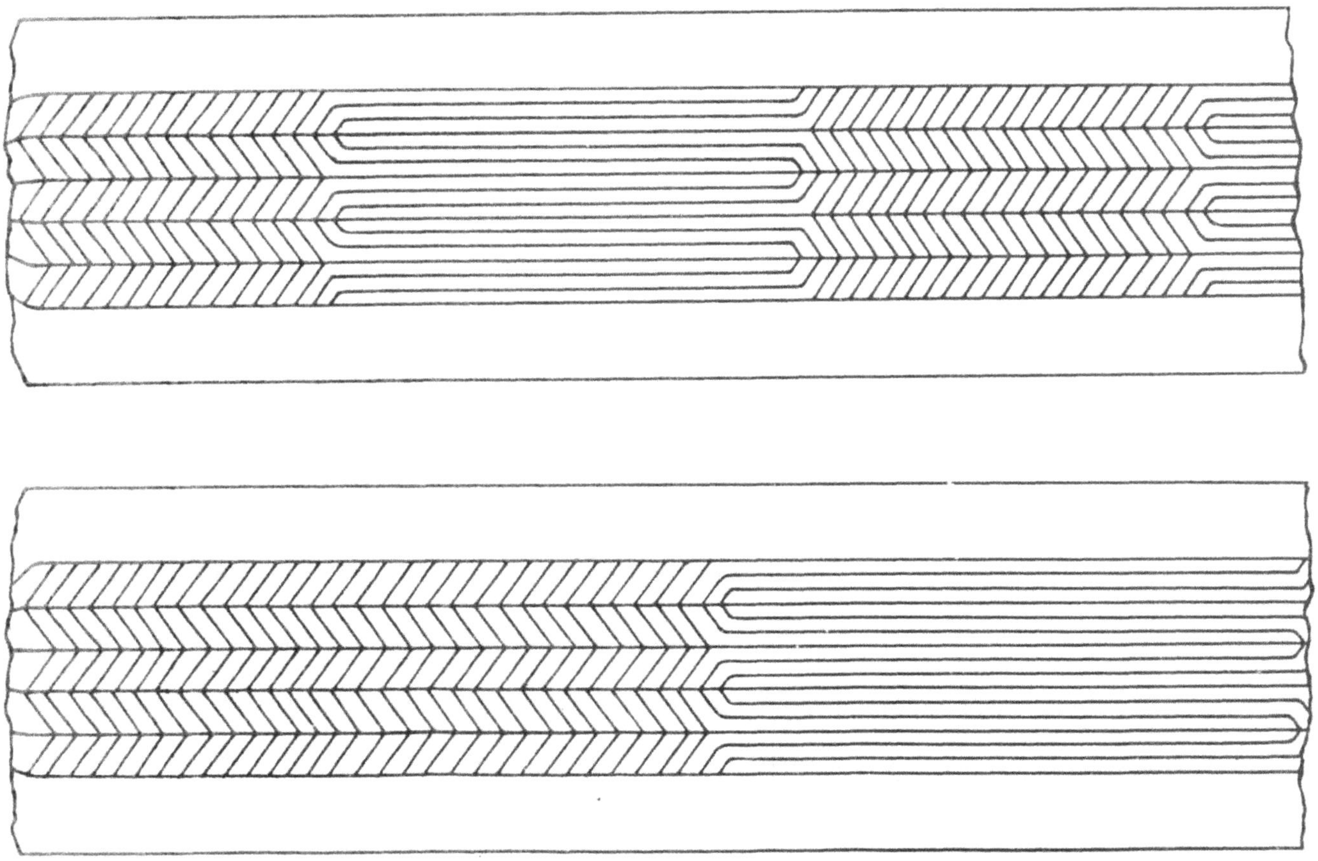

(d)

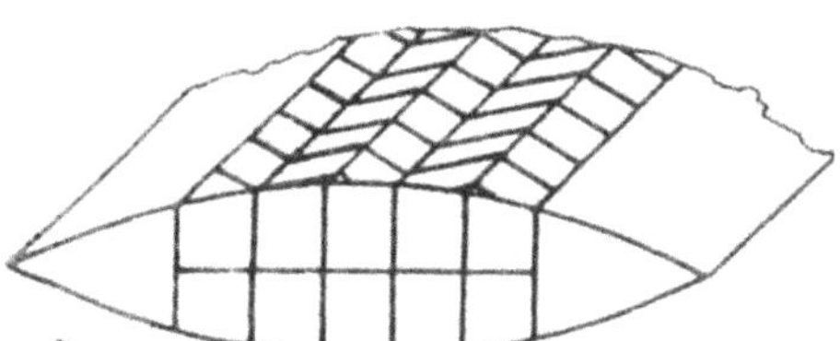

(e)

Fig. 79 (continued) Section X 19
+ surface detail drawn from X-ray
and three-dimensional view.

coarse-grained ferrite with some Neumann banding indicating some cold hammering after final forging. Nitride or carbide needles were visible in these pale bands. The darker bands consisted of fine to medium grained ferrite with some pearlite indicating a carbon content of about 0.1%. In some areas there was enough to have a Widmanstätten structure, indicating high forging temperatures.

Hardness values near the tip of the cutting edge (b) were in the range 164-165 HV. Values of only 85 and 104 HV were obtained near the inner weld of (b). Hardnesses in certain of the pale and darker bands of the pattern-welded part were as follows: pale, 154, 193, 202; 193, 175, 200; darker areas, 119, 133, 143 HV (Fig. 79a). Again the higher values of the paler ferritic areas suggest a higher phosphorus content although nitride would also raise the hardness. The blade had not been heat-treated after forging although it had been given some cold work.

S 40. Sword from Chesterton, Cambridge. (Peterborough Museum, 16/1979/1). A nearly complete sword blade minus end of tang and tip (Fig. 63) found in 1979 by chance in a ploughed field near Chesterton (TL 124 947) in Cambridgeshire). The exact circumstances of the find were not carefully recorded and no associated finds survive. It measured 82.8 cm long by 5.8 cm at the widest part of the blade which appeared to have had a flattened convex profile, 0.5 cm thick, although this had been obscured by the extensive corrosion of the surface. It was similar in size and shape to many swords found in Anglo-Saxon graves of about the 5th-7th centuries and this is probably the most likely attribution for this particular sword. When found it had probably already been disturbed from its original resting place as it was bent sharply in two places along the blade, the most likely result, when already weakened by corrosion, of having been struck by a heavy and sharp object such as a plough. The blade was heavily corroded and the edge was missing in places, but the fact that it had withstood being bent sharply without breaking showed that the metal of the core survived to some extent. No structural details could be seen on the corroded remains of the surface of the blade and no traces of scabbard, guard or pommel survived.

A radiograph of the sword showed the blade to consist primarily of a pattern-welded core zone running down the centre of the blade with a plain cutting edge on either side. The core zone appeared to consist of 2 superimposed sets of 4 adjacent composite strips. These strips exhibited a very pronounced light and dark 'grain' structure which alternated between being straight and twisted so that a herringbone or chevron pattern alternated with straight 'grained' areas. As far as could be seen the surface pattern was probably similar on both sides and a diagrammatic interpretation is given in Fig. 80c. The pattern-welding was visible continuing along the tang although much distorted in this area. The cutting edges did not exhibit any 'grainy' appearance.

A wedge-shaped transverse section extending just over half way across the width was cut from near the centre of the blade from a part which showed up as being straight 'grained' on the radiograph. When viewed unetched a markedly varied slag distribution was seen. The overall slag content was fairly low mostly consisting of small spots and ribbons aligned along various planes of forging. There were, however, several very large triangular shaped 2 phase slag inclusions at intervals along the central axis of the section of the core part of the blade.

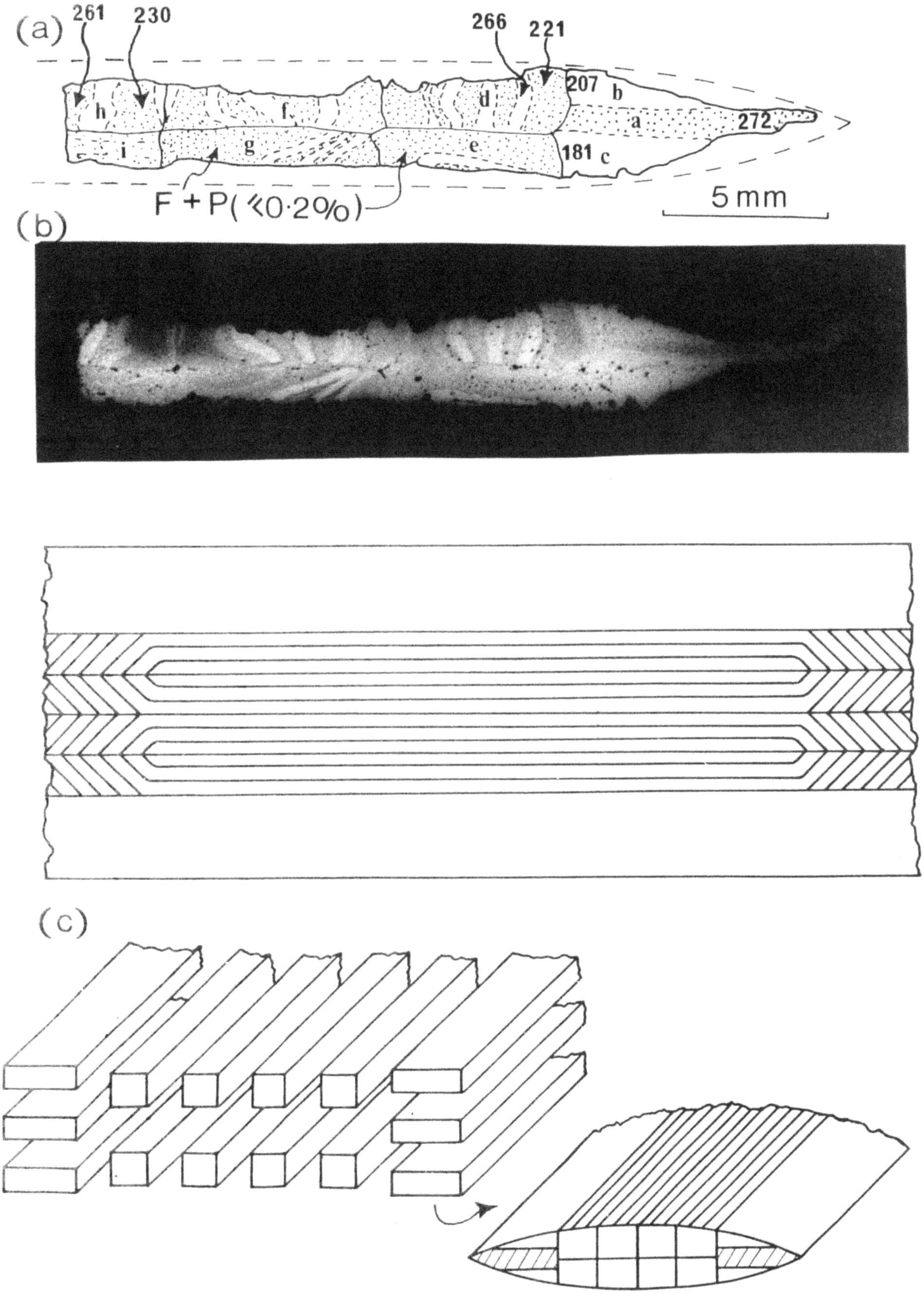


Fig. 80. Section of sword S 40 from Chesterton, Cambridge + surface detail drawn from X-ray and three-dimensional view.

When etched with nital it became clear that the core area was composed of several adjacent composite pieces welded back-to-back without any central core piece (Fig. 80(f). These are marked (d)-(i) on the diagram (Fig. 80a) taken from Fig. 80b. The cutting edge which appeared plain on the radiograph showed up as being a sandwich composed of three layers, marked (a)-(c) on Fig. 80a. The microstructure of the central part (a) of this sandwich consisted mainly of a Widmanstätten-like distribution of ferrite and an irresolvable or amorphous-looking constituent, probably fine grained or poorly formed pearlite. Towards the tip this became increasingly mixed with quite dark etching nodular troostite. This dark etching constituent faded out about two thirds of the way from the tip towards the weld junction with the core area. The proportion of pearlite in the ferrite/pearlite mixture also decreases towards this inner weld junction. This structure would appear to be the result of fairly rapid cooling although the critical cooling rate was too slow to produce any martensite. The fading out of the troostite and the apparent decrease in the proportion of pearlite present nearer the weld junction with the central core is probably just a reflection of the increasing thickness of the blade away from the tip of the cutting edge.

The outer parts (b) and (c) of the cutting edge sandwich, also consisted of a spiky looking mixture of ferrite and pearlite. The proportion of pearlite was generally quite low although it increased towards the middle of the section along the zone marking the junction with (a) in each case. This increase would appear to be the result of carbon diffusion outwards from the central piece (a) during the final heating and forging of the blade. The evidence of weld lines between these three parts of the cutting edge was difficult to see. Between (a) and (b) a narrow grey line of pearlite was fairly distinct, outlined by an adjacent line of ferrite whereas between (a) and (c) a faint grey line of pearlite could just be made out although it was not very clear. These lines appeared to mark the position of welds of a high standard to judge from the general absence of accompanying slag lines. The overall structure of the cutting edge was of a fairly even medium grain size.

Allowing for the heat treated nature of the microstructure it would seem that the outer pieces (b) and (c) of the cutting edge were very little more than wrought iron with a carbon content of probably no more than 0.1% and probably much less before diffusion from (a) took place. The central piece (a) was probably a low carbon steel with a carbon content of between 0.2 and 0.3%. The quenching rate may have been quite rapid but probably because of the low carbon content, too slow to give a fully quenched or martensitic structure. The cutting edge must have been assembled before being welded to the pattern-welded core part of the blade.

Most of the weld junctions between the composite strips, (d)-(h), of the pattern-welded core part of the section and between the core part and the cutting edge were marked by narrow grey lines of pearlite which in places were highlighted by adjacent white lines of ferrite and accompanied in places by a few spots and ribbons of slag (Fig. 80b). Fig. 80c clearly shows the rectangular shape of each composite rod in section as well as the very large, mainly triangular shaped, slag inclusions that were trapped at the corners of the composite bars when they were welded together.

The rods (d), (f) and (g) were each composed of seven alternate dark and pale zones which gave a distinctive banded appearance to the etched section (Fig. 80a and b). The darker zones consisted of a medium fine grained mixture of unresolved or amorphous looking pearlite and ferrite distributed mostly fairly evenly and in the same Widmanstätten-like formation seen in the cutting edge. These areas contained very little slag. The pale zones consisted of very large grained ferrite with occasional streaks of spiky pearlite and rather more slag than the darker pearlitic zones. Neumann banding was also visible in places in the large ferrite grains. The carbon content of all the darker pearlitic zones, allowing for the spiky nature of the ferrite and pearlite, probably varied between about 0.1 (or less) and 0.2% i.e. fairly similar to that of the central part (a) of the cutting edge.

The dividing lines between the dark and pale zones of the composite rods, were quite sharp with little evidence for carbon diffusion across these boundaries and instead the pearlite, in fact, tended to appear rather darker here. This thickening effect gave a slightly darker grey outline appearance to the pearlitic zones which sharpened the distinction between these and the pale ferrite bands. Little or no slag particles could be seen along these boundaries which would appear to represent near perfect welds.

The composite rod (e) mostly consisted of a thick grey band of spiky ferrite and pearlite with one narrow horizontal pale band of very large grained ferrite and the slight remains of a further dark band just showing on the corroded surface of the blade. The thick grey band appeared to be streaked in places with darker grey lines of pearlite suggesting that it might have been composed of several bands, or one piece folded and forged out several times (i.e. piled).

Only about half of the rods represented by (h) and (i) were included in the section. The part (i) was wholly composed of grey spiky pearlite and ferrite although in a similar way to those seen on (e) there were horizontal traces of darker pearlitic streaks possibly indicating weld lines between subdivisions. The piece (h) probably contained much the same seven alternate dark and light vertical bands as the adjacent pieces (d) and (f).

Hardnesses were measured on each of the three parts of the cutting edge, the steel central piece (a) near the tip of the cutting edge, and the two outer low carbon pieces near the welded junction with the core of the blade, (b) and (c). The results were as follows: (a), 272 HV; (b), 207 HV; (c), 181 HV. For each of the pattern-welded pieces, hardness was measured on the pale coarse-grained ferrite bands and on the darker pearlitic bands. The ferrite bands gave values of 266, and 261 HV and the darker bands 221 and 230 HV. This shows the usual likely high phosphorus content of the ferritic bands.

The two main constituents of the composite rods of the core of the blade were low carbon iron or mild steel of approximately 0.1–0.2% carbon and wrought iron, and unlike the cutting edge there had been little diffusion between these banded areas. The observed microstructures indicate that the final forging temperature was probably around 850–900 deg. C. from which it was finally quenched and the occurrence of Neumann banding suggests that some subsequent cold hammering took place.

<u>S 14. Sword from Eastrey (Kent)</u>: (Maidstone Museum, A.S. 121). Two fragments which together represented a nearly complete sword broken in the middle (Fig. 63). A small part of both the tip and tang were missing.

The total surviving length was 85.1 cm, thickness about 0.5 cm and with a width of 5.0 cm, it was a noticeably narrow blade with a fairly pronounced, slightly flattened, convex profile. The surface was very corroded with much pitting although it appeared to have a quite good surviving metallic core. No hint of any internal structure was preserved in the surface corrosion products. It was one of three very similar sword blades apparently found in graves at Buttsole, near Eastrey in Kent and donated by Mr. W. Cobb but no further details survive concerning the circumstances of the find. Date: approximately 6-7th century.

A radiograph showed a superimposed chevron-like grain pattern running down the central part of the blade. This pattern was not very clear but showed that along this part of the blade, a pair of composite twisted rods formed the surface on either side. There was no apparent indication of a separate central core. The criss-cross chevron effect did, however, vary and appeared to alternate with short straight-grained elements in places. The cutting edge showed little although some hints of straight graining were just visible in a few places.

A transverse section extending to just over half way across the width was cut from near the centre of the blade (Fig. 81). When viewed unetched the slag content was seen to be quite high in places although it was very variable and tended to be segregated into parallel zones giving a banded appearance to the section. A narrow broken line of slag spots ran across the section marking the weld position between the cutting edge and core of the blade. The slag inclusions across the blade were mostly quite small consisting of spots and well flattened ribbons except over the central part of the core area where an irregular distribution of large partly flattened inclusions occurred.

When etched with nital the structure shown in the macrograph (Fig. 81b) became visible. The cutting edge was divided mainly into two halves (a)/(b) and (c)/(d) (see Fig. 81a) and each of the two halves also appeared to consist of two parts although these subdivisions were much less clear. Part (a) consisted of very large grained ferrite with quite a lot of slag inclusions. There was no very clear dividing line with (b) although this was different in appearance, being a rather banded zone of medium-large grained ferrite with much slag (small spots and ribbons) and streaked with medium grained ferrite with a small amount of pearlite. All the slag seen in the cutting edge formed discontinuous lines running horizontally across the section.

Both (a) and (b) appeared quite pale on the macrograph whereas the other main part of the cutting edge, (c), was quite grey consisting of medium grained ferrite and pearlite. The pearlite was partly spheroidised and partly lamellar in form with mostly equiaxial grains and was lowest in density near the inner (or left hand) area where its carbon content was approximately 0.1% or less, whereas at the centre the carbon content was about 0.2%. Elsewhere across (c) the carbon content varied between these figures. The pearlite stopped abruptly along a straight horizontal line which gave the boundary with part (b). It was difficult to say whether this marked the exact position of a weld between (b) and (c) as it did not coincide with a line of slag inclusions. The overall slag content of part (c) was quite low although several broken lines of spots and ribbons showed up near the boundary with (b). It is possible that one of these, in fact, marked the true position of a weld between these two parts and that a small but even, amount of carbon diffusion masked this and appeared to show a weld

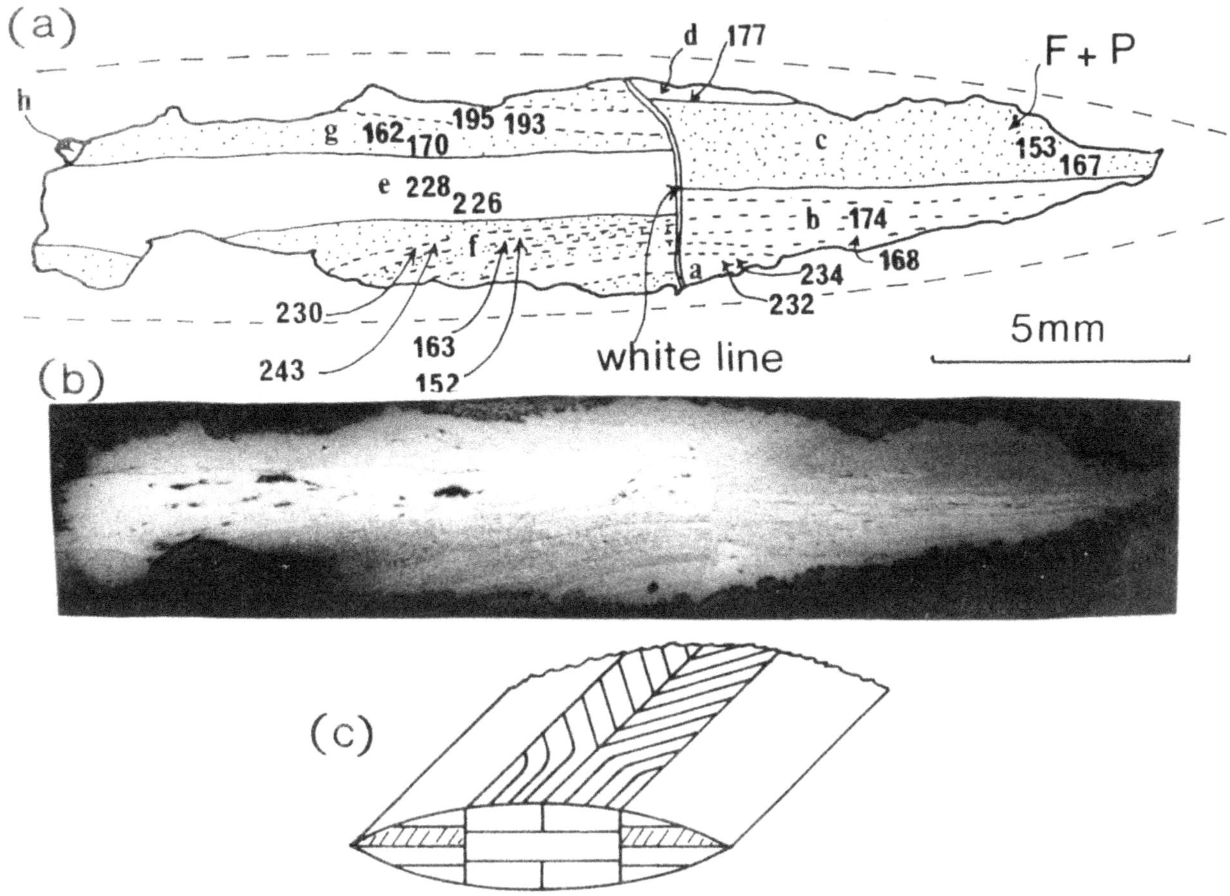


Fig. 81. Section of sword S 14 from Eastrey, Kent + three-dimensional view.

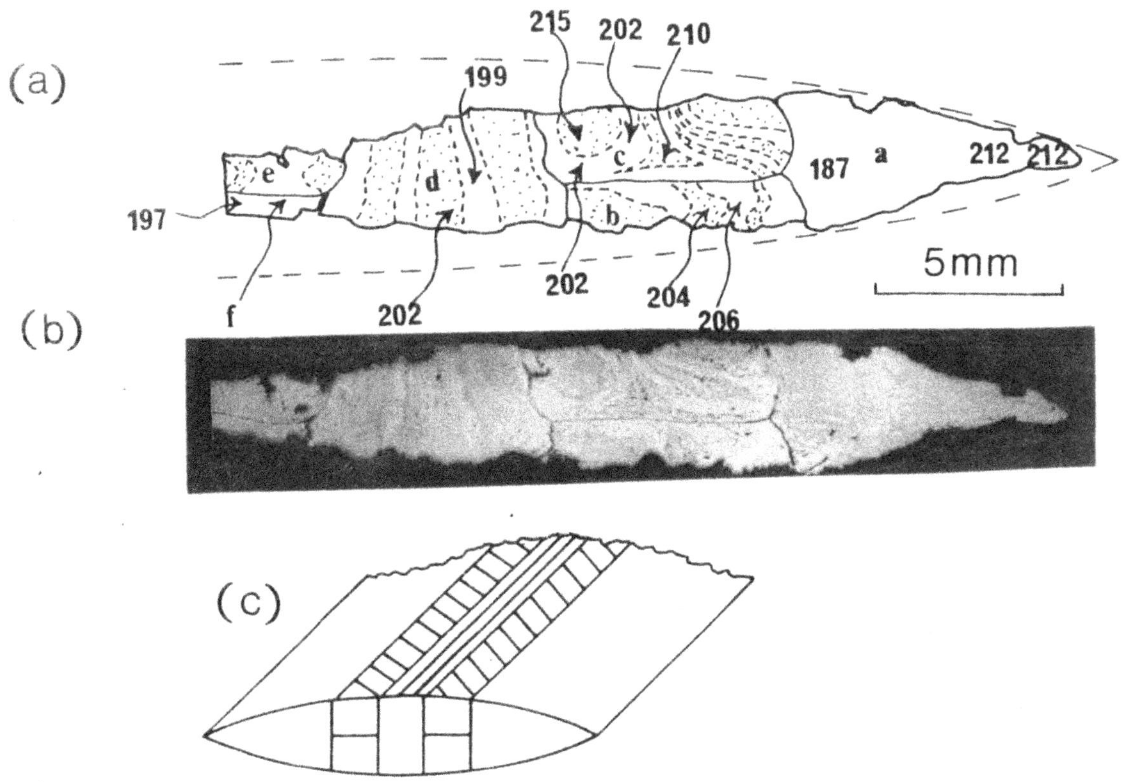


Fig. 82. Section of sword S 18 from Aylesford, Kent + three-dimensional view.

position a little to one side.

The small part (d) was very similar to (c) with a similar pearlite distribution giving a carbon content of about 0.1–0.2%, but it varied in that it contained much more slag.

The weld junction between the cutting edge and the core of the blade section was very clearly marked, both by a broken line of slag spots and a pale band of ferrite (Fig. 81a and b). The pearlite zones of the cutting edge and core stopped fairly abruptly at this line with little evidence of carbon diffusion across the weld into neighbouring ferrite areas where these were adjacent.

The core area of the blade was made up of a central strip (e) on to either side of which the composite rods which gave the blade its patterned appearance were welded. The composite pieces (f) and (g) appeared in the section with a very small part of a third composite piece (h) just appearing in the (upper left in Fig. 81) section divided from (g) by a few slag spots in a vertical line probably marking the position of a weld. The corresponding fourth composite piece on the other side of the blade did not appear within the area of the section.

The central core piece (e) consisted mostly of very large grained ferrite with occasional patches of medium grain size which were mixed with a little pearlite. Neumann banding occurred in a central patch near the weld junction with the cutting edge, indicating that some final cold hammering took place. Many of the large slag inclusions seen in this piece were 2 phase with voids where cracking had occurred.

The composite pieces (f) and (g) were very similar to one another, both being composed of alternate and roughly horizontal dark (pearlitic) and pale (ferritic) bands (Figs. 81, b). The inner grey pearlitic band in each case was rather wider than the others that survived. The pale bands consisted of medium grain size ferrite with some pearlite whereas the darker, grey bands consisted of fine grained ferrite and pearlite with a carbon content of approximately 0.1–0.2%. The pearlite was partly lamellar and partly spheroidised in form. The divisions between the bands were diffuse and there has evidently been fairly extensive carbon diffusion between what started out as alternate wrought iron (with very little carbon) and low carbon iron or mild steel bands welded together to form the composite rods. The slag content of these parts, (f) and (g), was inconsistent but mostly fairly low. The small part of (h) consisted wholly of ferrite and pearlite similar to the adjacent part of (g), with a carbon content approximating to 0.1–0.2%.

A series of hardness readings were taken for the parts (a)–(g) of the section and these were mostly done in pairs as a check on the consistency of the results. Three pairs of readings were taken for edge parts (a)–(c) and a single reading for part (d). The values obtained were: (a) 232 and 234 HV; (b) 168 and 174 HV; (c) 153 and 167 HV; (d) 177 HV. A pair of readings was taken for the core part (e) and for the pattern welded parts (f) and (g) in each case a pair of readings was taken firstly for one of the paler ferritic bands and secondly for one of the grey more pearlitic bands. The results were as follows: (e) 226 and 228 HV; (f) 230 and 243 HV, 152 AND 163 HV; (g) 193 and 195 HV, 162 and 170 HV. It is probable that the consistently higher readings for the ferrite areas are the result of a higher phosphorus content.

S 18. Sword, from Aylesford, (Kent). (Maidstone Museum). The corroded remains of a probable early Anglo-Saxon sword blade of which the lower or tip end, about half the tang and other smaller fragments were missing (Fig. 63). One main core fragment survived to which a number of flakes of totally mineralised surface incrustation adhered in places, and other similar flakes had become detached and in most cases it was not possible to work out where they had come from. The surface flakes looked as if they might contain the remains of a scabbard although this was barely recognisable as such. It was reportedly found in a gravel pit at Preston Hall, Aylesford in Kent, in March 1924 (Museum Records).

The main fragment was 58.3 cm in length by 5.2 cm in width. The thickness was difficult to estimate but appears to have been about 0.3-0.4 cm although it might have been slightly more, and its profile looks to have been either of a flattened or shallow but continuously convex form. The main core piece appeared to contain at least some network of metal over most of its length although it was clearly corroded right through in many places. Date: approximately 6-7th century.

A radiograph was taken of the main blade fragment plus the flakes adhering to it and those whose correct position could be worked out. This produced what seemed at first to be a surprisingly clear image considering the poor state of what survived and gave some good indications as to the internal structure of the blade. The core part of the blade was pattern-welded and the same triple banded design showed up running down this central area of the blade. This was composed of two outer twisted portions whose criss-crossed double image (i.e. of the two surfaces) showed up quite clearly in some places. Between these a straight-grained element ran right down the middle of the blade. Apart from the criss-crossing of the twisted parts there was little obvious sign that two images were superimposed and not enough to determine whether more than one set of composite rods were involved across the thickness of the blade. There was also no indication of any central core strip onto which the composite strip(s) might have been welded. The cutting edge appeared to be separate and exhibited some slight traces of a straight grained effect but nothing like as clear as that of the central portion of the composite core.

It is of some significance that the parts of the image(s) of the radiograph to show up most clearly were those devoid of surface incrustation whereas the areas particularly where this survived on both sides the radiograph was rather fuzzy and difficult to interpret.

A V-shaped section extending just over half way across the width was cut from a point 9.0 cm from the lower broken end of the sword. Unetched the section showed a very varied slag content and distribution. The weld lines between the main components of the blade showed up very clearly marked by lines of slag inclusions, both spots and ribbons and these were 2 phase with a pale (? wüstite) phase predominating. Inside the main components the slag content was mostly fairly low but was patchily high in places. Where it occurred it was noticeably different to that occurring at the main welds and it generally appeared as a dark (? fayalite) phase with only occasional paler spots visible within this.

When etched with nital the weld boundaries between the main components of the blade became even more heavily outlined than before as quite thick lines of pearlite appeared superimposed upon the lines of slag inclusions. The section was thus very clearly divided into a one-piece edge, (a), and the core area represented on the section by the composite pieces (b)-(f). The section was cut from a part of the blade where, judged from the radiograph, the metal survival appeared best but even then much had been lost to corrosion (Fig. 82), however, the main elements of the blade did appear to be present on the section.

The surviving part of the cutting edge (a) gave a rather blotchy or patchy appearance under low magnification (Fig. 82b). It mostly consisted of somewhat variable but mainly medium grain size ferrite which was mixed with some pearlite in irregular patches. Where it occurred the ferrite/pearlite mixture gave a rather spiky appearance typical of the Widmanstätten structure although this effect was not very pronounced. The pearlite was mostly irresolvable but did show a poorly formed lamellar structure in places. The carbon content at most was approximately 0.2% quickly fading to well below 0.1% over most of the area but in view of the rather spiky nature of the pearlitic areas this was probably an overestimate. Although the slag content of this area was mainly quite low, the distribution, orientation and size of these inclusions was fairly irregular. The overall impression of the cutting edge part (a) was of a piece essentially of wrought iron with some carbon, probably there by accident. It appeared to represent a piece of bloomery iron which had been subjected to some hammering after smelting to remove most of the slag but not subsequently forged to any great extent or piled which would have given a more banded and less irregular and patchy appearance in section. It is just possible that the cutting edge did originally have a small steel strip welded diagonally (in section) to one or both sides of piece (a) towards the tip of the cutting edge and subsequently lost to corrosion. There was, however, no evidence of this either from the section examined or from the radiograph of the surviving main fragment, therefore, this can probably be discounted.

Parts of all the composite or pattern-welded triple-banded core, observed on the radiograph, were represented in the section. The outer two twisted elements visible on the radiograph each consisted of two composite rods welded back-to-back, (b)/(c) and (e)/(f) (Fig. 82a) and between these was a single composite rod (d) which was subdivided into eight vertical (in section) bands corresponding to the straight grained central part of the radiograph. These vertical bands were highlighted as alternate paler and darker etched areas. The folded appearance of the composite parts (b), (c), (e) and (f) was very marked, parts (b) and (c) also consisting of eight alternate paler and darker bands. Parts (e) and (f) looked similar but only half of them appeared in the section so the total number of bands could not be counted.

The microstructure of the paler and darker areas of composite parts (b), (c) and (d) were very similar. The paler bands consisted of very large grained ferrite with a quite high content of slag inclusions in patches although it was mostly fairly low. The darker bands consisted mainly of medium grain size ferrite mixed with some pearlite in the same rather Widmanstätten type of distribution seen in the cutting edge area (a). The pearlite was either partially unresolved or granular in appearance or it exhibited a poorly formed lamallar structure. Even making no allowance for the nature of the pearlite the carbon content of these darker bands is clearly very low, a visual estimate giving no more than about 0.1% and often less. The composite pieces (e) and (f) were similar but here the carbon content was even lower. The bands within the composite pieces mostly showed because of the contrast between the paler and darker area on etching and there were no particular indications of weld lines such as lines of slag inclusions or grey or white segregations of pearlite and ferrite.

Three hardness readings were taken for the cutting edge piece (a), two nearer the cutting edge and one near the inner weld and the results were: 212, 212 and 187 HV. From the twisted composite piece (b) hardness values were obtained for one of the pale very large grained ferrite bands and one of the pearlitic bands and these were 204 and 206 HV. For the adjacent twisted composite part (c) two readings were taken for one of the pale very large grained ferrite bands and one each for the grey pearlitic bands and in this order the results were: 202 and 202 HV, 210, 215 HV. Readings from one pale very large grained ferritic band and from one grey pearlitic band gave hardnesses of 199 and 202 HV. The large grained part (f) gave a hardness of 197 HV.

At most the darker bands of the composite rods could only be described as a very low carbon steel, little more than wrought iron, although it had clearly been carefully combined with alternate strips of carbon free iron. It was noticeable how, if anything, the highest carbon content of the blade occurred in places along the weld boundaries between the main parts of the blade where the grey pearlitic lines were at their densest although even here it was not much higher than elsewhere.

The blade was probably quenched after final forging although the carbon content was too low for this to have had much effect on the microstructure or hardness of the blade.

S 19. Sword, from Wickhambreux, Kent. (Maidstone Museum). One surviving piece of a sword blade with the barely recognisable and flaky remains of part of a scabbard attached to it (Fig. 63). It was found in a gravel pit at Wickhambreux, Kent in 1886[73]. It came from part of an Anglo-Saxon cemetery disturbed by gravel quarrying and associated finds included at least two (fragmentary) swords, a large bronze bowl and a blue glass vessel. Of these other swords one (S.12) was metallographically examined and the other was X-radiographed but no metallographic examination was carried out as it appeared to be completely mineralised - although it showed the vaguest hints of what may have been a twisted composite core. Date: approximately 6-7th century.

The sword fragment was 75.2 cm long by approximately 5.0 cm in width and of uncertain thickness - roughly 0.5 cm. The tang, the tip end and much of the edges of the blade were missing and what was left was very badly corroded with some metallic core surviving.

A radiograph of the sword showed a rather diffuse image and very little structural detail could be made out although a criss-cross pattern, indicating the double image effect of twisted composite elements in the core of the blade, was visible in places. In one place this criss-cross effect appeared to be superimposed upon a straight grained element to the same core area. The superimposition of a criss-crossed and straight grained effect is probably indicative of a separate central core piece. From what was left it appeared that the patterned part of the blade probably consisted of three twisted composite pieces on either side welded on to a single central core piece (Fig. 83c). Separate edge pieces could be seen in places but these bore no discernible structural traces.

A transverse section across the whole width was taken from near one broken end of the blade. Much slag was visible across the unetched section especially in the central area but no individual welds were discernible as specific lines of slag inclusions.

When etched with nital it became clear that the blade was composed of two cutting edges which had been welded to a core made up of two sets of three composite strips welded on to either side of a central core piece (Fig. 83). The metal survival was patchy (Figs. 83a and b) but despite this parts of all the main constituents of the blade were represented; part of one cutting edge, (a) (in Fig. 83a), parts of four , (b)-(e), of the original six outer composite pieces of the core and about half of the central core piece, (f). The edge piece (a) survived as an 'island', isolated by corrosion from the core part of the blade at this point, the weld zone having been lost. The slag content of the edge piece (a) was quite high, consisting of small mostly well flattened inclusions. This area otherwise consisted mainly of large grained pearlite with ferrite dispersed amongst it in a distinctive Widmanstätten form. Much of the pearlite appears lamellar although somewhat granular in some areas. The Widmanstätten formation must represent fairly fast cooling although probably not quenching. In an equiaxed grain structure pearlite of the proportion that appears present here would represent a carbon content of about 0.5% but an estimate of about 0.3-0.4% may be nearer in view of the spiky form. The pearlite content dropped away sharply towards the sides of this edge piece and more gradually towards the tip and at the sides the carbon content is less than 0.1%. In these lower carbon areas the grain structure was more nearly equiaxed. This sudden change was not accompanied by obvious signs of a weld (slag etc., lines) and may have been the result of fairly extensive decarburisation of a single, at first fairly homogeneous steel edge piece. It is possible, however, that the edge piece may have been composed of a three part, iron-steel-iron sandwich skillfully welded together and the welds subsequently masked by carbon diffusion outwards. There was not sufficient evidence to resolve this problem. A single sandwich construction is suggested in Fig. 83d.

The composite core pieces (b)-(e) were all similar and all the fairly small areas surviving gave a folded appearance typical of twisted and flattened composite rods in section. They consisted mostly of wrought iron, the folded appearances being partly the result of slag segregation and partly by a banded appearance created by alternating slightly darker and paler bands. The darker bands consisted of equiaxed medium grain ferrite with a little pearlite at the grain boundaries, the carbon content being mostly well under 0.1%. The paler bands consisted of large grain ferrite. Slag content of the composite pieces was fairly even as small inclusions, although the overall quantity was somewhat less than in the cutting edge area (a). The weld lines between the composite pieces (b)-(c) and the central core piece (b) were marked on each side with a broken line of slag inclusions upon which was superimposed a white line of segregated ferrite.

The core area (b) consisted of wrought iron which had a high slag content, much more than the cutting edge (a) or composite pieces (b)-(c), and it was present both as small and large partially flattened inclusions. This was mostly aligned along the main axis of the section and was fairly evenly distributed. Apart from slag this area consisted of mainly large grain ferrite with occasional grain boundary carbide present.

Three hardness readings were taken for the edge part (a) giving a value

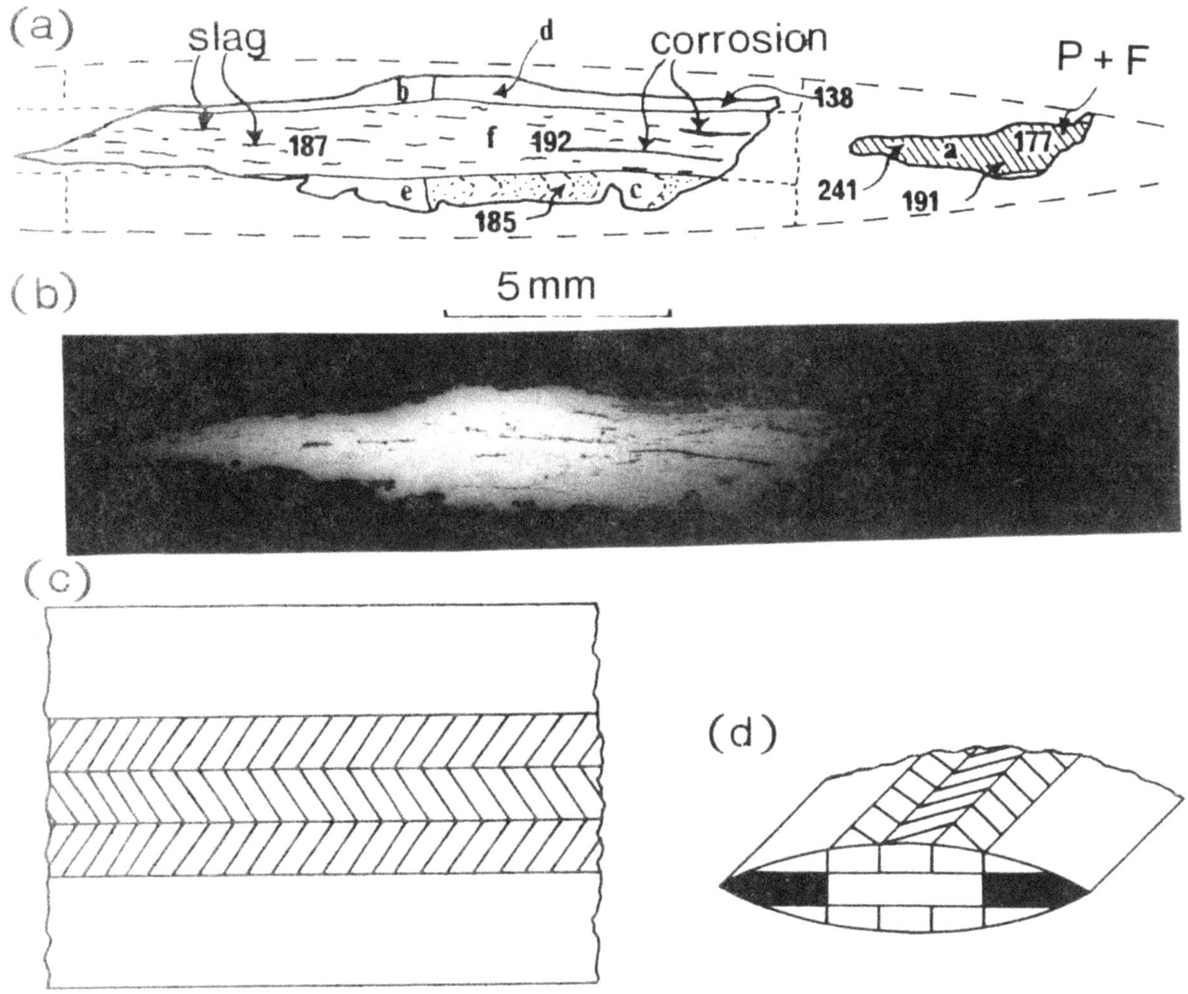


Fig. 83. Section of sword S 19 from Wickhambreaux, Kent + surface detail drawn from X-ray and three-dimensional view.

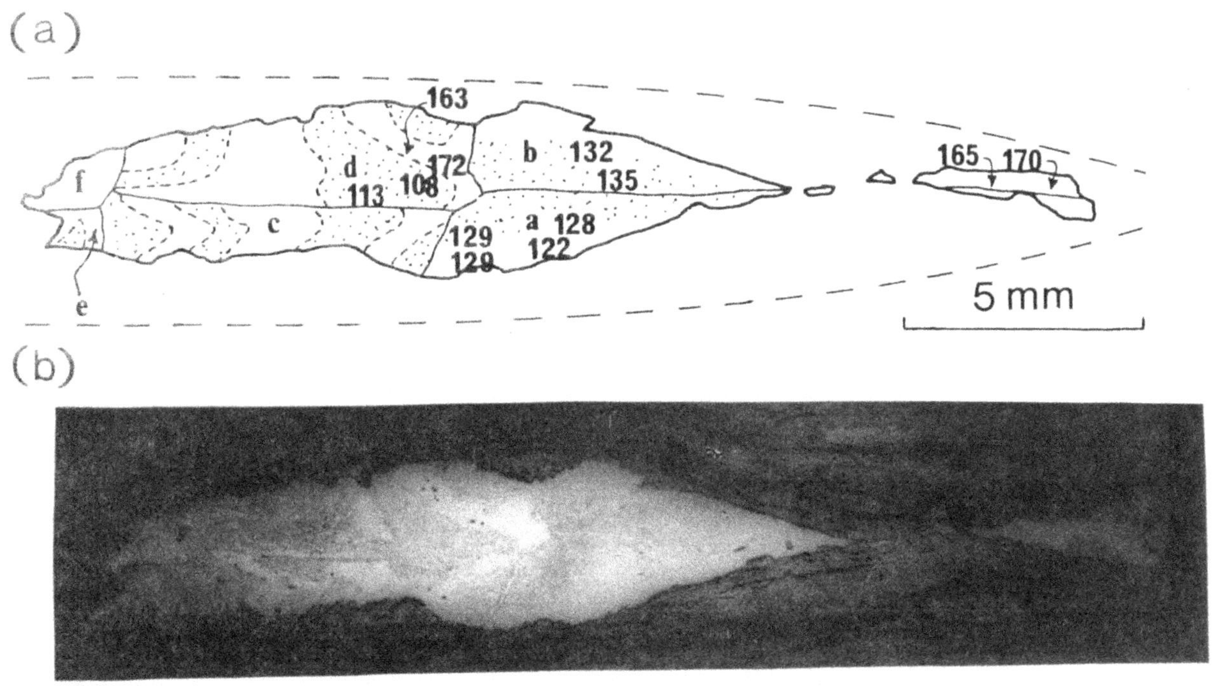


Fig. 84. Section of sword S 16 from Sarre, Kent.

(c)

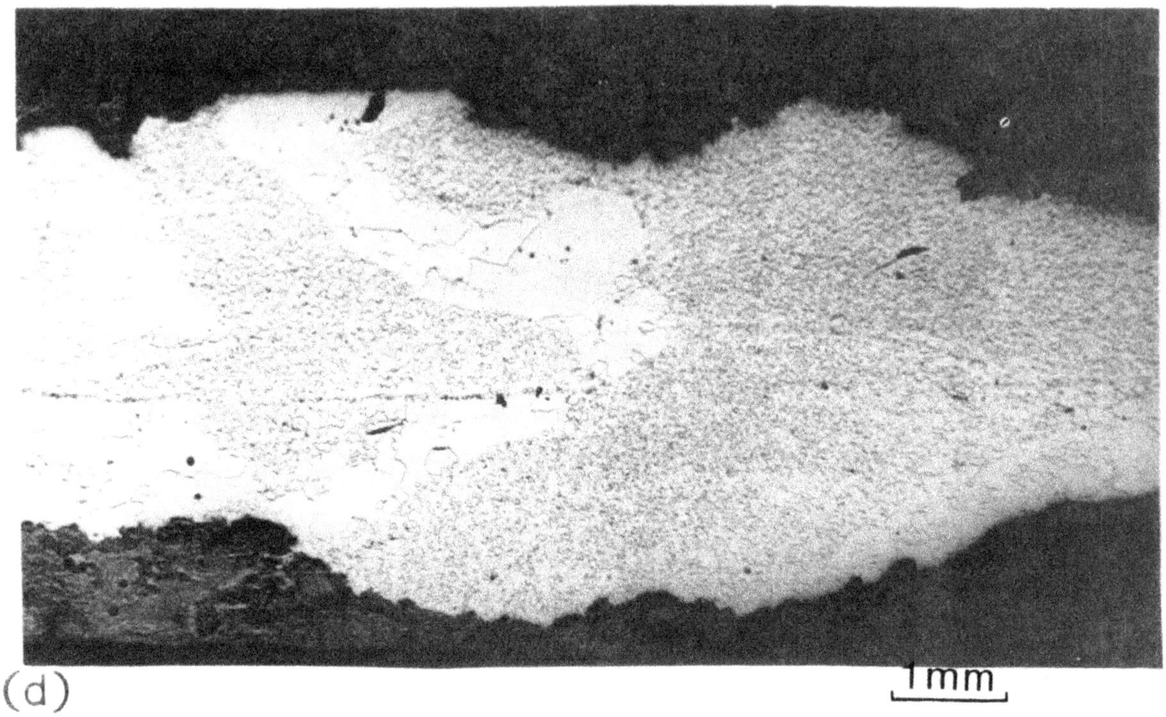


(d)

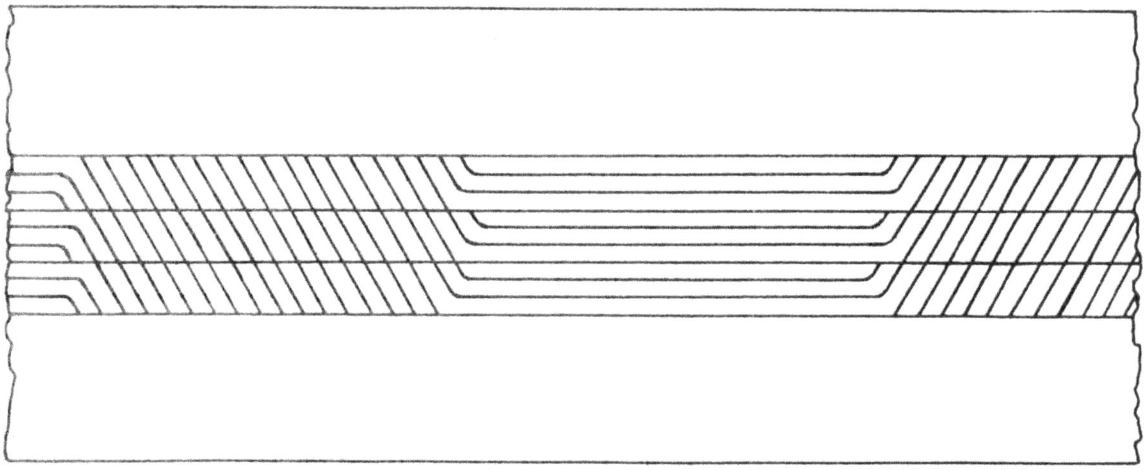

(e)

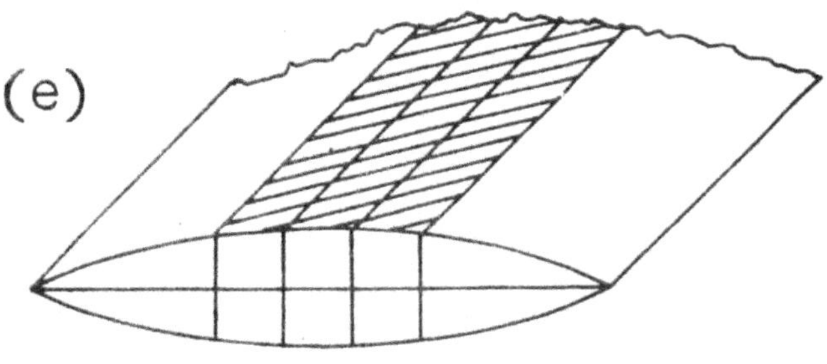

Fig. 84 (continued) Section X 21 + surface detail drawn from X-ray and three-dimensional view.

of 241 HV for the most pearlitic part and values of 177 and 191 HV in the areas nearby which showed some ferrite. Hardness values of 187 and 192 HV were obtained for the central core piece (f). One of the larger grained paler ferrite bands of the composite pattern welded part (c) gave a hardness of 185 HV and one of the darker smaller grain ferrite bands of pattern welded part (d) gave 138 HV.

The blade was mainly of wrought iron with the exception of the steel edge piece (a) which was probably sandwiched between two pieces of wrought iron in a similar way to S 17 and S 20.

S 16. Sword, from Sarre, Kent. (Maidstone Museum, KAS 818). A nearly complete sword blade consisting of four fragments (Fig. 63). It was excavated (?) from grave 264, of the Anglo-Saxon cemetery at Sarre in Kent in 1864[72a]. The end of the tang and several small fragments were missing from the blade which was in a very poor state, very corroded and flaky and rusted right through in places. Its length was 85.7 cm; width 6.3 cm and thickness difficult to gauge but looked, allowing for expansion during corrosion, from the breaks to have been about 0.5 cm. The width included the thick uneven crust of rusty corrosion products round the blade which appeared to represent the remains of a scabbard, although hardly recognisable as such. Date: approximately 6-7th century.

A radiograph showed some structural detail to the sword although it was mostly not very clear. A core with three adjacent composite rods on either side of the blade, running down the centre, was indicated and these mostly appeared to give an alternating straight and twisted pattern although in individual rods this appeared to alternate in some places with straight grained lengths (Fig. 84d). Some very slight indications of a straight graining effect could be seen along the cutting edge. Several cracks were visible on the radiograph between both the cutting edge and the core and between composite rods in the core. This is probably the result of corrosion penetrating and proceeding along the welds between these parts, and where this was noticed between the composite rods it was probably a fairly good indication that there was no separate central core piece to the blade.

A V-shaped section extending approximately halfway across the blade was cut from an area near the hilt end of the sword where, on the radiograph, metallic survival appeared to be better. When viewed unetched the slag content appeared variable but mostly quite low but there were a few very large inclusions which along with broken lines of much smaller inclusions appeared to mark weld positions between components.

When etched with nital the structure shown in Fig. 84a, b and c became visible. The cutting edge was composed of two halves (a) and (b) which were divided by a horizontal (Fig. 84a) pale ferrite line which coincided with a line of flattened slag inclusions. This line stood out in contrast to the rest of (a) and (b) which in each case showed up as a grey pearlitic background near the weld with the core zone becoming progressively paler towards the tip of the cutting edge. The line also showed up as a raised ridge under oblique illumination. Even in the darkest parts of the grey areas of (a) and (b) the proportion of pearlite was low and the carbon content here approximated to 0.1%. The grain size was quite small and the pearlite appeared partially lamellar and partially granular or irresolvable in form. The pearlite constituent of (a) and (b) mostly faded out towards the cutting edge tip and the grain size increased and was mostly quite large in these ferrite areas. In the largest grained ferrite areas of the cutting edge the short rod-like nitride or carbide formation showed up in places. The weld between the cutting edge and the core part of the section showed up as a clear V-shaped line whose point coincided with the weld line between the two halves of the cutting edge (Fig. 84a). This V-shaped weld line showed up partly as a result of slag inclusions along it but mainly because of the superimposed white line which contrasted sharply with the grey pearlite areas of the edge parts (a) and (b) and the adjoining paler ferrite areas of the core. There had been little carbon diffusion across this weld junction and consequently the line remained clear. Parts (c)-(f) of four composite rods were represented on the section although only a small area of each of (e) and (f) were included. These composite pieces were welded back-to-back with no intermediate central core piece. The weld lines between them were fairly clear, marked by lines of slag inclusions highlighted, in places, by the contrast between grey pearlitic and paler ferritic parts of adjacent composite pieces. The composite pieces were all similar in composition each consisting of alternate pale and darker or grey bands. The bands in each case bore a distorted or folded appearance which showed the twisted nature of the composite rods in section. The pale areas of the composite core pieces consisted of large grained ferrite which in places also showed up more of the short rod-like nitride or carbide formations seen in the ferrite areas of the cutting edge. The grey bands contained a fairly even but small proportion of fairly fine grained pearlite which mostly represented a carbon content of about 0.1%. The boundaries between these ferritic and pearlitic zones within the composite pieces were mostly marked by the contrast between the paler and darker bands. Little carbon diffusion appears to have occurred between these areas and the boundaries between them are also not clearly marked by lines of slag inclusions. The boundaries, however, must more-or-less mark the positions of welds between the parts that together formed these composite rods.

Four pairs of hardness readings were taken across the edge of the blade, one pair near the tip of the cutting edge part (b) and a pair each for parts (a) and (b), further in towards the inner weld and a final pair for part (a) near the inner weld. For these the values obtained were 165 and 170 HV; 132 and 135 HV; 122 and 128 HV, 129 and 129 HV. Pairs of hardness readings were taken for one of the large grained ferrite bands of pattern welded parts (c) and (d) and a pair of readings for one of the grey slightly pearlitic bands in part (d). The values obtained were: 124 and 129 HV; 163 and 172 HV, 108 and 113 HV (Fig. 84a).

Each of the composite rods (c) and (d) were made up of six individual strips and it is probable that the other composite rods were similar - certainly the small areas of (e) and (f) appeared similar in section and the overall appearance of the composite rods is much the same on the radiograph. The metal of both the two halves of the cutting edge and the core areas of the blade appears to vary between wrought iron and very low carbon steel itself little more than wrought iron.

<u>S 17. Sword, from Sarre, Kent.</u> (Maidstone Museum, KAS 830). A sword nearly complete except for the tip and still within the encrusted and barely recognisable remains of its scabbard. It measured 82.6 cm in length by 5.3 cm in width. It was in very poor condition and looked to be mostly mineralised although the fact that it held together suggested that at least a network of metal survived in the core of the weapon. It was excavated in

the 1860's from the Anglo-Saxon cemetery at Sarre in Kent,[71, 72] but it is not known whether the break occurred in antiquity or whether the missing tip fragment was lost subsequently. It could be seen from the break that the profile of the blade approximated to a flattened convex shape but it was so corroded that only a very rough estimate of the original thickness could be made. This would appear to have been about 0.5 cm. Date: approximately 6-7th century. (Fig. 63).

A radiograph was taken which showed the blade to have a pattern-welded core. The detail was not very clear, probably to a large extent due to the interference effects caused by the encrusted and mineralised remains of the scabbard in the path of the X-rays. The general form of the pattern, however, was fairly easy to make out and this appeared as a herringbone design running uninterrupted from the hilt down the surviving length of the blade. The design was formed by three adjacent twisted composite rods welded side by side. It was very difficult from the radiograph to say whether or not two sets of three adjacent composite rods were present on either side of the blade and no sign of any separate central core piece could be seen. The herringbone pattern shows up as being criss-crossed on the radiograph which could either have resulted from composite rods with opposing (i.e. right and left hand) twists having been welded behind one another or, by a single layer of rods whose twists would have to point in the opposite direction on either side of the blade. No sign of any structural detail could be seen along the cutting edges of the weapon. Fig. 85c shows how the pattern may have appeared on the surface on either side of the weapon.

A V-shaped section, extending approximately half way across the width, was cut from a point 8.0 cm from the broken end of the blade. When viewed unetched the slag content over most of the section appeared quite low. Some quite large ribbon-like slag formations together with broken lines of small spots ran horizontally (i.e. parallel to the main axis) across the section suggesting that both the core area and the cutting edge might consist (in section) of three horizontal layers. The areas near the surface of the core part contained more slag than the rest of the section.

When etched with nital the composition of the section became clearer (Fig. 85a and b). The cutting edge appeared to consist of three main horizontal bands, (a)-(c) in Fig. 85a. A fourth piece may have been welded on to piece (a) to form the outside of the cutting edge but the evidence for this was largely lacking. Part (a) varied a great deal in appearance from one end to the other. Near the tip of the cutting edge it consisted of a mixture of fairly pale etched coarse martensite and a very dark etched irresolvable constituent, somewhat nodular in appearance (probably troostite). The martensite was restricted to the tip end and the very dark etched constituent was mixed with increasing amounts of a Widmanstätten-like distribution of ferrite with grey irresolvable pearlite. The proportion of ferrite appeared to increase and it became paler towards the weld line between (a) and the core part of the blade. A very narrow surviving paler area along the outside part of part (a) consisted mostly of fairly large grained ferrite with some acicular ferrite/pearlite mixture although much less than the adjoining band (a) and the contrast between them is fairly pronounced. There was no line of slag inclusions or any other sign of a weld along this zone between the darker band (a) and the adjoining paler area. This and the proximity to the surface of the blade suggests that the paler area may have been the result of the decarburisation, possibly just localised to the surface of this part of the cutting edge.

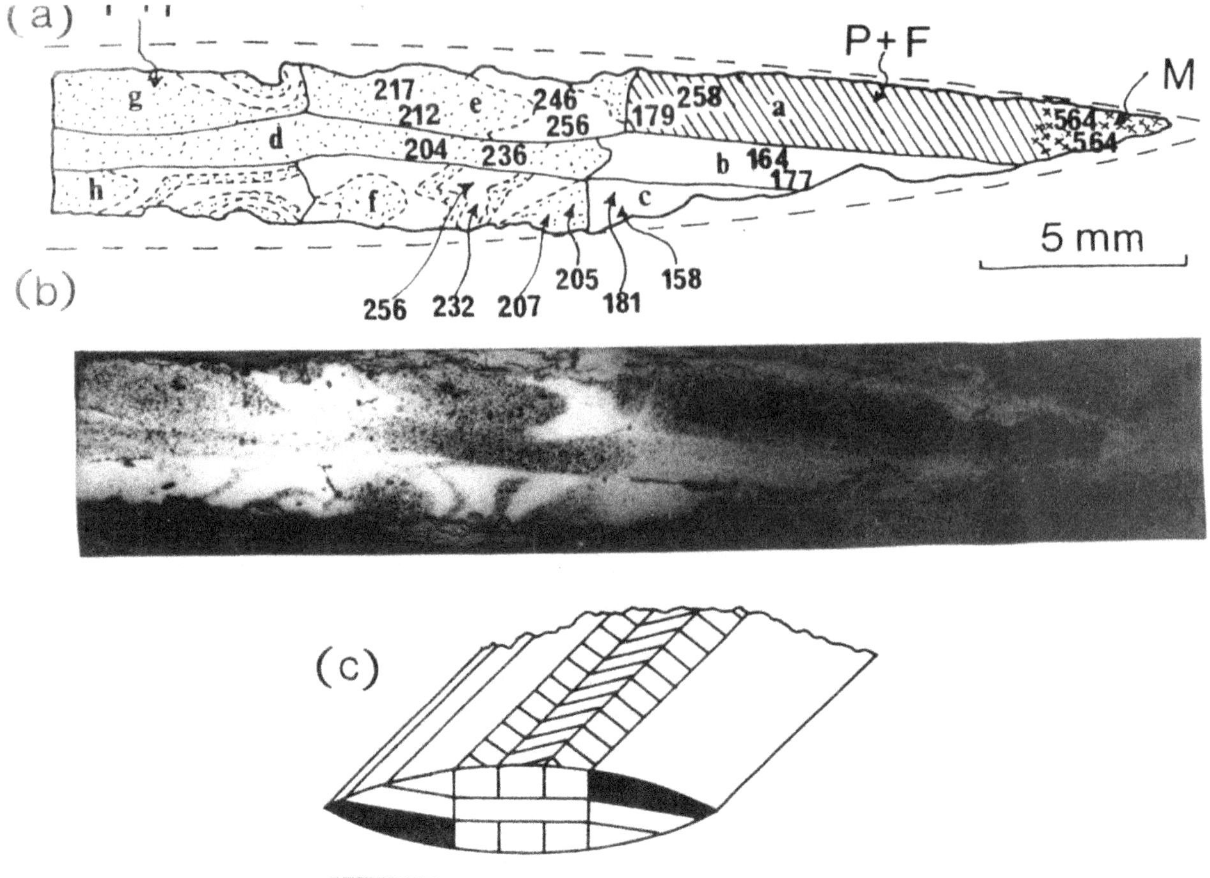


Fig. 85. Section of sword S 17 from Sarre, Kent + three-dimensional view.

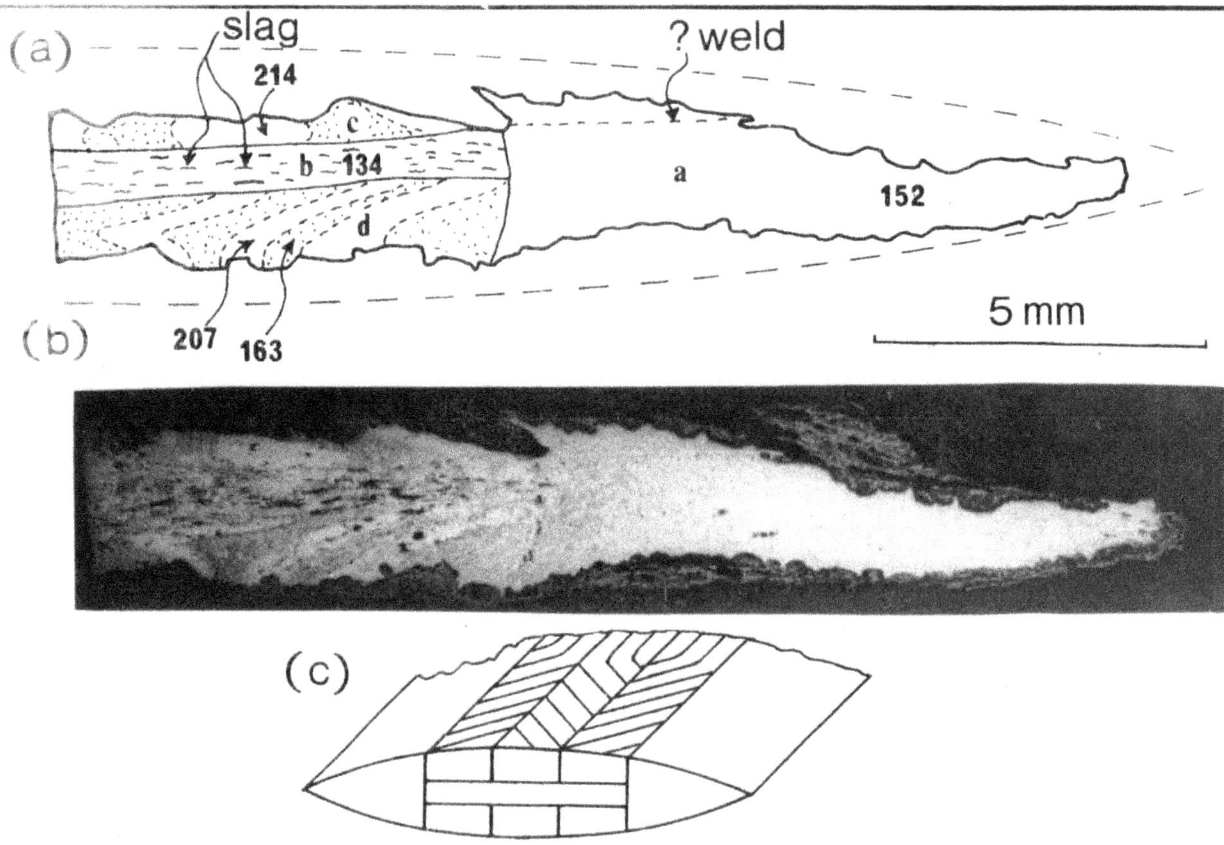


Fig. 86. Section of sword S 56 from Bifrons, Kent + three-dimensional view.

An intermittent line of slag spots indicated the position of a weld line between strips (a) and (b) of the cutting edge. Extensive carbon diffusion appears to have taken place across the weld line between (a) and (b) which otherwise consisted mainly of fairly large grained ferrite with a low proportion of Widmanstätten ferrite and unresolved pearlite scattered across it. Part (c) also consisted mainly of fairly large grained ferrite with a low proportion of spiky Widmanstätten ferrite and unresolved pearlite but was separated from (b) by a clear line of slag inclusions indicating a weld between these components.

The different microstructures observed in part (a) of the cutting edge were clearly the result of the quenching of the blade after final forging. The progressive nature of the quenching effect was very noticeable; the martensite forming near the tip of the cutting edge which cooled first and fastest, and the progressive change in the microstructure, towards the weld junction with the core of the weapon, to a mixture of Widmanstätten ferrite and unresolved pearlite was seen. This inner part of (a) would have cooled slowest and this change in the microstructure probably reflects the differences in the cooling rate between the inner and outer edges (as seen in section) of strip (a) rather than any appreciable difference in its carbon content. Unfortunately the quenched nature of the microconstituents precluded any reliable visual estimate of carbon content although at the inner end a very rough estimate (making some allowance for the acicular Widmanstätten distribution) of 0.2–0.4% carbon may be made.

The carbon contents of parts (b) (ignoring the effects of diffusion from (a)) and (c) were clearly very low, probably less than 0.1% so that the cutting edge appears to have been made by welding together three strips, one of (probably) fairly low carbon steel, and the other two of little more than wrought iron. The cutting edge must then have been welded to the already assembled core part of the blade. The weld junction was mostly not very clear, being marked along most of its length by a faint line of slag spots. Large slag inclusions did occur however within the core part of the blade where more than two components met at the same place.

The core part of the blade consisted of composite twisted pieces welded onto a central core piece, (d) on Fig. 85a. Of the total of six composite pieces, four were represented on the section, (e)–(h) on Fig. 85a although parts (g) and (h) were not quite complete.

The central core part (d) showed a fairly even Widmanstätten distribution of ferrite and unresolved pearlite of a medium grain size and with a low slag content. The structure appeared very similar to the inner or core side of the cutting edge piece (a) and the two parts may have had similar carbon contents i.e., very approximately 0.2–0.4%. The central core strip (d), therefore, appeared to be a similar low carbon steel to piece (a).

The folded appearance of the twisted composite rods (e)–(h) can be seen in 85b. They were all similarly composed of alternating bands of very large grained ferrite and medium grained sized ferrite and unresolved pearlite in a Widmanstätten formation. The pearlitic bands looked very similar to the central core part (d) and thus appear to have been a low carbon steel of a similar carbon content. In (d) and the pearlitic parts of (e)–(h) a few small grained patches occurred where the carbon content appeared to be somewhat lower than the rest. The ferrite bands of (e)–(h) contained rather more slag and a solitary pair of Neumann bands were also visible.

A pair of hardness readings was taken for each of the low carbon iron edge parts (b) and (c) and two pairs of readings taken for the steel edge part (a) one pair near the inner weld and one pair near the tip of the cutting edge. In this order the results were as follows: (b) 164 and 177 HV; (c) 158 and 181 HV; (a) 258 and 279 HV, 564 and 564 HV. One pair of readings were taken from the central core piece (d) which gave values 204 and 236 HV. For the pattern-welded parts (e) and (f) two pairs of readings were taken in each case firstly for one of the very large grained ferrite bands and, secondly for one of the grey pearlitic bands. The values obtained were (e) 246 and 256 HV, 212 and 217 HV; (f) 232 and 256 HV; 205 and 207 HV.

It would appear that the cutting edge of the blade was made from 3 or possibly 4 pieces of which one (a) was of low-medium carbon steel with the rest of fairly low carbon iron. The central core and pearlitic bands of the pattern-welded parts were of a similar low-medium carbon steel to the edge part (a). The relatively higher hardness values for the paler ferrite bands of the pattern welded pieces probably indicates that a high phosphorus iron was used for these. The blade was quenched although slightly imperfect results were obtained probably indicating that the carbon content of the steel was too low for the particular quench. The blade appears to have been subjected to some final cold hammering.

S 56. Sword from Bifrons, Kent. (Maidstone Museum KAS 837). A blade in two fragments without the tang and the end of the tip. In poor condition, mostly mineralised with the remains of the scabbard. It was found in 1863 in grave 39 of the Anglo-Saxon cemetery at Sarre[72a] and is approximately of 6-7th century date. The two fragments together measured 70.4 cm in length with a max. width of about 6 cm. (Fig. 63).

Radiography showed the blade to be pattern-welded with three sets of adjacent composite and partly twisted rods on each side. The pattern alternated between a herringbone and straight-grained design (Fig. 86c).

A wedge shaped section extending about half-way across the width was cut from the blade 12 cm down from the shoulder and showed the usual slag content which was very high in places (Fig. 86). It was clear that the pattern-welded composites had been welded to a central core piece before the cutting edge was welded on to either side. The cutting edge consisted of medium to coarse-grained ferrite with a little carbide in the finer grained areas. The central core piece was also ferrite. The composite pieces showed the usual banded structure, the paler bands showing nitride or carbide needles.

The hardness of the cutting edge, part (a), was 152 HV and the central core-piece, 134 HV. One of the lower carbon darker bands had a hardness of 164 HV while the paler ferritic bands had hardnesses of 207-214 HV indicating higher phosphorus.

Clearly this blade was made of nine main components (Fig. 86c) all of wrought iron but essentially of two types, with and without a high phosphorus content. This would have ensured that the pattern showed up after suitable treatment.

S 15. Sword from Holborough, Kent. (Maidstone Museum). The main part of a sword blade of which the tip end is missing (Fig. 63). It was one of two swords found in the Anglo-Saxon cemetery at Holborough in Kent. Associated finds indicate the cemetery to be 7th century in date[75]. The surviving blade fragment was 76 cm long by 5.6 cm wide and it was badly corroded with a very uneven flaky and pitted surface although there was at least some metal left underneath the corroded surface. The original thickness of the blade could not be measured but would appear to have been about 0.5 cm and it appears to have had a flattened convex profile. Much of the edges of the weapon had corroded right away.

A radiograph produced a clear image showing a pattern-welded core to the blade as well as other grain effect detail. The pattern-welded part consisted of three adjacent composite rods forming the surface on either side of the blade. The pattern was the same on either side and on the radiograph the images overlapped only slightly and produced an alternating herringbone and straight grained effect in panels each of which measured about 5 cm in length except for the uppermost, a herringbone panel next to the tang which was about 10 cm in length. The rods continued through the tang and showed here a straight grained effect. Fig. 87c shows a diagrammatic reconstruction of how part of the pattern showed. The cutting edge also shows a clear straight grained effect in places.

A V-shaped transverse section was cut from the blade approximately 10 cm from the tang and just within the first and longest of the herringbone 'panels' which showed on the radiograph. The section extended to rather under half way across the blade. When viewed unetched the visible slag content varied and was mostly quite high. A few slag spots formed a faint intermittent line vertically across the section, marking the weld position between the cutting edge and the core part of the blade. On the core side of this weld swirling slag patterns mark the folded aspect of the twisted composite rods that shows in section. The slag content of the cutting edge was especially varied, very high in patches with small-medium sized spots, partially flattened ribbons and quite large irregular inclusions and it was mainly 2 phase, a pale phase in generally predominant darker background (? wüstite in a glassy matrix). Some areas, however, contained very little slag.

When etched with nital the structure became clearer although it did not show up very well in the macrograph (Fig. 87b). The main reason for this was that the structure which showed up was entirely one of wrought iron with no steel present at all and so there was not the contrast between pale and darker (i.e. ferritic and pearlitic zones) that has been observed so clearly on other sections.

The cutting edge appeared to consist of one piece (a) although two very large slag ribbons, the dark corroded outlines of which can be seen on Fig. 87b, running at roughly 45 degrees across the cutting edge suggest the possibility that two pieces of poorly prepared wrought iron were welded together to form this part of the weapon. These large slag inclusions do not however, line up very well and this and the generally rather variable

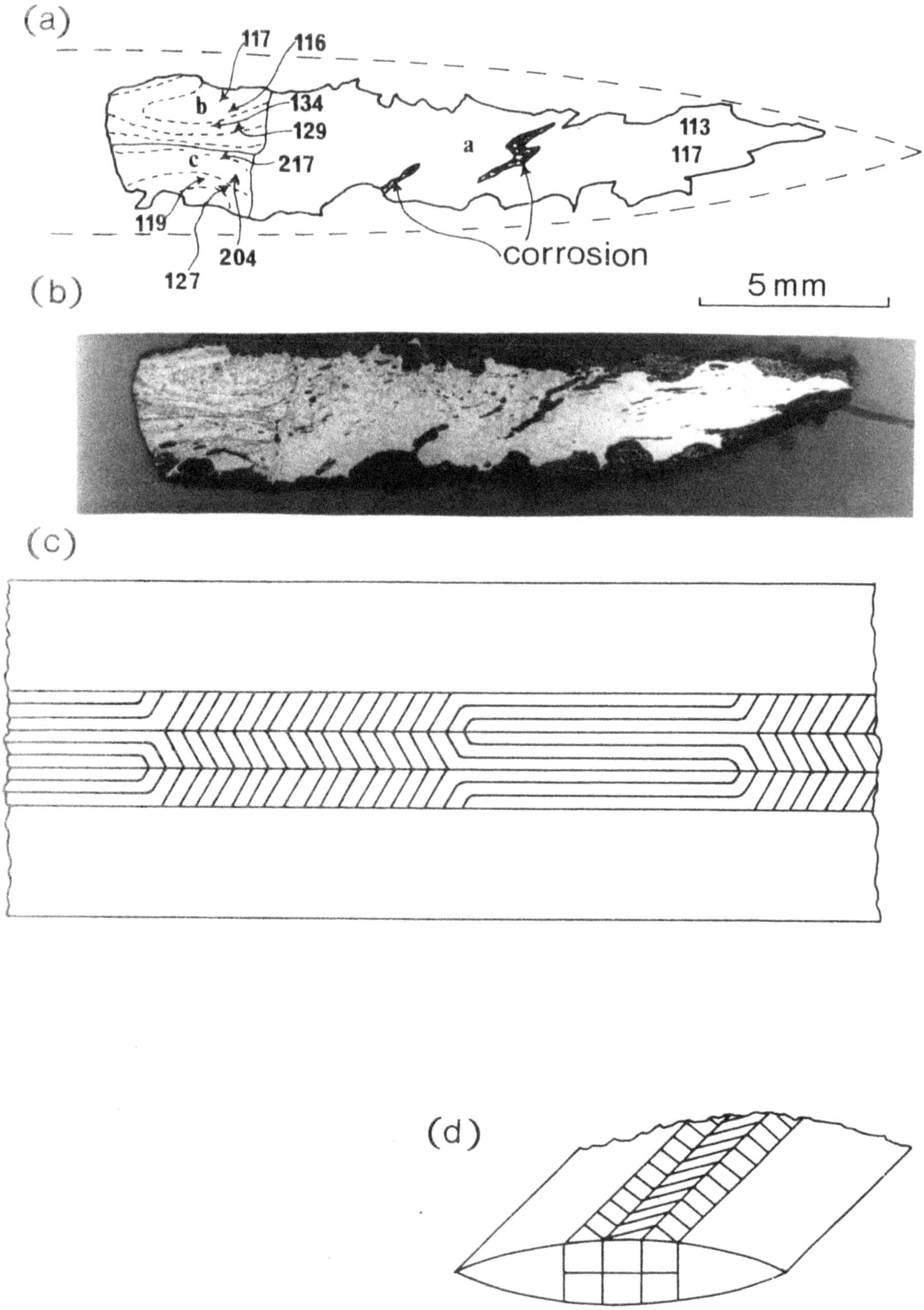


Fig. 87. Section of sword S 15 from Holborough, Kent + surface deta
drawn from X-ray and three-dimensional view.

orientations and distribution of the slag may just indicate that the cutting edge part (a) consisted of one piece of poorly prepared bloomery iron forged out to the desired shape before being welded to the core. The grain size of the ferrite of part (a) varied between very large nearer the tip of the cutting edge and medium-large nearer the core area where a very little pearlite was visible at the grain boundaries in places. The very large grained area near the tip also contained a patchy distribution of rod-like nitride or carbide 'needles'.

In the core area of the section only two of the composite strips were represented on the section. These, (b) and (c) (in Fig. 87a) had been welded back-to-back without any intermediate central core piece. The weld line between (b) and (c) showed up as a narrow intermittent line of small slag spots and ribbons. These parts consisted almost entirely (slag, excepting) of ferrite and the folded appearance being slightly accentuated by alternating bands of very large grained ferrite, and a rather smaller grained ferrite with a very little pearlite visible at the grain boundaries in places.

A pair of hardness readings were taken for the cutting edge (part (a) and, for the pattern-welded core parts (b) and (c) pairs of readings were taken, firstly in one of the paler larger grain ferrite bands and secondly in one of the finer grained ferrite bands. In this order the results were as follows: (a) 113 and 115 HV; (b) 129 and 134 HV; 116 and 117 HV; (c) 204 and 217 HV, 119 and 127 HV (Fig. 87a). Only the paler band in part (c) gave a markedly higher value and this probably indicates a higher phosphorus content here.

<u>S 31. Sword from Caversham, Reading</u>. (Reading Museum, 51:80). The remaining lower half of a sword blade (Fig. 63) found near Caversham (SU 7386 7511), in the flood plain of the Thames. No further details were recorded and the circumstances of its finding are not known. Its condition and the way it had corroded suggested that it was a river find. It had corroded very unevenly with some parts corroded right through while in some places, as well as the metal core, were well preserved as was the wide shallow fuller that had run down on either side of the blade. What did remain of this blade fragment survived mostly as metal although deep etching by corrosion had revealed some of the internal structure of the blade. Parts of what appeared to be a triple banded pattern-welded design were visible in the central fullered part of the blade and, in places, the cutting edge had a coarse straight-grained appearance. The broken end of the blade was very irregular and badly corroded and there was no indication as to whether the sword had been already broken or bent when it was lost or buried.

The blade profile with the wide shallow fuller is very similar to that of S 46 below which has been given an approximate 7-9th century date. This blade is unlikely to be earlier and a similar date range is probably the safest estimate. Its length was 53.7 cm and maximum width 4.8 cm.

A radiograph of the blade confirmed the observations described above but provided a little more detail. A triple banded pattern-welded element to the blade could clearly be seen. The herringbone pattern appeared to overlap in places which is probably indicative that two separate sets of triple twisted composite strips form the surface of the central part of the blade on either side. The herringbone pattern given by adjacent twisted strips alternated with straight grain variations to the pattern (Fig. 88c). Where the cutting edge parts of the blade were visible on the radiograph they showed a fairly pronounced straight grain effect.

A transverse wedge shaped section extending approximately half way across the blade was cut from an area where the surface on one side of the blade survived quite well (Fig. 63).

A very variable slag content was visible before etching although no very clear weld lines were discernible. After etching (Fig. 88b) the horizontal weld lines marking the division between central plain core of the blade and the outer twisted composite parts showed up as dark lines which can be clearly seen on the left side of the macrograph. The weld junction between the upper two composite surface strips (the junction between the lower two having been corroded away at this point) and the weld position between the cutting edge portion of the blade and the central part also became clear as dark grey lines. These lines consisted mainly of pearlite which coincided in places with broken and rather discontinuous lines of small slag spots and ribbons. Although it is not quite so clear on the macrograph it was possible to separate the cutting edge area of the section into four different areas from which it appears that the cutting edge was assembled from four separate pieces. A simplified interpretative diagram of the macrograph is shown in Fig. 88d.

The central part (d) of the cutting edge consisted of a piece of steel with a carbon content of approximately 0.3-0.4% near the centre which decreased fairly evenly towards the boundaries with (b), (c) and (e) where it varied between about 0.1 and 0.2%. The slag content of (d) was mostly fairly low - mainly fairly evenly distributed small spots and ribbons with a few large inclusions. There has clearly been much carbon diffusion outwards across the weld boundaries between (d) and (b), and (d) and (c). The welds between these zones, as was the case elsewhere on the section, were of very high quality with little slag marking the boundaries. Part of the weld boundary between (d) and (c) was also, and much more clearly, marked by a grey line of pearlite. This was not the case between (a) and (b) whose weld boundary had been marked by carbon diffusion between (a) and (b). Zone (b) was, however, of a quite different character to (a) consisting for the most part of very large grained ferrite with much slag mostly as fairly small ribbons with some large more amorphous inclusions. The slag was aligned along the section and gave this zone (b), a rather streaky appearance. Area (c) also consisted mainly of very large grained ferrite but in contrast to (b) contained little slag. Zone (a) appears in the macrograph as a very dark wedge-shaped area bounded on the inside by two rather faint and diffuse white lines which are themselves highlighted by further dark diffuse area. The wedge-shaped area (a) appears to represent a narrow higher carbon steel strip that has been welded between the two outer edges of strips (b) and (c) during the assembly of the cutting edge. It consisted of a mixture of very fine martensite with some very dark etching nodular troostite. The grain structures became much coarser towards the welds and the proportion of nodular troostite increased. The paler diffuse lines coincide with a rather indistinct line of slag spots and appear to mark the position of welds between (a) and (c), and (a) and (b). The further dark zone round the inside of (d) is indicative of extensive

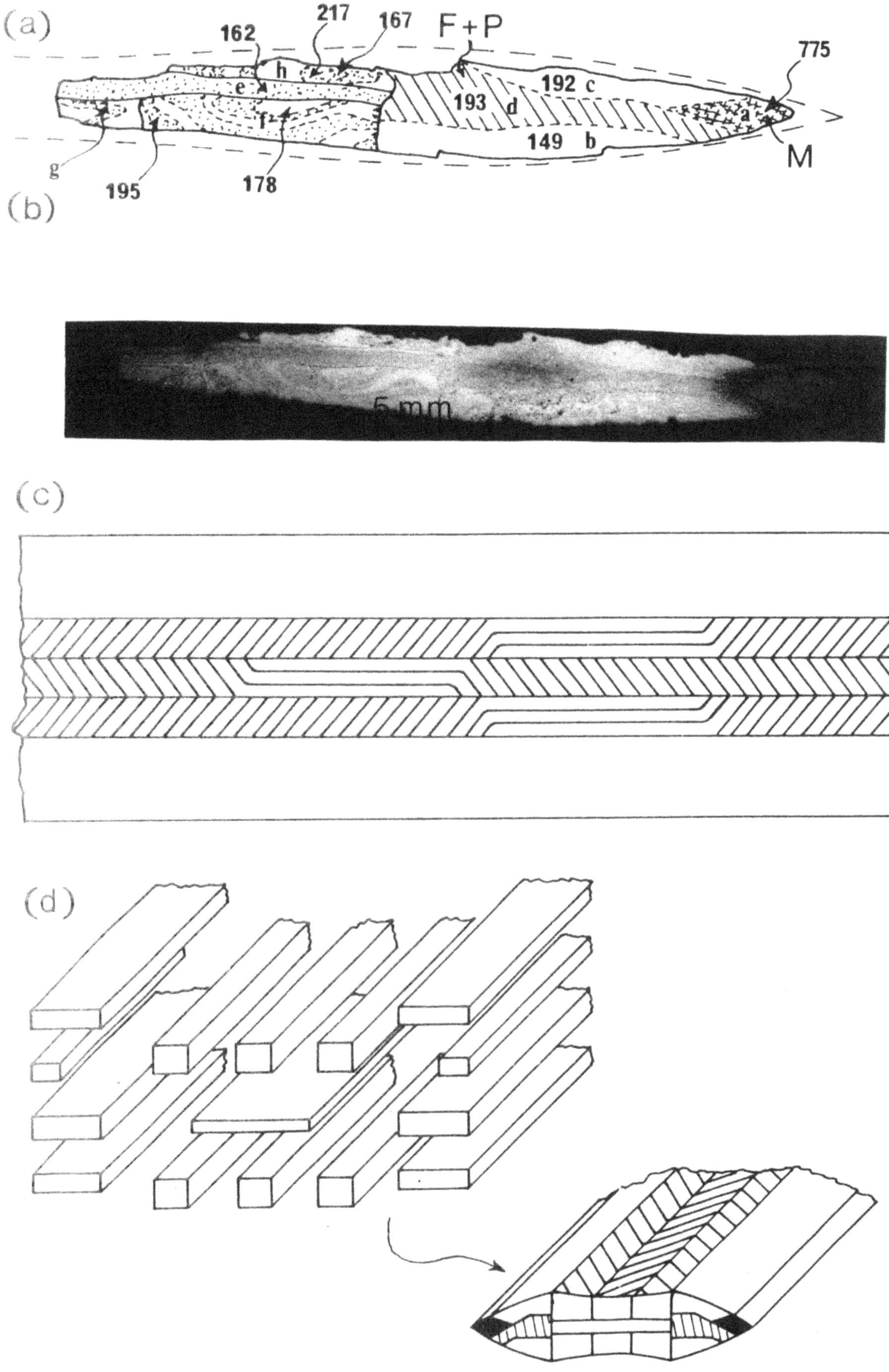


Fig. 88. Section of sword S 31 from Caverhsam, Reading + surface detail drawn from X-ray and three-dimensional view.

carbon diffusion outwards from (a). The structure of (a) clearly shows that the blade was quenched but the quenched structure means that the carbon content is difficult to estimate although a proportion of approximately 0.5-0.8% C may be likely.

The grainy appearance on the radiograph of the cutting edge of the blade is most probably a result of the high slag content of strip (b) and the effect has probably been magnified by corrosion, around and along the slag stringers.

The central part of the blade did prove to consist of two sets of three composite strips welded side by side on to either side of a central plain core. The central core (e) consisted of a fairly even low carbon steel, carbon content approximately 0.1-0.2%. The pearlite here was difficult to resolve but appeared mostly fine grained and rather amorphous or granular in places. Slag content was fairly low with some ribbons, spots and the occasional larger inclusions. The composite strips (f), (g) and (h) consisted mostly of ferrite. The alternate light and dark folded bands of each strip consisted of interleaved (light) areas of very large grained ferrite with little slag, and (darker) areas of medium grain size ferrite with a little pearlite around the grain boundaries and much slag which occurred as large and small two-phase inclusions, some fairly well flattened and some not. Although these composite strips consist in each case of, probably, six interleaved strips welded together (piled) then twisted and forged flat, there is almost no trace left of any weld between the component bands. What slag there was occurred within each band and not at the weld boundary which was only marked by the contrasting structural appearance of each band. These welds were clearly of a very high standard.

The martensitic tip part (a) of the cutting edge gave a hardness of 775 HV. The pearlite and ferrite of the central part (d) of the cutting edge gave 193 HV whereas the outer ferrite parts (b) and (c) gave 149 and 192 HV. The low carbon central core piece (e) gave 162 HV. Two hardness readings were taken from the pattern welded part (h) from one of the very large grained ferrite bands and from one of the smaller grained ferrite bands with much slag. These gave results of 217 and 167 HV and one of the medium grained ferrite bands with much slag gave 195 HV. The rather high value for some of the ferrite areas may indicate a higher phosphorus content for these parts.

The blade was probably quenched in water but does not appear to have been tempered. The hardened steel of the cutting edge insert would have made this a very effective weapon and was a very efficient and economical use of steel.

S 46. A Sword, probably from the Thames, at Brentford, Middlesex. (Museum of London, 0.2112). The greater part of a sword blade with tang and an iron pommel in place. It was probably found in the Thames at 'Old England', Brentford. The effects of corrosion were very uneven with a lightly patinated surface surviving in places typical of burial in mud. The surviving blade was 79 cm long and 6.4 cm wide with a maximum thickness of 0.4 cm (Fig. 63). The pommel is similar to that shown by Wheeler on a sword from Walthamstow (Wheeler, 1927, Plate XIII, p. 178). The blade was also similar in that it had a fairly shallow fuller running down the centre on both sides, a feature commonly found in blades of the 9th century or later. But Wheeler suggests a 7-8th century date for Walthamstow, derived from the pommel; but the fuller might indicate a somewhat later date for the Brentford sword, possibly 8-9th century although this may be an earlier example of a fullered blade.

Surface evidence showed that the blade was pattern-welded with both straightforward and herringbone elements to the pattern (Fig. 89c). Radiography confirmed this structure but there was no indication of more than one twisted composite rod occupying the full thickness of the blade. A section was taken about 1 cm below the 'bite' near the broken end (Fig. 63). It had a low slag content and consisted of seven separate pieces. The cutting edge was composed of a sandwich of three parts. The central part (b) contained areas of tempered martensite in a matrix of pearlite. The carbon content was probably in the range 0.3-0.4%. The weld lines on either side of the central part of the sandwich were light and perhaps high in arsenic and were also marked by slag inclusions. The three parts of the cutting edge had been hammer welded together before being attached to the blade (Fig. 89).

The twisted composites which gave the herringbone patterns occupied the full thickness of the blade. They were made of the usual alternate bands of low carbon and slightly higher carbon iron. The central zone which gave a straight-grained effect on the radiograph consisted of two halves welded together before being welded to the twisted composite pieces. More 'white lines' emphasised the welds. The two central strips consisted almost entirely of lightly tempered martensite and represent a steel with a carbon content of about 0.4-0.8%.

The hardness of the martensitic area of the edge was 647 HV and that of the lower carbon areas 212 and 215 HV. The martensitic central strips reached 737 and 788 HV.

The central part of the blade consisted of 6 pieces, two central strips of fairly high carbon steel welded back-to-back, on either side of which were welded, side-by-side, two twisted composite rods of wrought iron or low carbon steel. The cutting edge consisted of a central core of medium carbon steel sandwiched between two outer pieces of low carbon iron.

After forging, the blade had been quenched and then lightly tempered. The blade was thus provided with a hardened steel spine as well as a similarly hardened cutting edge.

S 41. Sword, from the Thames, at Vauxhall, London. (Museum of London, A 13591). Two fragments of a sword blade. The tang and lower end of the blade were missing although a straight iron lower cross guard survived in place (Fig. 63). The total length of the fragments was 50.0 cm and width of blade 5.8 cm and its thickness approximately 0.45 cm. The blade had undergone fairly heavy but mostly even surface corrosion and beneath this the metal survival looked to be quite good. In profile there appeared to be a shallow fuller running down the centre on either side of the blade and in places along this part the corrosion products showed up a chevron pattern indicating this central area to be pattern-welded. The sword is difficult to date from surviving pieces, although it would appear to belong to the

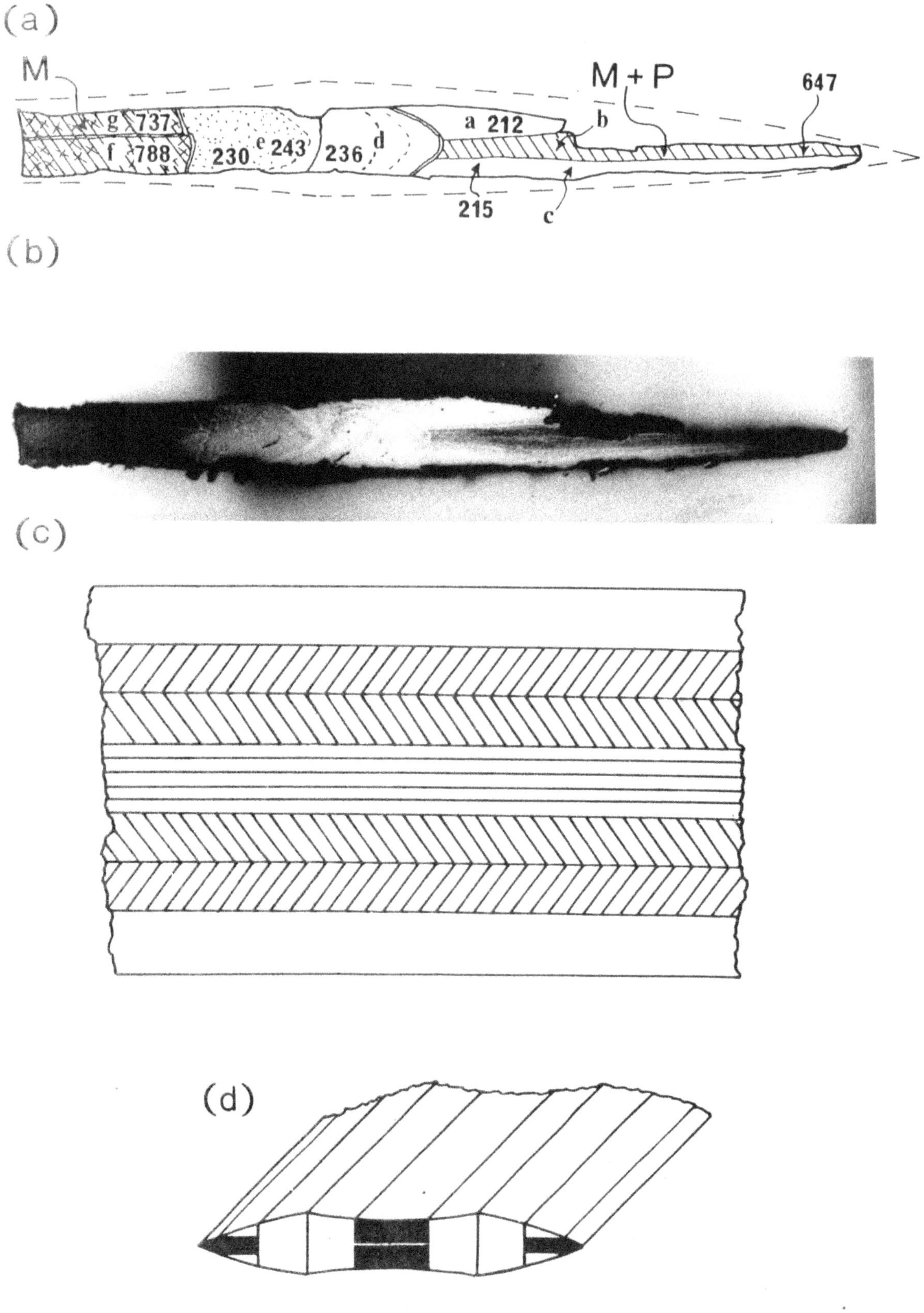


Fig. 89. Section of sword S 46, probably from the Thames at Brentford, Middlesex + surface detail drawn from X-ray and three-dimensional view.

late Saxon period. The straight form of the lower cross guard appears in the Wheeler Types I and IV and also occur in various of the Petersen Types, and from these an approximate 8-10th century date is suggested here, although it may be later. It may have been broken before deposition in the river, but this is not clear. The breaks were clearly not recent and may have been mainly the result of localised corrosion whilst submerged. At the lower break the core of the blade appeared to have corroded in preference to the cutting edges.

A radiograph confirmed that the blade had a pattern-welded central area. The pattern on each side consisted of two twisted composite strips which side-by-side gave a chevron pattern which ran uninterrupted down this central part of the surviving blade fragments (Fig. 90c). This pattern was fairly clear on the radiograph which also showed the double surface image up as a criss-cross effect along the composite area. The images overlapped to a large extent, especially along the central axis of the blade; enough to be fairly sure that there were two pairs of twisted composite rods on either side of the blade. There was no (straight grained) sign of a central core although this could not be ruled out at this stage. The edges of the blade were separate and bore fairly distinct traces of a straight grained effect in places.

A V-shaped section extending half-way across the width was cut approximately 10.0 cm from the lower broken end of the lower blade fragment, i.e. from what was about the middle of the original sword. When viewed unetched the overall slag content appeared quite low, although it was much higher in a few patches over the core area where it appeared mostly as small spots. A few quite large ribbon-like formations of slag on the central part of the section were indicative of welds in this area.

When etched with nital the structure shown in Fig. 90 became visible. The cutting edge appeared to consist of three parts (Fig. 90b); one main part (a) (in Fig. 90a) which mostly etched very dark but a paler zone (b) possibly representing a separate piece welded on to one side, although this was not clear. A further, rather curious circular area (c) was quite clearly outlined by slag ribbons. This appears to have been a separate small piece welded on to the main part (a) of the cutting edge, before the cutting edge was welded to the core of the blade. Also the more obvious part of the straight graining visible in the cutting edge on the radiograph occurred along the boundary with the core. The roughly circular area (c) would, therefore, appear to be the view in section of a piece of wire welded on to the inner side of the cutting edge before the final assembly of the blade.

The greater part of the cutting edge, piece (a) consisted mainly of a dark and fast etching constituent. This was partially lamellar and feathery looking, although the structure was difficult to resolve it appeared to have broken down somewhat and become spheroidised. Near the tip this was mixed with a dark etched acicular constituent, tempered martensite. The mixed lamellar or feathery looking constituent was probably a pearlite or carbide distribution resembling upper bainite. The way this predominates with a small martensitic area near the tip of the cutting edge means that the blade was inefficiently quenched, quite probably deliberately so. The tempered nature of the martensite and somewhat spheroidised nature of the very dark pearlitic constituent, show that the

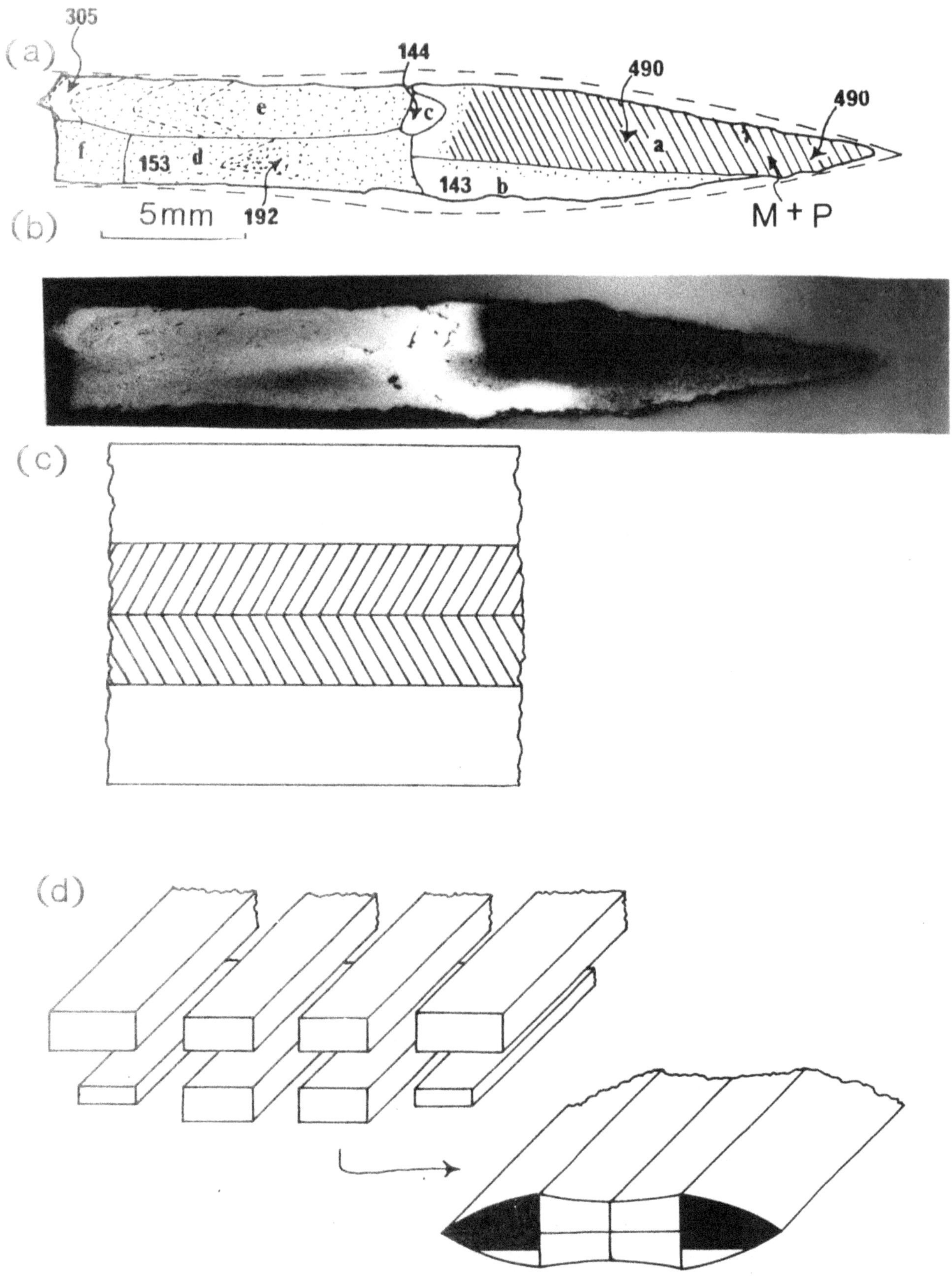


Fig. 90. Section of sword S 41 from the Thames at Vauxhall, London + surface detail drawn from X-ray and three-dimensional view.

blade has undergone tempering to some extent. The heat treated nature of the structure meant that the carbon content was difficult to estimate optically but it probably represents a fairly high carbon steel of between about 0.5 and 0.8% carbon.

Near the weld junction with (c), area (a) was much paler looking and here contained a high proportion of ferrite. This was probably the result of localised decarburisation while piece (c) was being welded to (a). The paler area (b) visible to one side of (a) quite probably was also the result of the decarburisation of part of the surface of (a), and would most likely have taken place at the same time, i.e. when (c) was welded to (a). There was no evidence (in the form of slag inclusions etc.) that (b) was a separate piece and in this case the partial decarburisation of the cutting edge piece (a) appears the most likely explanation, although the possibility of a separate piece (b) should not be ruled out completely. Some further decarburisation may have taken place after the cutting edge was welded to the composite core, but there was no particular indication of this. Piece (c) consisted mainly of a very low carbon steel, really no more than wrought iron with a little carbon - probably less than 0.1% - in a medium grain size structure with mostly irresolvable pearlite. Near the weld boundaries (c) had become decarburised and here considerable grain growth appeared to have taken place, particularly around the larger slag inclusions.

The core of the blade consisted of two pairs of composite rods welded back to back without any separate intermediate core piece. Two of these (d) and (e) and a small part of a third (f) came within the section (Fig. 90a and b). The weld between (e) and the adjoining piece on the same side must have been only just outside the edge of the section. The weld junction between (d) and (f) was well inside the section which accounts for the overlap effect visible along the central part of the radiograph. The core pieces did not give a particularly banded appearance when etched. The slag content was low and well scattered, although where visible, as small flattened inclusions, it tended to congregate along lines which outlined the folded appearance of these twisted pieces in section (d) (f) and most of (e), which appeared to consist of a low carbon steel which looked somewhat paler and darker in places, although the differences were slight, diffused and rather patchy. The microstructure of these areas consisted mostly of fairly equiaxial small-medium grain ferrite and unresolved pearlite. A visual estimate of the content varied between about 0.1 and 0.3%.

There was however, a very pale band clearly visible at the inner end of area (e), (i.e. near the centre of the blade) and this consisted of large grain ferrite with little or no pearlite. A small part of a further darker band (similar to most of the rest of (e)) was also clearly visible by contrast on the extreme inner edge of area (e). The composite rods forming the core of the blade appear to have been made from a somewhat varying low carbon steel with some wrought iron included.

Two hardness readings were taken in the cutting edge part (a) near the tip where a very dark partly acicular tempered martensitic structure showed and in the centre of part (a). For both the hardness was 490 HV. The lower carbon iron area (b) gave 143 HV and the similar area (c) gave 144 HV. Two of the darker grey ferrite and pearlite areas of parts (d) and (e) gave values of (d) 192 HV; (d) 153 HV; (e) very large grain ferrite 305 HV. This very high hardness value for the ferrite probably indicates a very high phosphorous content for this band.

S 23. Sword from Reading Museum, ? from the Thames. (Reading Museum 112.66/1). A complete sword in two parts, broken about a quarter of the way along the blade from the tip (Fig. 63). It was 94.6 cm long and 5.4 cm wide. A thick straight iron guard and similar pommel still survived in place. The surface of the sword was badly pitted by corrosion but traces of a wide central fuller could still be seen running along the central part of the blade on both sides. The straight guard and pommel are similar to some of Petersen's Type H and the Wheeler Type I[65,66] which indicates the sword to be a Late-Saxon period weapon probably of the 8-10th century.

A radiograph of the sword revealed that a lattice or similar design had been inlaid into the surface of the blade on both sides near the hilt. The inlaid design appears to have been contained within the fuller of the blade and consisted of a series of uprights (at right angles to the cutting edge) and diagonal pieces of twisted or pattern-welded iron rods forged into the surface (Fig. 91e). An interpretation of the form of the design is given in Fig. 91d. The type of design is quite common and has been found in a number of Viking age sword blades in both Britain and Scandinavia[77]. The radiograph was difficult to interpret because of the profusion of corrosion pits on both sides of the blade but it was just possible to see hints of a straightish 'wood grain' effect in places. It is doubtful, however, had the corrosion occurred much more evenly whether the 'wood grain' would ever have been more than fairly faint on the radiograph.

A section was cut from the tip fragment of the sword, along the edge of the break with the main part of the blade (Fig. 63). It was thus an oblique transverse section at an angle of about 60° with the cutting edge and it also extended about two thirds of the way across the blade.

Before etching, the section proved to have only a fairly small amount of slag inclusions of mostly fairly uneven size and distribution. A line of slag particles running diagonally across the centre of the section showed the position of a scarf weld between two main components. The cutting edge area appears to be divisible into 3 parts separated by discontinous lines of slag inclusions along which much corrosion had taken place. Corrosion was also visible along the scarf weld and around some of the other larger slag inclusions.

The section etched rapidly and the result was rather patchy and uneven looking (Fig. 91b). The central scarf weld and corroded slag lines of the cutting edge are visible. A paler band can be seen between the two corroded slag lines which appears fairly distinct from the darker area towards the central scarf weld and the narrow rather patchy area towards the tip of the cutting edge. The cutting edge thus appeared to consist of a sandwich of three parts (a), (b) and (c).

Examination under higher magnification revealed that the somewhat patchy outer part of the sandwich consisted of mainly (paler etching) martensite with some amorphous or irresolvable (darker etching) pearlite. The middle paler band appears to consist of a mixture of irresolvable pearlite and ferrite. The inner darker part consisted mainly of dark etching unresolved or amorphous pearlite with small patches of martensite. The martensite itself etched rather darker than might be expected probably indicating some degree of tempering.

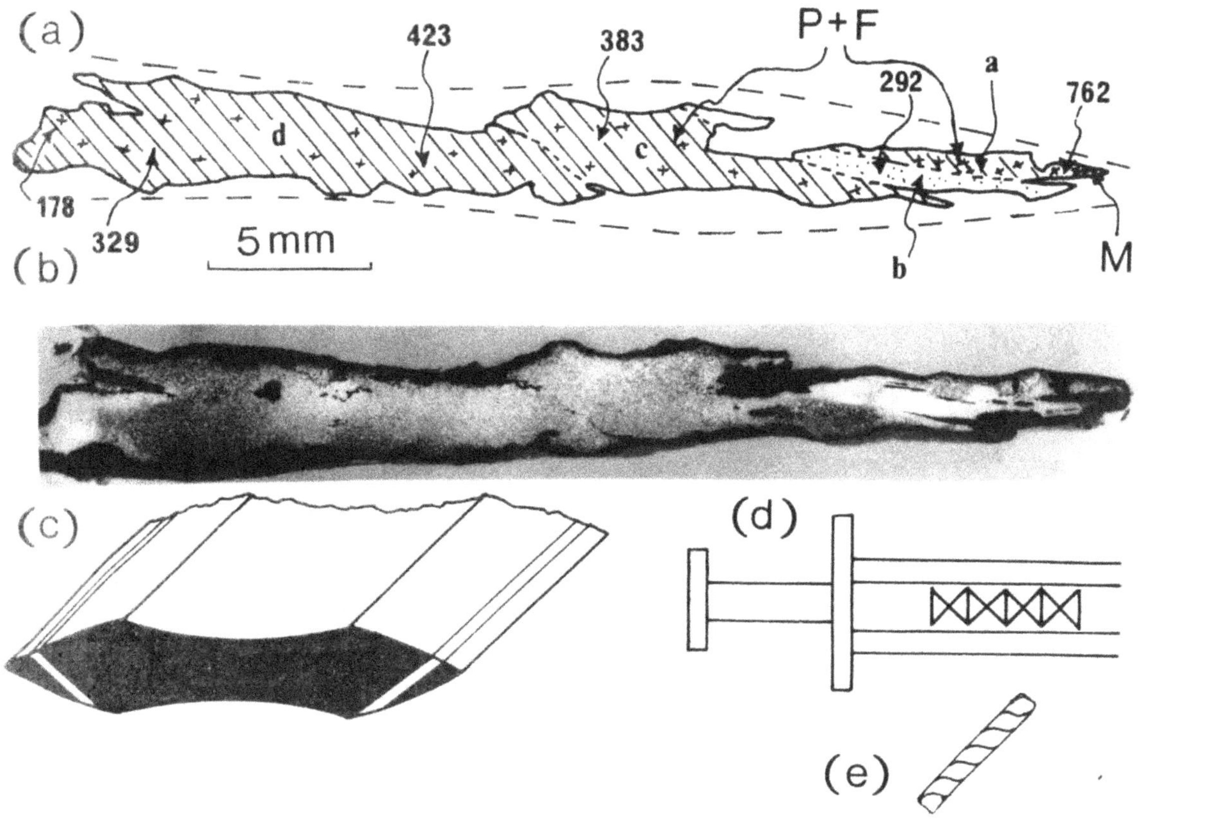


Fig. 91. Section of sword S 23 probably from the Thames near Reading + three-dimensional view, inlaid design drawn from X-ray and twisted inlaid piece.

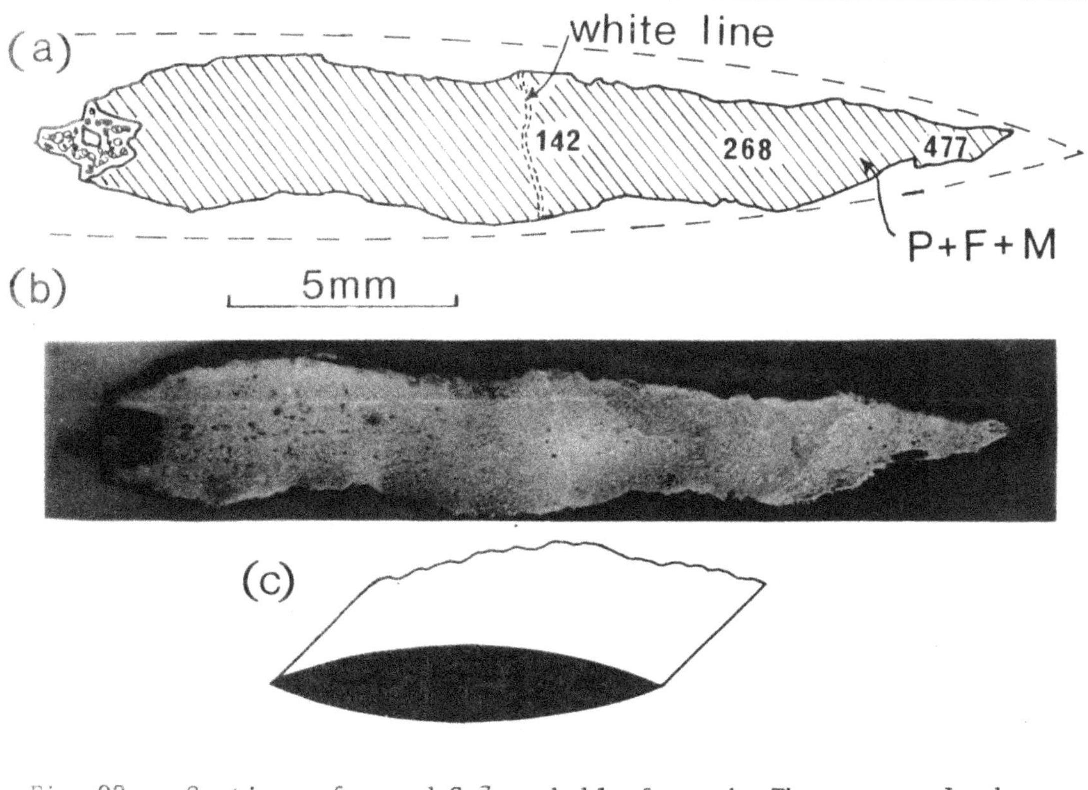


Fig. 92. Section of sword S 7 probably from the Thames near London. + three-dimensional view.

The central fullered part of the blade, against which the sandwich cutting edge had been scarf welded consisted mostly of a fairly even distribution of dark etching amorphous pearlite mixed with some ferrite and sporadic patches of paler etching martensite. Also visible was a fine white network across much of this part of the section. This was an outline, in ferrite, of a large grain structure infilled by the generally dark fine grain structure of the amorphous pearlite/ferrite/martensite mixture.

This middle part of the blade is clearly a fairly homogeneous piece of steel. The sandwich construction of the cutting edge has been highlighted by etching indicating a variation of carbon content between the three parts although a good quality steel still appeared to form the greater proportion. the presence of dark etching pearlite and martensite with a particular concentration of martensite in the outer part of the cutting edge shows that the blade had been quenched fairly rapidly from a temperature above the lower critical point of 723°C, although not fast enough to give a fully quenched martensitic structure.

Six hardness readings were taken across the section. The martensitic area near the tip of the cutting edge part (a) gave a value of 762 HV. The paler band (b) and higher carbon part of (c) gave hardnesses of 292 and 383 HV. Two of the high carbon non-martensitic areas of the central part (d) gave 423 and 329 HV whereas a mostly ferrite patch of part (d) near the blunt end of the section gave 178 HV.

The white ferrite network visible in part (d) was most probably a 'ghost' image of a former large austenite grain outline. Its survival may show that the sword was not heated very far above the lower critical point before quenching. The quenched structure of the steel meant that the carbon content was difficult to estimate but the high proportion of amorphous pearlite was probably indicative of a carbon content, over much of the blade, in the region of 0.6–0.8%. The surviving ghost image of a former grain outline would not be possible had the temperature risen above the upper critical point which would have probably been between about 750 and 800°C. In this case the sword blade was probably quenched from a temperature of 725–750°C.

<u>S 7. Sword ? from the Thames in London)</u>. (Museum of London, C 2260). The upper part of a sword blade with a straight iron guard and a flattish semi-circular iron pommel (Fig. 63). Its surface was heavily pitted by corrosion but underneath this the metal survival appeared to be quite good. The total surviving length was 62.5 cm with a maximum blade width of 5.2 cm but it was too corroded to determine the original profile. The find site is uncertain but its condition suggests that it came from a river and the Thames in or near London would seem most likely. It does not readily fall into one of the Wheeler or Petersen types but also the short straight guard is typical of Wheeler's Type I, c 9–10th century[66], and is similar to Petersen's Type X[65]. Date: 9–10th century.

There was no indication of it having been broken or bent before submersion and the loss of the lower part may be the result of differential corrosion and water movement. The projecting part of the broken end of the blade was cut off to give a section representing about half the width of the blade. Examination of the unetched section revealed a sparse, but even, distribution of slag inclusions with no obvious linear concentrations that might indicate a weld line between different components. The inclusions were mostly small and of one or two darker grey phases (? fayalite and a glassy constituent).

Etching with nital revealed the structure visible in Fig. 92b which shows a rather curiously patchy looking appearance in which the white speckles are ferrite forming a matrix around areas of pearlite interspersed with paler patches of martensite and a very dark, fast etching, nodular constituent, troostite (Fig. 92b). This structure was fairly even across most of the section although it was more predominantly martensitic near the tip of the cutting edge. A rather irregular and faint white (ferritic) line just visible running vertically across the centre of the section (Fig. 92) might suggest a weld position between the cutting edge and core parts of the blade although no corroborative proof such as a (faint) line of slag inclusions could be seen. The structure visible on either side of this line shows that a fairly homogeneous steel was used across the whole section and therefore, probably the whole blade. Weld lines can sometimes leave virtually no trace and so this white line might indicate where a single piece of steel was folded and skillfully welded together during the preparation of the metal for the blade.

It is concluded here that a single piece of homogeneous steel was used to make the blade, however it was prepared beforehand. It was quenched but not tempered, resulting in the structure shown. The carbon content is difficult to estimate in this state but a medium carbon steel of between about 0.3 and 0.6% carbon can probably be assumed. The incomplete (i.e. only partially martensitic) nature of the quenching may be the result of 'slack' quenching, possibly in oil or brine, as opposed to water quenching, or it might be the result of a carbon content towards the lower end of the range suggested above (i.e. about 0.3%).

Hardness values averaging 477 HV were obtained for the area near the tip of the cutting edge, about the value to be expected for the not fully martensitic structure encountered here. The dark etched area nearer the centre of the blade gave values averaging 268 HV and the palest finer grained areas of ferrite and pearlite, also near the centre of the section, gave an average value of 142 HV. This last value is fairly typical of a low carbon steel which may indicate that the estimate of 0.3% carbon for this piece is approximately correct, the other variations in values indicating the variable effectiveness of the quenching to which the blade was subjected.

S 8. Sword ? from the Thames. (Museum of London, C 2258). The upper part of a sword with short straight upper and lower iron guards still in place (Fig. 63). The surface was badly but evenly pitted by corrosion, although beneath this the metal survival appeared to be quite good. The total surviving length was 37.9 cm and the maximum blade width was 5.5 cm. The find site is uncertain but the state of the preservation suggests a river bed and the most likely possibility would appear to be the Thames in or near London. The short straight upper and lower guards are of the Wheeler Type I[66] and similar to Petersen's Type M[65] and therefore the sword may be of a 9-10th century date. The original profile of the blade had been removed by corrosion but a faint hint of a shallow fuller could just about be made out running down the centre of the blade. There was no evidence that the blade had been broken or bent before submersion and the loss of the lower end may be largely the result of differential corrosion.

A wedge shaped section extending about half way across the width was cut from one side of the blade approximately 10.0 cm down from the lower guard. Unetched, the section exhibited a rather uneven distribution of slag inclusions of one or two darker grey phases (? fayalite and a glassy constituent). When etched with nital one half (horizontally) of the section became a dark grey while the rest remained pale (Fig. 93) and this suggested that the blade was made from two pieces one of steel and one of wrought iron welded together and forged out to give the bi-metallic structure shown in Fig. 93c.

The pale half was of large grained ferrite, whereas the darker half consisted of spheroidised pearlite - a distribution of cementite globules in a ferrite matrix. At its densest this appears to correspond to a carbon content nearing the eutectoid i.e. about 0.8% carbon. There was no discernible sign of slag particles indicating the position of a weld between the two halves and there has clearly been much carbon diffusion from the steel part to the wrought iron part resulting in a rather diffuse junction between the two halves. It seems unlikely that this structure could have been achieved any other way than by the welding together of a piece of wrought iron and a piece of steel despite the apparent absence of a visible weld line. The extensive carbon diffusion was probably the result of prolonged heating during or after the final forging and the spheroidised nature of the pearlite indicates that a further and final fairly prolonged period of heating took place but this time it must have been below the critical temperature of about 700°C at which pearlite re-dissolves.

A series of micro-hardness readings were taken from points running across the section. The results of these varied between 81 HV in the ferrite area to 125 HV in part of the intermediate zone containing approximately 0.4-0.5% C and rose to maximum value of 172 HV in the higher carbon (0.8%) area. The lower value is typical of a fairly pure wrought iron and the higher value consistent with a higher carbon steel that has been subjected to sub-critical annealing. This final heat treatment would have toughened the steel of the blade thereby rendering it much less liable to cracking when flexed.

<u>S 48. A Sword from Rochester, Kent.</u> (Maidstone Museum, 13, 1956). A blade including part of the tang with a straight lower iron hilt guard, 78.8 cm long and 5.3 cm wide (Fig. 63). It was found by chance near the junction of Watts Ave and Margaret St. in Rochester and evidently came from a grave. Corrosion was uneven and some edges were missing but in general there was plenty of remaining metal. There was a shallow fuller running down the centre of both sides of the blade. The guard was similar to those shown by Wheeler as Type I, or Type IV (Wheeler,[66] 1927) which suggests a 9-10th century date but in any case a fullered blade of this kind is typical of swords of the Late Saxon period.

Radiography showed that the blade had a pattern-welded zone running down the centre. It was clear that more than one twisted composite bar occupied the thickness of the blade and that the surface of the central zone on either side consisted of a pair of composite rods twisted in alternate directions and then welded side by side. Each pair was welded back-to-back

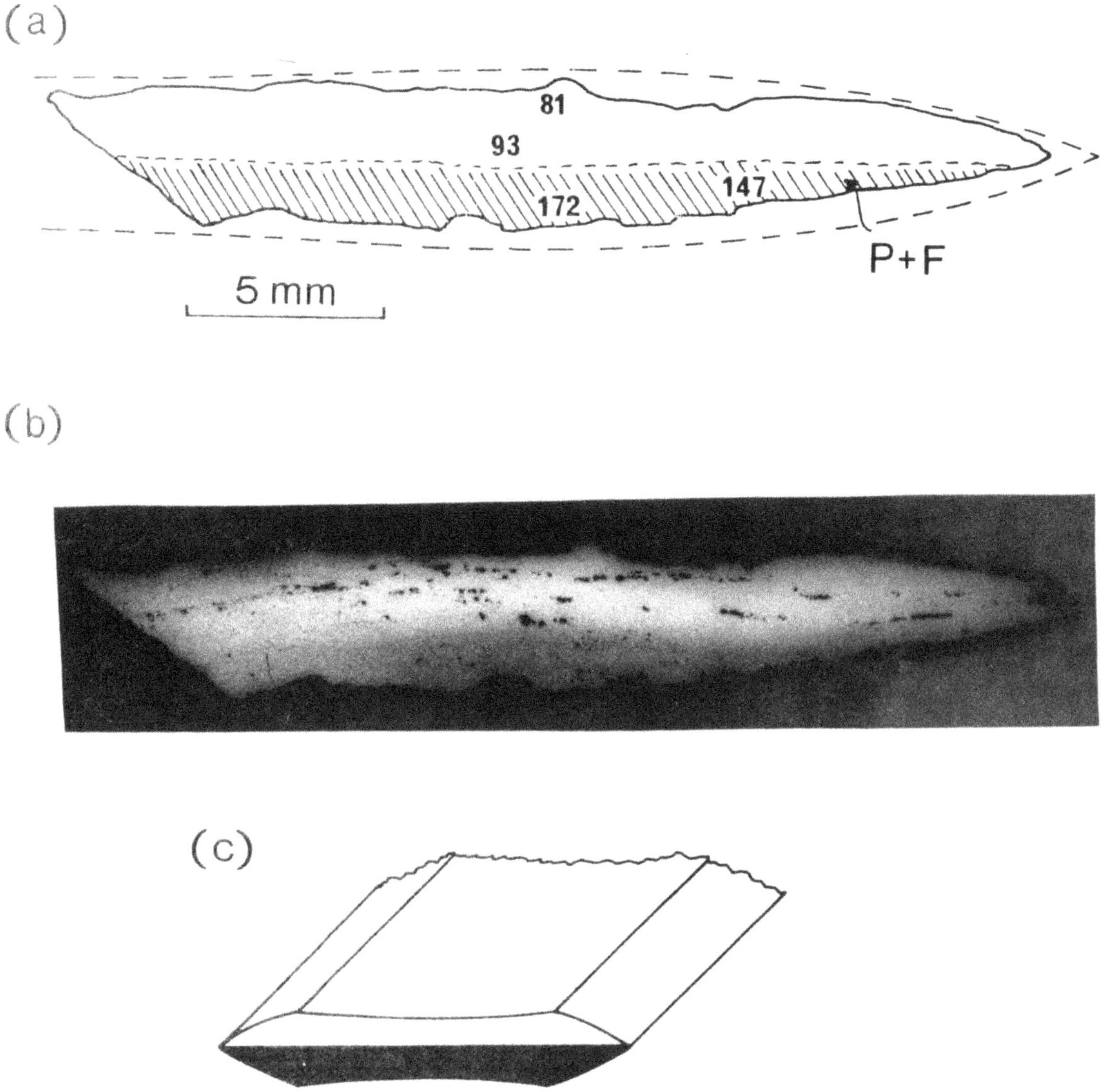


Fig. 93. Section of sword S 8 probably from the Thames near London.

to form the central part of the blade to which the cutting edges were attached.

A section was cut through half the width of the blade 13.5 cm up from the tip. The section comprised the cutting edge and two composite core parts. A large two-phase slag inclusion was entrapped between the parts. The edge consisted of three zones of variable carbon content and near the tip this had a coarse martensitic structure. The composite core pieces had the usual banded structure (Fig. 94).

The hardness of the martensitic regions of the cutting edge was 564 HV maximum decreasing to 305 away from the tip. Ferrite and pearlite structures had hardnesses in the range 179-226 HV. The hardness of the composites was 200 HV in the ferrite + pearlite bands and 163 in the coarse-grained ferrite bands.

The blade had been finished by quenching but this was not fast enough to produce any martensite in the central regions.

S 47. Sword from the Thames at Waterloo Bridge. (Museum of London, A 3670). The remains of the upper part with a straight iron guard and an upper pommel guard surviving in place (Fig. 63). The lower guard also had the remains of a copper-base alloy cladding adhering to it. According to Wheeler it falls within the Type I with a date of about 900 A.D. (Wheeler, 1927, p. 36). The remains were 37.7 cm long with a maximum width of 6 cm and a thickness of 0.45 cm. The surface was very corroded and pitted but a central fuller could be seen.

Radiography showed a series of inlaid pieces along the central part of the blade near the hilt. This seemed to take the form of a lattice design very similar to that found on sword S23 from Reading. A suggested interpretation of the design is shown in Fig. 95c. This was made of short pieces of twisted iron rod hammered into groves cut into the surface of the blade. Lattice designs similar to this have been found in conjunction with lettered inscriptions such as in the well-known ULFBERHT or INGELRI sword blades although only the lattice designs seem to have been employed on the three inlayed blades in this study.

A section was cut from near the broken end of the blade, in such a position as to avoid the inlaid design. It consisted of two parts separated by a diagonal line marking the position of a scarf weld (Fig. 95a, b and e).

This was very difficult to see under low magnification but it was delineated by a line of small slag inclusions. Part (a) was nearing a eutectoid composition (0.8% carbon) and consisted of rather spheroidised pearlite and a little ferrite which increased as the weld was approached where it had a Widmanstätten form. Part (b) had a variable carbon content although the mean was lower than (a).

Hardness readings from the higher and lower carbon areas of Part (a) were 232 and 144 HV respectively; similarly, those on Part (b) were 239 and 138 HV.

The structure would suggest that the blade spent a considerable time below or near 700°C before final cooling. It was made from two pieces of

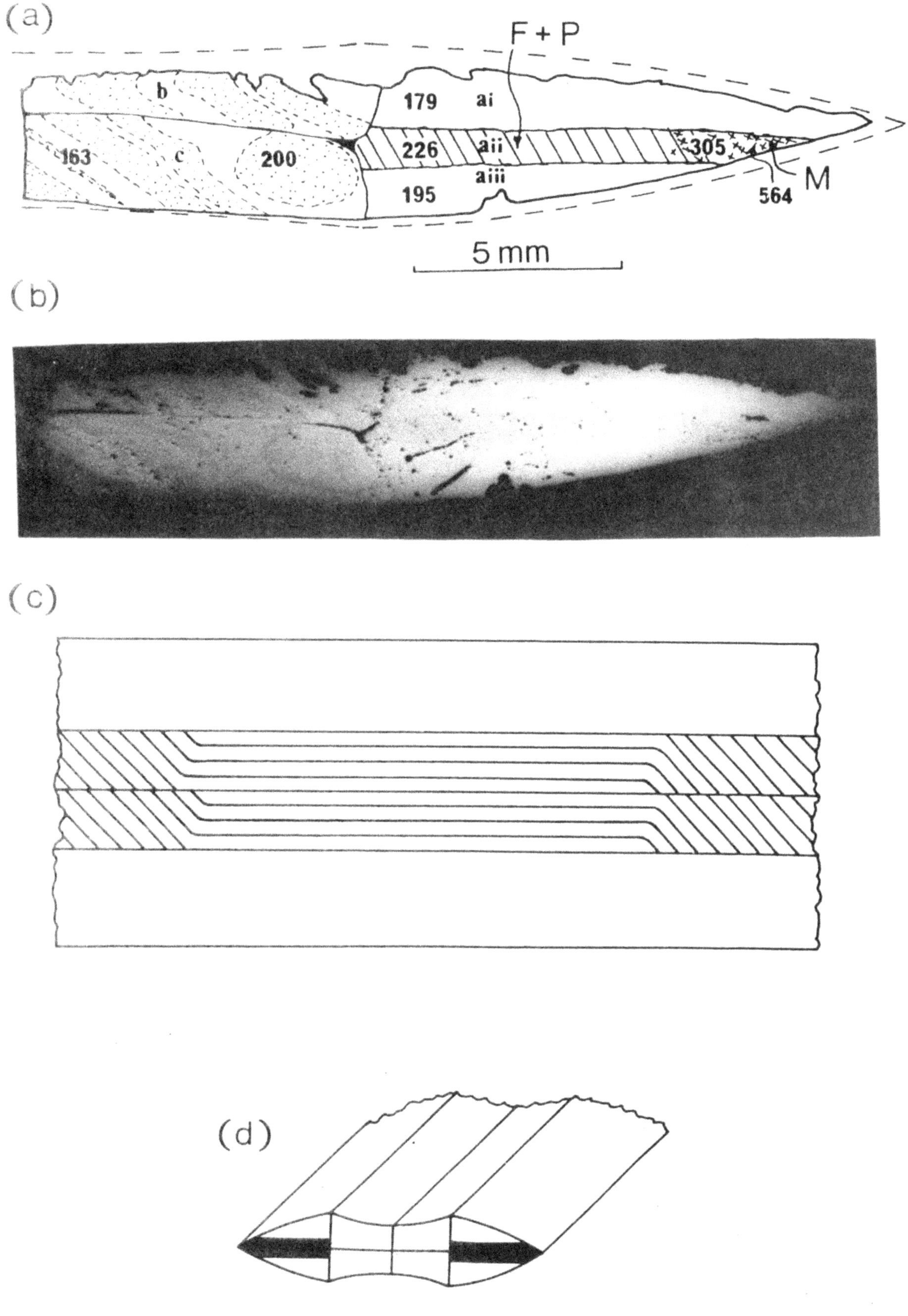


Fig. 94. Section of sword S 48 from Rochester, Kent + surface detail drawn from X-ray and three-dimensional view.

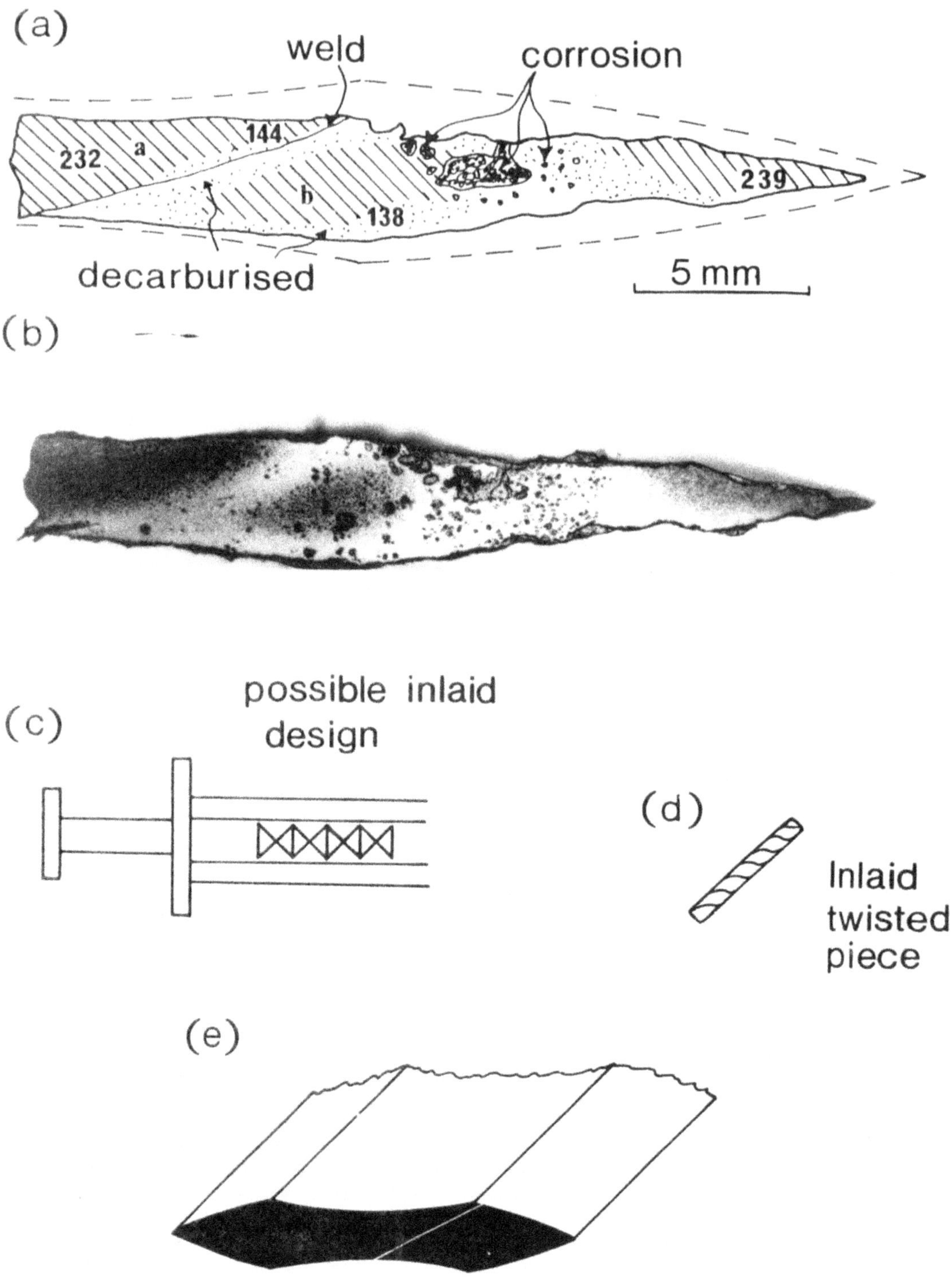


Fig. 95. Section of sword S 47 from the Thames at Waterloo Bridge + inlaid design drawn from X-ray, twisted inlaid piece and three-dimensional view.

steel with a medium to high carbon content quite unlike any pattern-welded blades. The final annealling toughened the steel of the blade.

S 24. Sword from the Thames at Cleeve, Goring on Thames, Oxon. (Reading Museum, 256:63a and b (T.C.B.)). This sword was found in the River Thames at Cleeve, and survives in two parts (Fig. 63) which represent most of the weapon, the tip and pommel being missing. The total length was 82.5 cm and the width 4.8 cm. The breaks were clearly not recent, having been subject to fairly extensive corrosion over a long time but it was not possible to say whether or not the blade was broken before deposition in the river. A slightly curved iron guard survived in place. The effects of corrosion were uneven, some parts of the blade having corroded right through whilst others appeared to survive in relatively good condition with only some slight but even surface corrosion. In these areas the etching effect of the corrosion had highlighted a pattern-welded core to the blade. This appeared to consist of three parallel narrow bands but the detail was not very clear. The blade was not fullered but had a fairly pronounced convex profile which was somewhat flattened in the area of the pattern-welded core. This weapon is probably Late-Saxon and although most swords of this period appear to have had fullered blades, the slightly curved iron guard appears similar to Wheeler's Type VI and VII.[66] The very narrow strips used in the pattern-welding of the core is a form of decoration common to Late-Saxon scramasaxes but not to earlier Saxon weapons. Date: approximately 9-11th century.

X-radiography confirmed the presence of a pattern-welded core to the blade, consisting of three very narrow parallel twisted strips. Some straight wood grain effect was also faintly visible in places along the cutting edges of the blade. The pattern-welded core and the cutting edge parts could be clearly seen to continue up through the tang.

A transverse V-shaped section was cut from the lower surviving blade fragment. The cut was taken so that the section would extend right across the pattern-welded area. Unetched, the section showed up as having a generally fairly low slag content distributed fairly evenly across it, mostly appearing as small spots and flattened ribbons. Various orientations and concentrations were visible which indicated the probable presence of welds and piling.

The structure of the section became much clearer upon etching. The macrograph of Fig. 96b shows the pattern-welded central core to consist of three main parts (d, e, f) each separated from the other and from the cutting edges by a slightly diffuse grey line. Further grey lines show the distorted, folded composite nature (of the transverse view) of each of the three strips forming the core. The grey lines that separate the main elements of the section clearly mark the position of welds. The cutting edge can also be seen to consist of three main parts (a, b, c). The upper and lower parts (b and c) are zones of parallel grey and white bands while the central zone (a) is visible as a thicker grey band. This represents a sandwich consisting of two outer parts of piled wrought iron or low carbon welded to a central, fairly uniform steel strip. Examination under higher magnification also showed that the grey lines consisted of narrow bands of pearlite with a carbon content of approx. 0.1-0.3% C, often coinciding with lines of slag spots or ribbons which marked the positions of welds. The pearlite of the edge parts (b and c) was most probably introduced, not as thin steel strips or partial carburisation of thin wrought iron strips -

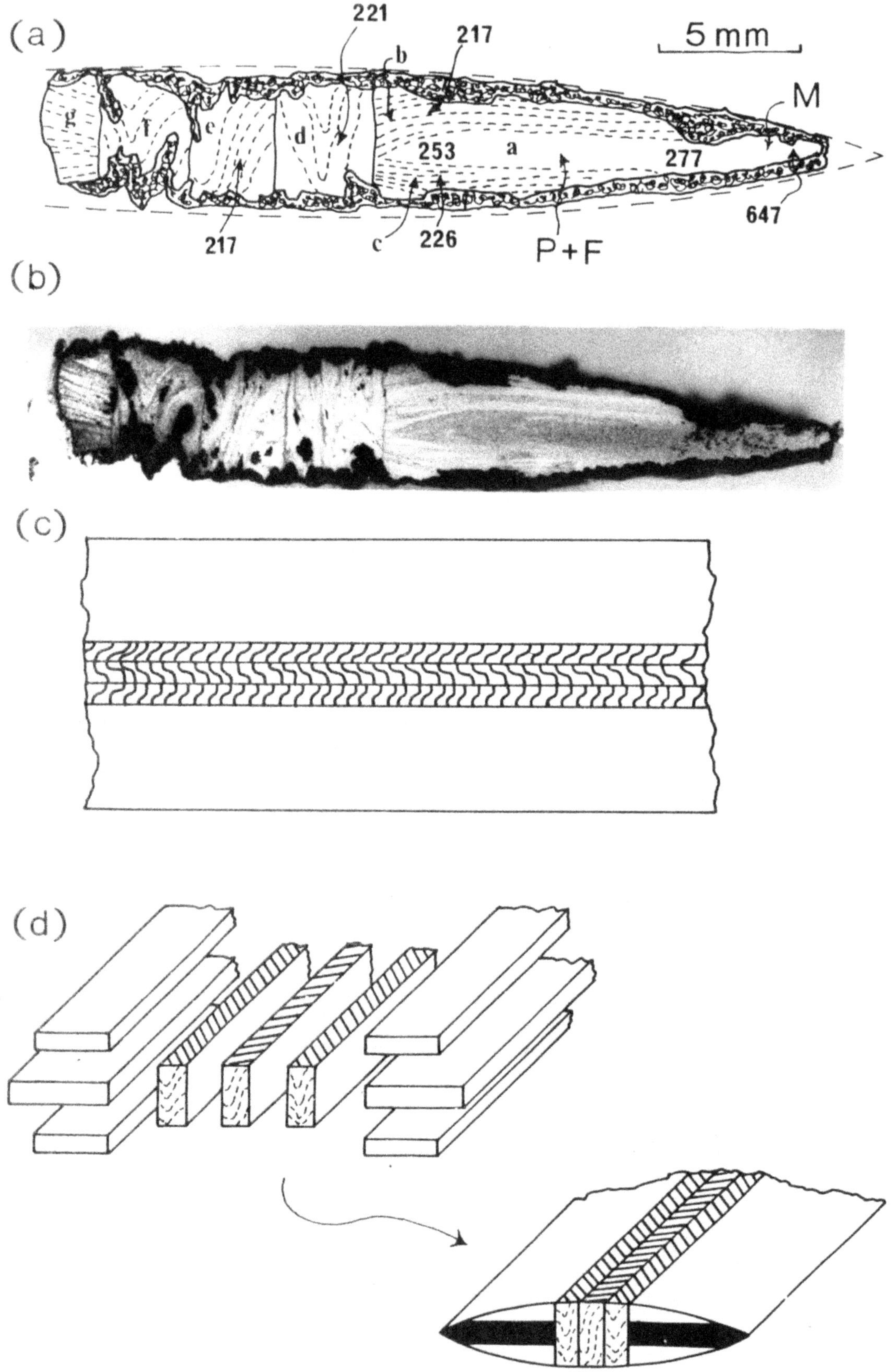


Fig. 96. Section of sword S 24 from the Thames at Cleeve, Goring on Thames, Oxon + surface detail drawn from X-ray and three-dimensional view.

neither of which would have been feasible - but as carbon in a fairly finely divided form, possibly as a constituent of a flux used during the forge welding or piling of these outer strips and when the cutting edge was welded on to the core. The same explanation, if correct, must apply to the other diffuse grey lines visible at the weld junctions between the various parts of the blade (Fig. 96) and may also account for the distorted lines of pearlite visible in the central pattern-welded strips.

The wider central grey band (a) of the cutting edge. It consists of a piece of steel with a fairly even carbon content of about 0.4-0.5%. Towards the tip of the cutting edge this central strip showed up mainly as martensite patches mixed with some dark etching irresolvable pearlite (troostite ?). Going inwards towards the central core the proportion of martensite decreased while that of the darker etching constituent increased. Where the carbon content was lower, at the junction with the ferrite areas and in the diffuse grey lines, the pearlite occurred in a Widmanstätten-type distribution.

Three hardness readings were obtained along the steel central part (a) of the cutting edge sandwich, firstly 647 HV in the mainly martensitic area near the tip, then 277 HV in the centre in a pearlitic area between martensitic patches, and lastly 253 HV near the inner weld where the structure showed as pearlite + ferrite. The low carbon iron of the outer edge parts (b) and (c) gave hardnesses of 217 and 226 HV. The very similar low carbon iron of the pattern welded core pieces (d) and (e) gave 221 and 217 HV.

It would appear that the blade was quenched from within the critical range, about 750-800°C. The quenching was not rapid enough to give a fully martensitic structure even at the tip of the cutting edge.

S 43. Sword, from the Thames. (Museum of London, A 17923). The upper part of a sword blade with tang and grip found in the Thames near London (no more specific findspot known). The remains of a wooden grip were quite well preserved, the remains of a fabric covering or binding could also be seen. A distinctive downward curving lower cross-guard survived in place (Fig. 63) as did half of an upper straight cross-piece, although there was no sign that any additional pommel had been attached to this.

It was classified by Wheeler as being one of his Type I swords of c. A.D. 900[66]. The total length of the sword fragment was 32.5 cm, the maximum blade width approximately 5.1 cm. The blade appeared to have a roughly flat backed profile and at its thickest appeared to have been about 0.5 cm, although this was difficult to gauge owing to the poor state of preservation of the surface which was quite heavily pitted by corrosion, which had also removed much of the edges. The broken end was heavily corroded and the break was clearly not recent but there were signs that it had been bent or broken before deposition in the river. The age and cause of the break are uncertain but are probably the result of differential corrosion during immersion. The surface corrosion had, in places, left traces of a pattern-welded area visible down the central part of either side of the blade. As far as could be seen this appeared to show up as a fairly continuous herringbone design on either side. In one or two patches, where the corrosion was deeper, the herringbone pattern was missing and here hints of a straight grain survived in the corrosion products underneath, suggesting that the blade might have had a central core piece.

X-radiography indicated a herringbone pattern composed of three parallel twisted composite bars welded side-by-side (Fig. 97c). The cutting edges on either side of the blade did not exhibit any visible structural effects.

A wedge-shaped transverse section was cut from half of the blade (Fig. 63). When examined unetched a rather uneven slag content could be seen. No clear signs of weld lines identified by slag could be seen.

When etched with 2% nital the structure shown in Fig. 97b became visible. Fig. 97a shows this more clearly. The cutting edge, which had shown very little structure on X-ray, appeared to consist of two parts, an outer steel part, (a), folded round an inner wrought iron part (b). The carbon content of part (a) varied between about 0.8% in places near the tip, where an almost eutectoid structure of partially spheroidised lamellar pearlite was visible, down to about 0.1% near the weld junction with part (b). The weld junction between (a) and (b) was quite clearly marked by a thin grey line of pearlite superimposed on a white line of ferrite but not accompanied by any obvious lines of entrapped slag particles. Part (b), by contrast, consisted of ferrite with a higher slag content than (a). Fairly extensive carbon diffusion had taken place across the weld boundary between (a) and (b) - showing as an even margin of pearlite along, but steadily decreasing away from the weld.

The central part of the blade was composed of several components of which the largest part was the core piece (c) with thin composite strips forming the surface on either side. Parts of two of these, (d) and (e) survived on the section (Fig. 97). The core piece (c) mostly consisted of ferrite with a varying slag content and a few patches of nitride or carbide needles. A narrow horizontal pearlite band across the centre part of core piece (c) suggests that a low carbon steel strip (of no more than 0.1-0.2% carbon) formed the centre around which the larger ferrite piece was welded.

The surviving parts of the two composite twisted pieces, (d) and (e), were visible on either side of the central core piece (c). Although the herringbone pattern from these showed up fairly well on X-ray the banded nature of these strips did not show up very well on the section. The carbon content of these strips was fairly even but low - mostly below about 0.1% and the pearlite of this was quite well spheroidised. Slag content was also low and consisted mostly of small well flattened inclusions mainly of one dark phase (? fayalite). These composite strips (d) and (e) were quite thin, partly a result of surface corrosion, and the weld lines between each of them and the central core piece (c) was fairly clearly marked by a narrow grey line of pearlite.

The combined results of X-radiography and metallography suggest that the blade was made from 12 separate parts in approximately the following order: a central core part was made by folding and welding a piece of wrought iron round a thin strip of low carbon steel. Side-by-side on either side of this were welded three parallel thin composite strips consisting of low carbon iron. Each of these may have been formed from several wrought iron and low carbon steel rods twisted and forged together and then forged out to produce these quite thin strips. The cutting edges with their steel outer parts welded on to wrought iron inner pieces must have been produced separately, then hammer-welded on to the central part once this had already been assembled. The blade must then have been forged out to produce its

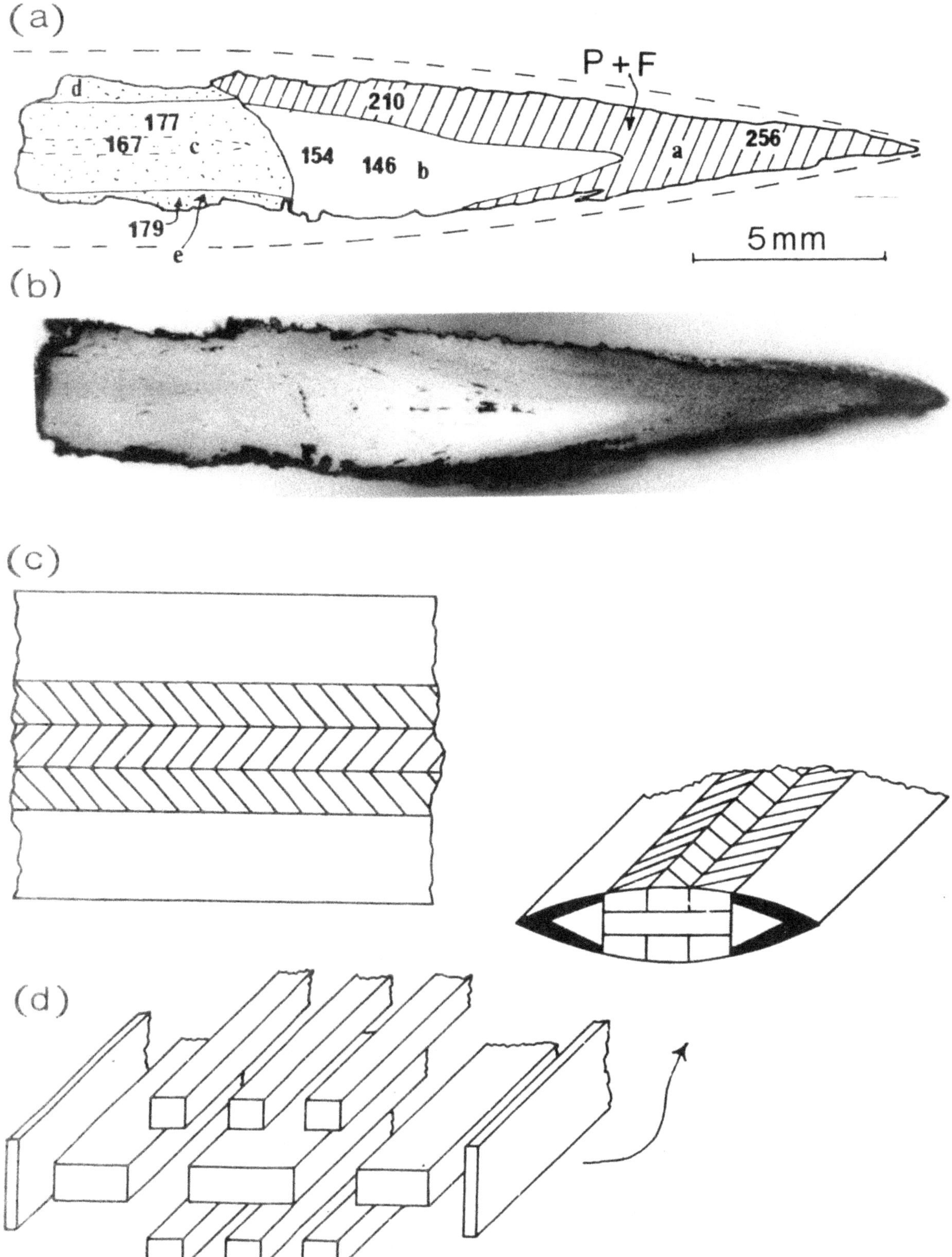


Fig. 97. Section of sword S 43 from the Thames near London + surface detail drawn from X-ray and three-dimensional view.

final shape. The fairly high degree of carbon diffusion observed in section and the partially spheroidised nature of the pearlite indicates that the blade underwent fairly prolonged heating at about 700°C, to toughen the steel edge, during the final forging of the complete blade. The spheroidised structure of the pearlite also indicates that no further heat treatment took place.

A hardness value of 256 HV was obtained for the high carbon (app. 0.8%) area of part (a) near the tip of the cutting edge. The hardness of the inner edge part (b) averaged 150 HV. The central core piece (c) gave values of 167 and 177 HV and the pattern-welded piece (e) a value of 179 HV. These values are rather high for what is predominantly ferrite and may indicate the presence of a small proportion of an alloying element such as phosphorus.

S 25. Sword from Oxford Road, Reading. (Reading Museum, 16.36). This sword consisted of two fragments (Fig. 63 showing the lower of the two) representing the upper two-thirds of the weapon, the tip end being missing. The total length was 92.7 cm and the width 5.0 cm. It was badly corroded with at best a heavily pitted surface, although a shallowly curved iron guard and pommel survived in place. It was found in 1936 in Reading in the Oxford Road at Tilehurst Street, but no further burial circumstances are known. The form of guard and pommel are similar to Petersen Type Y and Wheeler Type VI (Wheeler, 1927, p. 32, 35), which suggests a 10-11th century date for the weapon. Although very heavily corroded the remains of a fuller could just be seen running down the centre of each side of the blade. The blade fragments were bent as well as broken and there is a strong possibility that this damage was inflicted on the blade before burial, but unfortunately, because the circumstances of the find were not recorded, it is not possible to be more definite.

X-radiography showed the blade to be pattern-welded along the fullered central zone and this showed up fairly clearly in places as a criss-cross pattern resulting from superimposed herringbone patterns (Fig. 98c and d). It was uncertain on X-ray whether this represented one or two pairs of twisted composite rods across the thickness of the blade.

A transverse section was cut from the lower blade fragment from near the break with the upper blade fragment. The section extended from one cutting edge approximately half way across the blade (Fig. 98a).

Before etching, a very contrasting, non-uniform slag content could be seen across the section. In the cutting edge lines of flattened slag inclusions gave a banded appearance suggestive of a piled structure in which the slag content was generally fairly low. The core area, particularly the central part (d), contained much slag of varying size which had only been partially flattened. Extensive corrosion had taken place around the slag in this area. The central core part (d) had corroded away preferentially to the outer pieces (e) and (f).

Etching revealed the structure shown in Fig. 98a in which the cutting edge appeared to consist of three parts, (a), (b) and (c), divided from each other and the core by thin, clear white lines marking the position of welds. The upper part c (as seen in Fig. 98) consisted of a more or less homogeneous piece of fairly high carbon steel, about 0.6% C, in which pearlite and ferrite were mostly distributed in a coarse-grained

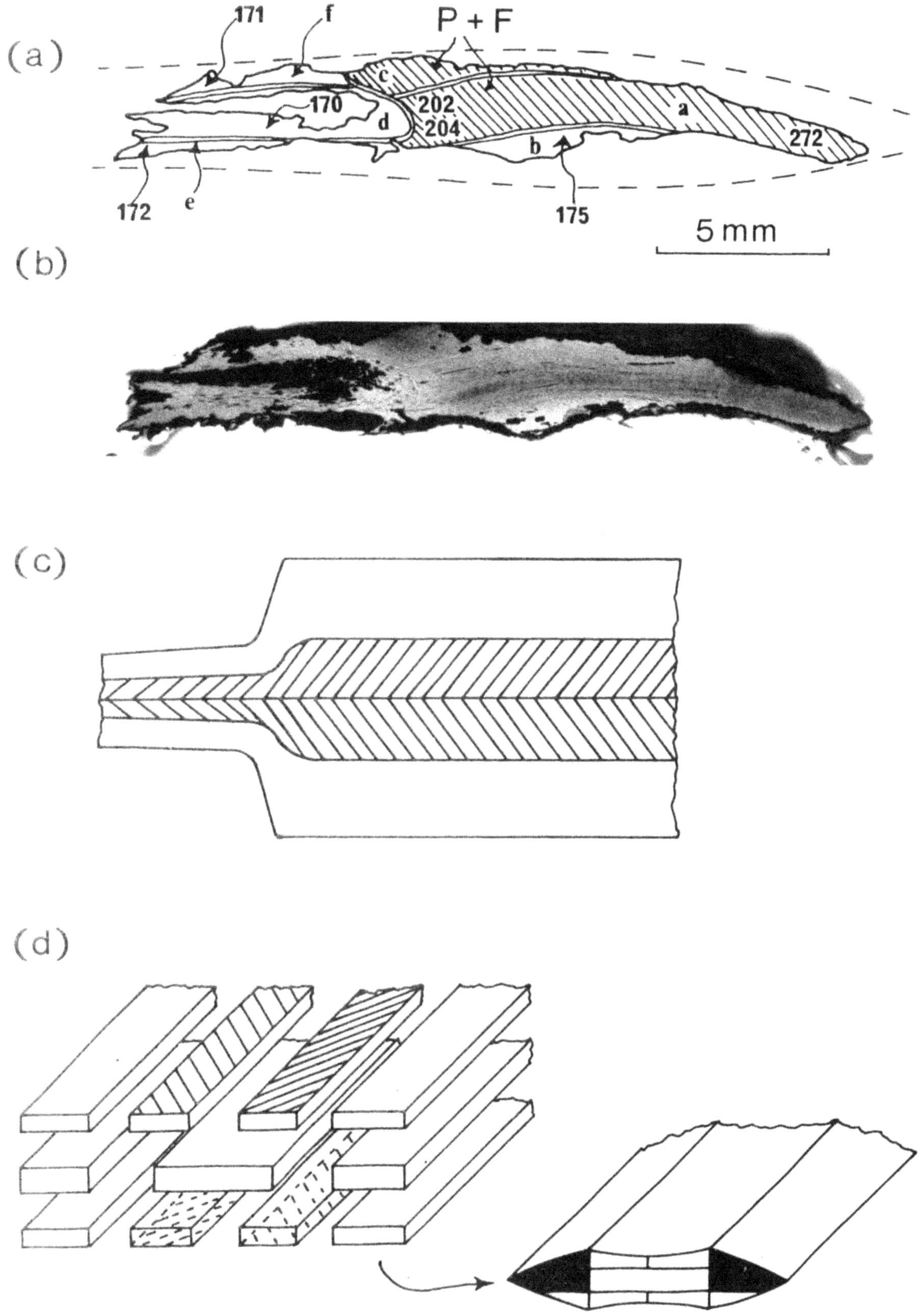


Fig. 98. Section of sword S 25 from Oxford Road, Reading + surface detail drawn from X-ray and three-dimensional view.

Widmanstätten formation. Within this formation the pearlite was partially lamellar and partially spheroidised in structure. The central and main part of (a) the cutting edge gave a distinctly banded appearance which appeared to consist of 6 alternately light and dark bands. Both were pearlitic, the carbon content of the darker bands being about 0.5–0.6% falling progressively to about 0.3% in the lighter bands which gave a diffused appearance. Towards the tip of the cutting edge the bands gradually merged to an area of fairly even, approximately 0.6%, carbon. This central component of the cutting edge at one stage must have been a piece of steel or steeled wrought iron of uneven carbon content which has been piled and heated for long enough for extensive if incomplete carbon diffusion to have taken place. The third part (b) (lowermost in Fig. 98) of the cutting edge was mostly ferrite and appears to have been a piece of well-forged (i.e. low slag) wrought iron that has absorbed a certain amount of carbon after being welded on to the central piled part (a). A fairly even zone of pearlite can be seen where it has diffused across the white line of the weld. As might be expected, the proportion of pearlite steadily decreases away from the weld until it fades out altogether.

The core of the blade was also divided into three parts by weld lines marked by narrow white lines (Fig. 98b). This clearly shows that the pattern-welding of the blade consisted of two pairs of parallel twisted composite strips which were welded onto a single plain central core strip which itself did not show up at all on the radiograph. Only one of each pair of pattern-welded strips was represented on the section and these were both badly corroded. In both cases the strips had a fairly high slag content which outlined the folded nature of these components of the blade. The lower strip (e) (Fig. 98) has partially corroded along the folds highlighting the non-corroded folded metal. The uncorroded metal of these pattern-welded strips consisted mostly of ferrite with some pearlite in places. They were made from fairly coarse (high slag) wrought iron with a very low carbon content that may be accidental. The central, plain core strip (d) had been made from very coarse wrought iron that contained so much slag as to be hardly more than a piece of poorly prepared iron bloom.

Three hardness readings were taken along the central piled steel part (a) of the cutting edge, firstly near the tip where the carbon content was about 0.6% C and in two places nearer the inner weld where the carbon content was less, about 0.3–0.4% C. The values obtained were 272, 202 and 204 HV. The mainly wrought iron part of (b) gave hardnesses of 175 HV. The mainly wrought iron central core part (d) and pattern-welded surface part (e) and (f) gave 170, 172 and 171 HV (Fig. 98a).

The coarse grained Widmanstätten form of much of the pearlite in the cutting edge may have resulted from overheating during forging, causing excessive grain growth in these areas. The partially spheroidised lamellar nature of the pearlite shows that after forging the blade must have cooled fairly slowly (probably in air) and then been reheated for a time at about 700°C, allowing some of the pearlite lamellae to 'ball up' or spheroidise, a process known as sub-critical annealing.

The reconstruction diagrams of Fig. 98d shows how the blade appears to have been assembled.

S 44. A Sword from the Thames, at Brentford. (Museum of London, A 24419). This was the upper part of a sword with a down-curving iron cross guard and

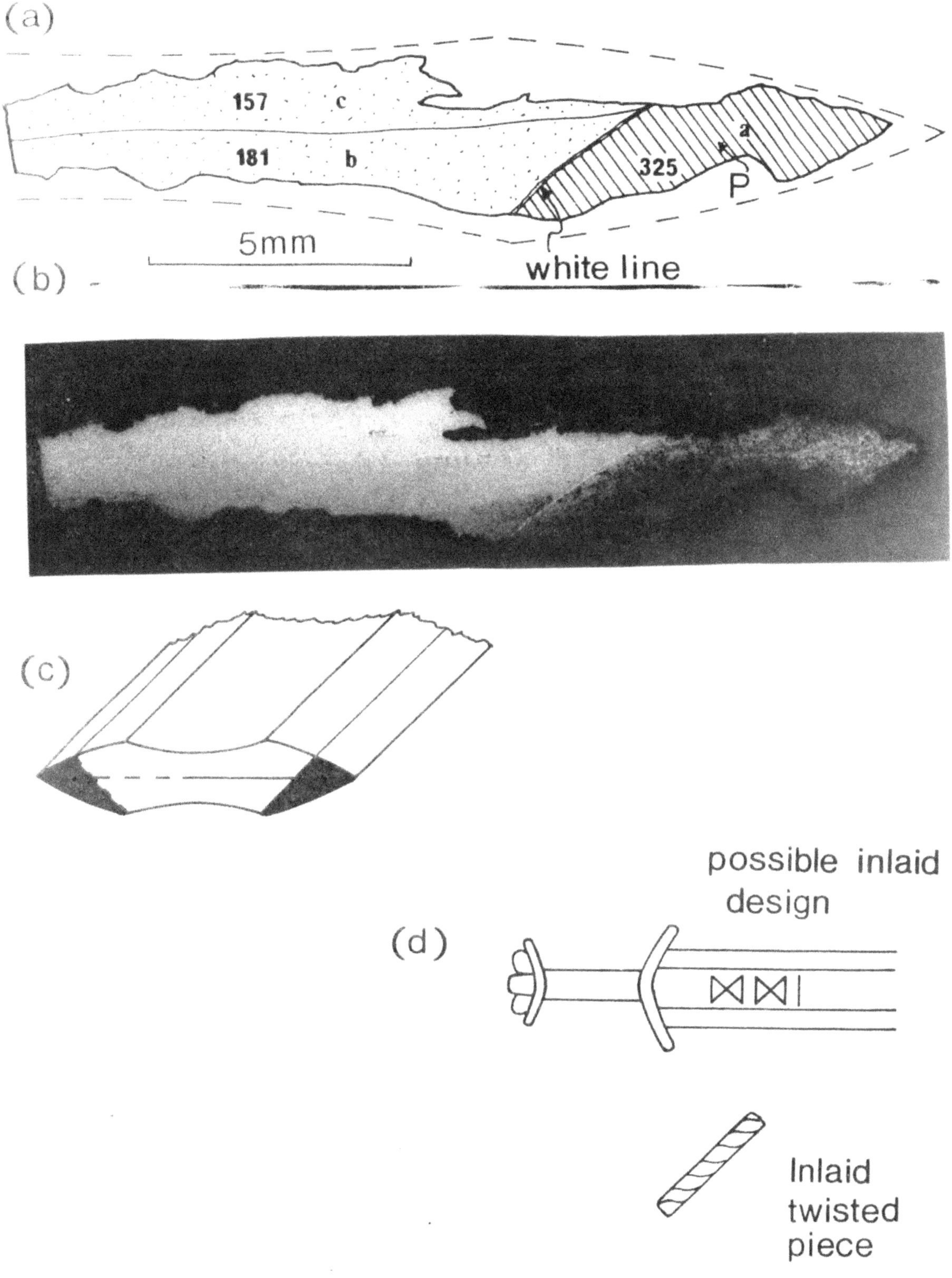


Fig. 99. Section of sword S 44 from the Thames at Brentford + three-dimensional view. inlaid design drawn from X-ray and twisted inlaid piece.

curved three-lobed pommel surviving in place. Both of these were inlaid with a covering of copper-base alloy incised with fine lines. Wheeler classified the weapon as Type VI of probable 11th century date[66]. The surface was very corroded but the blade showed no sign of having been bent. The total length remaining was 48.7 cm, the max. blade width 5.1 cm and the thickness about 0.5 cm. It had a central fuller 2.2. cm wide (Fig. 63).

Radiography revealed traces of a lattice or similar design on both sides of the blade. Each straight part of the design consisted of a twisted (?composite) piece of iron forged into the surface of the blade. A wedge-shaped section was cut 4 cm from the broken end. This consisted of three different elements, a cutting edge and a central zone of two halves separated by weld lines (Fig. 99). The cutting edge etched darkly and consisted entirely of coarse lamellar pearlite (0.8% carbon). It had been scarf-welded to the central core. One part of the core consisted of ferrite and pearlite with a carbon content of about 0.2%. The other half consisted entirely of coarse-grained ferrite with a little pearlite visible at the grain boundaries.

The hardness of each of the three parts was as follows: the high carbon cutting edge, 325 HV; the low carbon steel portion of the core, 181 HV and the ferrite part, 157 HV.

The core of the blade was made by welding a rod of low carbon steel to one of wrought iron and scarf welding to this core cutting edges of high carbon steel (Fig. 99c). The blade was air cooled after final forging, and no attempt was made to obtain maximum hardness from the cutting edge.

S 26. Sword from the Thames, at Wallingford Bridge. (Reading Museum, 1608.64). The bent remains of a blade fragment 44 cm long and 4.6 cm wide (Fig. 63). The condition varied between heavily corroded at the irregular, broken end where little metal survived, and fairly good at the other end where corrosion appeared to have been restricted to fairly light surface pitting. This end had been cut at right angles across the blade fairly recently but no record of this having been done could be found. The blade had a very pronounced convex profile which had a tendency to be rib- or ridge-like along the centre of each side. An even approximate date is difficult to assign without further details of form or provenance; however, the thickness and almost ridge-like form of the blade profile is not commonly found amongst swords of the Saxon and earlier periods but is a commoner form in later medieval weapons. On this basis, a tentative date range of 12-14th century may be suggested for this sword blade.

A section was taken extending across the blade parallel to the existing cut. Unetched, several lines of flattened slag ribbons and spots were visible suggesting possible weld lines running diagonally across the blade. Corrosion had also taken place along some of these as well as around some larger slag inclusions.

Fig. 100a shows diagrammatically the view seen in the macrograph of Fig. 100b. On either side of the blade steel strips (a and h) both of approximately 0.6% carbon formed the tip parts of the cutting edges. The approximate carbon contents of the other strips were as follows: (b) nil, (c) 0.8%, (d) nil, (e) 0.2-0.3%, (f) 0.4-0.5%, (g) nil. These estimates do not allow for the fact that a certain amount of carbon diffusion had occurred in places across steel/iron welds, especially in the area of

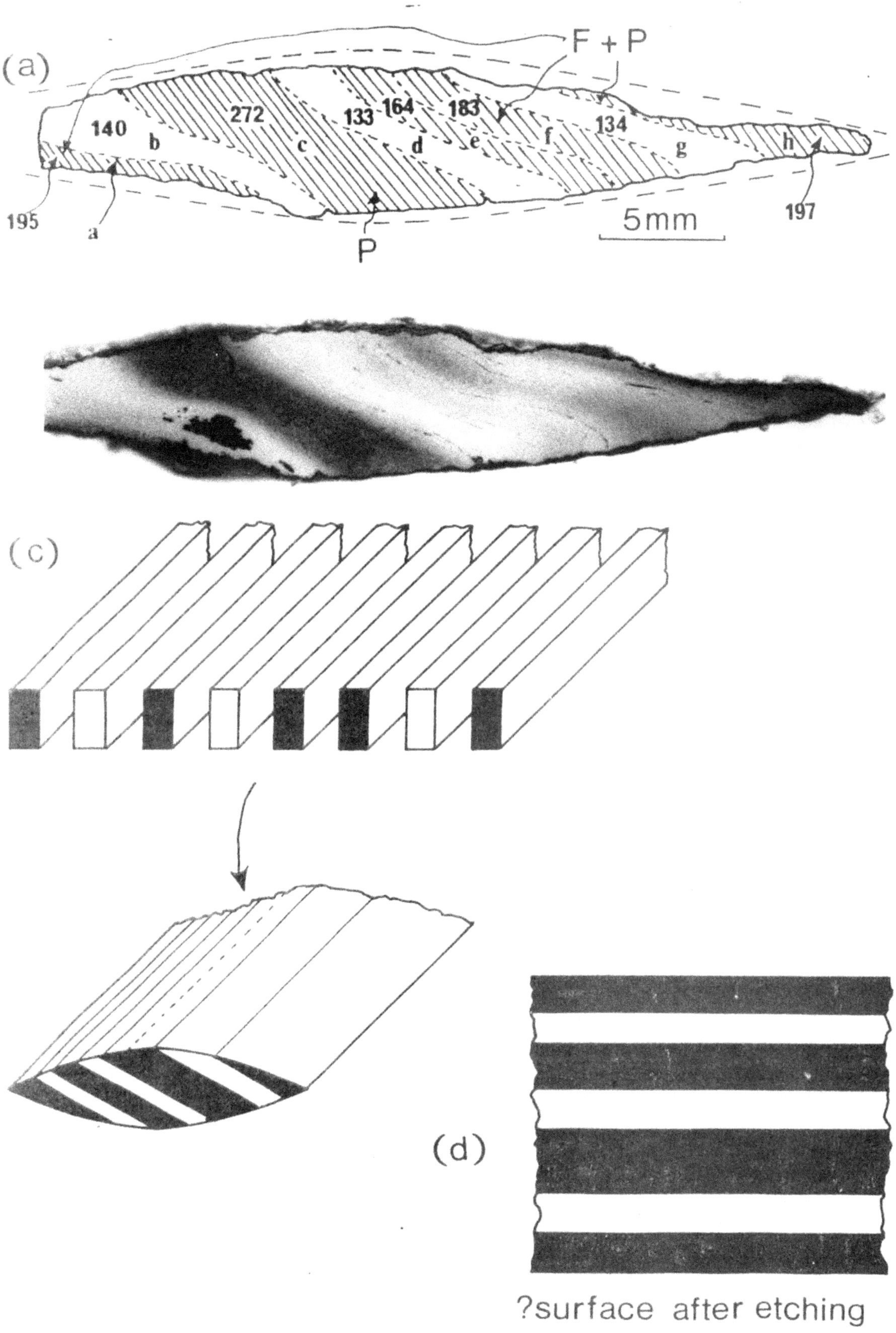


Fig. 100. Section of sword S 26 from the Thames at Wallingford Bridge + surface detail drawn from X-ray and three-dimensional view.

(f)/(g). Also considerable decarburisation was evident in places along the sides of the steel strips, most noticeably between the steel strips (e) and (f) and along the high carbon steel strip (c).

The slag content of the ferritic strips varied between (b) which appeared only to have the occasional spot or ribbon and (g) which contained many slag inclusions of one or two dark phases (? fayalite and a glassy constituent). The pearlite of the steel strips was in a mainly spheroidised form which indicates that after forging the blade was subjected to sub-critical annealing - fairly prolonged heating at below about 700°C which has allowed the pearlite lamellae to spheroidise.

A series of 8 hardness readings were taken one from each of the parts of the blade (a)-(h) running horizontally across the middle of the section. The results were as follows: (a) 195 HV; (b) 140 HV; (c) 272 HV; (d) 133 HV; (e) 164 HV; (f) 183 HV; (g) 134 HV; (h) 197 HV.

<u>S 28. Sword, no provenance.</u> (Reading Museum, 50.80 (H.L.)). The upper half of a sword blade 43.6 cm long and 4.4. cm wide (Fig. 63) with a long narrow iron guard which widened to a knob at each end. The pommel was missing and the surface of the blade was fairly heavily, but evenly, corroded and the metallic core appeared to have survived quite well. A narrow fuller ran down the centre of the blade on either side. The provenance of this sword fragment is not known but the distinctive long, narrow form of the guard is typical of sword guards (or quillons) of later medieval weapons of the 13th or 14th centuries. The narrow form of fuller is also common amongst sword blades of this period.

A transverse section was taken from the broken end extending across the full width of the blade. Unetched, the section showed a fairly low content of slag, as spots and ribbons running horizontally across the section which suggested a piled or sandwich structure to the blade. This was confirmed on etching (Fig. 101) from which it appeared that the blade consisted of a sandwich of 5 inter-leaved steel, or steel and iron strips, of alternating higher and lower carbon content, the latter possibly the result of carbon diffusion from the former. The outer dark-etching strips, (c) and (e) only survived on the less heavily corroded part of the section. These consisted of a mixture of troostite (rapid, very dark etching, nodular or amorphous looking pearlite) and paler martensite patches. The thick central dark band (a) consisted of a similar structure in addition to which could be seen a number of parallel fine white lines which appeared to occur at the probable weld boundaries with the paler zones on either side and also within the strip. Towards one end (Fig. 101) most of these inner white lines converged, which suggested the possible presence of a fold here. This area was fairly slag free but some slag spots did also appear to form a convergent pattern which would also tend to suggest the presence of a fold. These indications suggest that the central strip has been piled - in this case folded and forged out a possible two or three times before the final assembly of the blade. The clear narrow white lines may be the result of localised arsenic enrichment caused by the migration of arsenic towards surfaces being welded.

Hardness values of 496 HV and 503 HV were obtained for areas of tempered martensite in part (a) firstly near the tip on one side and secondly near the centre of the blade on the other half of the section indicating that the blade was tempered after quenching. The small grain ferrite and pearlite

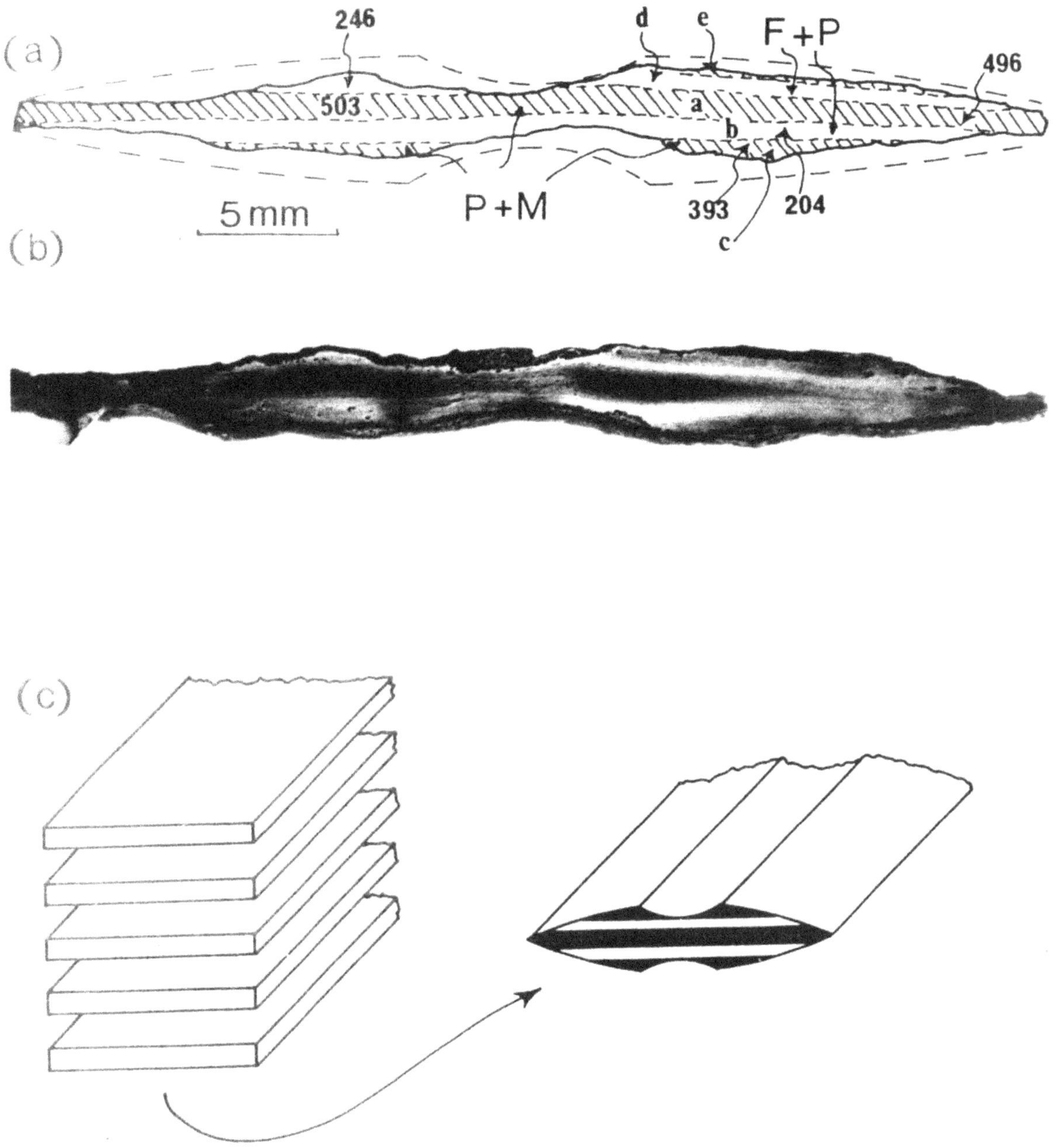


Fig. 101. Section of sword S 28 from Reading Museum (no provenance) + three-dimensional view.

of the intermediate part (b) and (d) gave 204 and 246 HV while the martensite of the outer part (c) gave 393 HV where the effects of tempering were more marked.

S 30. Sword from Sunbury Weir Stream, Middlesex. (Reading Museum). Fragment of sword blade (Fig. 63) broken approximately half way along towards the tip and also at the shoulder of the blade where it narrows to become the tang. It was found near the Thames in the Sunbury Weir stream. It was in quite good condition, the corrosion effects having mostly been limited to fairly even and shallow etching of the surface, although some deeper 'pitting' had occurred. In profile the blade thickened to a central rib on either side giving an ogival section. The fragment was 46 cm long with a maximum width of 4.6 cm and a thickness of 6 mm at the central rib. An old label attached to the blade said '? 17th century', but apart from its fairly distinctive profile it bore no particular distinguishing characteristics from which its date or type might be deduced. Date: Late Medieval/post Medieval.

A V-shaped section extending a little over halfway across the width was cut from the blade. The section showed a small amount of slag fairly evenly distributed as fairly small spots and ribbons. There were no lines of slag suggesting a weld position.

When etched, a macrograph of the section (Fig. 102b) showed the blade to consist predominantly of steel. It was mainly of a sandwich construction with two similarly high carbon steel - 0.6-0.8%C outer parts (a) and (c), with a low-medium carbon steel 0.3-0.4% band (b) running across the section from the centre of the core between the two outer parts, meeting the surface a little before the tip of the cutting edge. Following the central band in the opposite direction (leftwards in Fig. 102b) from the core of the blade, it can be seen to curve fairly sharply towards the surface of the blade near where the rib would have been had it not been corroded away here. It curved over another high carbon steel portion similar to the other outer two parts. Fig. 102a is a schematic diagram to show in a simplified form the structure visible on the macrograph of Fig. 102b. Part of a fourth higher carbon steel piece, (d), was also visible next to (a) on the 'blunt' end of the section. The structure was similar to that of (a) being one of mainly coarse lamellar pearlite.

There were no clear weld lines between these zones but a certain amount of slag segregation gives a definite hint of weld lines between (b) and (c) and (a) with similar but less definite hints of welds between (b) and (c); and (c) and (d).

The higher carbon zones (a), (b) and (d) were predominantly pearlitic with only small areas of ferrite visible (Fig. 102b). Much of the pearlite showed a fairly large grain structure and was mainly in a fairly coarse lamellar form.

The structure shown in Figs. 102b and c indicate that the blade was not of a true sandwich construction (i.e. a series of parallel inter-leaved bands) but mainly consisted of two halves; one a fairly homogeneous piece of high carbon steel (a) in Fig. 102a), and the other consisting of pieces of alternately high and lower carbon steel (b), (c) and (d) in Fig. 102a. These parts appear to have been welded together leaving little traces in the

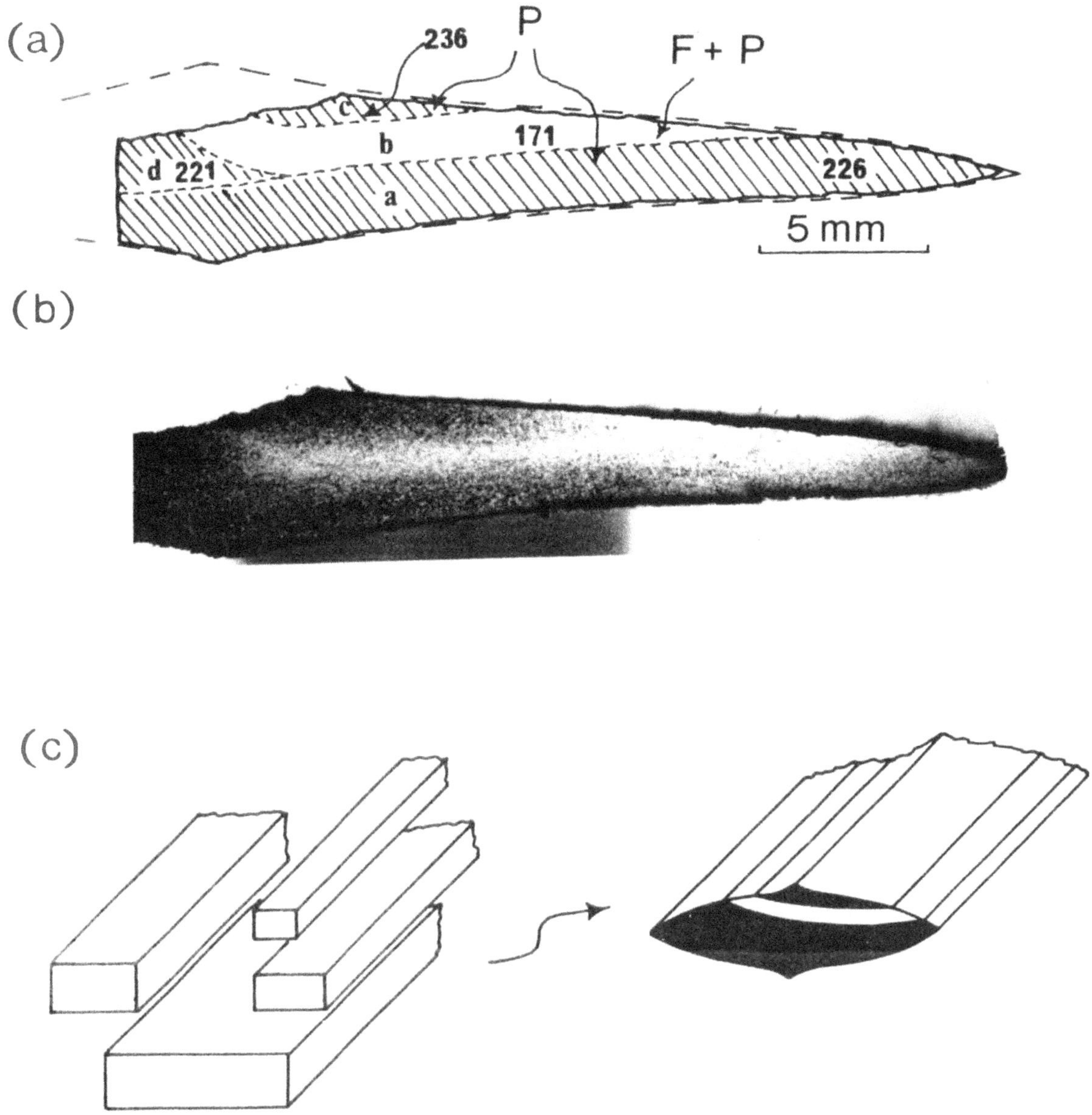


Fig. 102. Section of sword S 30 from Sunbury Weir stream + three-dimensional view.

form of entrapped slag. The divisions between the different parts have become very blurred as a result of extensive carbon diffusion across the likely weld junction. This is probably the result of prolonged heating of the blade above the upper critical point - in this case (for the higher carbon zones) about 750°C. This would explain the fairly extensive occurrence of large grained pearlite.

The lower carbon zone is subdivided in places by some localised small bands or 'lenses' of lower carbon content, about 0.2%, showing a much smaller grain size. There were contrasting pearlite grain sizes within zone (c). This may indicate that the upper critical temperature for these localised lower carbon areas, i.e. approximately 800°C, was not exceeded during this prolonged re-heating or annealing of the blade. The fairly coarse lamellar nature of the pearlite is indicative that the blade was finally cooled slowly.

A hardness reading was taken for each of the four parts of the blade visible on the section and the results were as follow: (a) 226 HV; (b) 171 HV; (c) 236 HV; (d) 221 HV.

## Discussion on Edged Weapons.

### Arrows and Spearheads.

The number of arrowheads examined here are too few to draw much in the way of conclusions. Roman spearheads are nothing more than wrought iron and must have been quite adequate for their purpose. The later spearheads do not seem to be much harder but two are pattern-welded. One of these has an edge hardness of 496 HV as a result of heat treatment, while the others have a moderate hardness with a ferrite-pearlite edge. Earlier Anglo-Saxon spearheads have been fully discussed by Swanton[78] and the two pattern welded ones figured here are more typical of the later period. Both of these would have looked impressive when polished and etched.

The 301 pattern-welded spearheads seen by Anteins[59] fell into the length range 19 - 59.5 cm which is considerable. Anteins claims that they date from the 11th - 14th centuries which is later than for pattern-welded swords. Most of these came from cremation sites.

Two blades had quench-hardened edges in the range 260 - 478 HV. Most had a 'serrated' join of the edge to the core and this sort of join, together with the pattern welded core, must have made a very imposing weapon when polished and etched. Composition (% Ni) and decoration suggest that many were imports, probably from Sweden, possibly Gotland.

A leaf-shaped spearhead from Clifton on Trent was examined by Brewer;[80] it was not pattern-welded and had a ferrite core with a hardness of only 114 HV1. Metal near the tip of the blade seemed to have been carburised slightly and gave hardnesses in the range 178 - 214 HV1. A Viking spearhead examined by Coghlan,[52] which was designed to penetrate armour, was slim, heavy and had two strong ribs at right angles to the blade. The total length was 30 cm. The metal was piled wrought iron with a carburised cutting edge and a hardness of 160 HV. A small medieval spearhead had a slim shaft with a wide socket and leaf-shaped blade. This was a low carbon wrought iron.

## Scramasaxes

The single-edged weapon known both as seax and scramasax was popular during the Anglo-Saxon period. These weapons vary greatly in size ranging from single-edge swords down to small daggers. Size and burial circumstances can be used as very general criteria to divide smaller examples into weapons, and knives used for other purposes. The two classes merge into one another, the weapons being mostly larger and the knives smaller. The term scramasax is used here to cover the whole range of these single-edged weapons.

It is not known to what extent knife-smithing was practised separately from sword-smithing. As scramasaxes are single-edged weapons not unlike large knives, it is possible that they were made by knife-smiths rather than sword-smiths. Although only eight scramasaxes were examined in this study, the comparisons with the main structures found in the large numbers of both swords and knives are interesting.

The Romano-British period dagger from Brancaster, included here as a broadly similar weapon, consisted only of ferrite. The scramasax from Barham Down was the only Early Saxon weapon of this kind examined. It is a clear example of a Type 4 blade with a slaggy iron core around which a piece of higher carbon steel has been welded. The finished blade was not quenched but given an alternative final heat-treatment in which the steel jacket was toughened by fairly prolonged heating at about 700°C after initial air cooling from forging temperature. This would have been very much a functional rather than decorative blade, and would have remained very effective so long as the steel jacket remained intact. The abrading of the steel jacket of the Type 4 blades by the repeated sharpening to which most of the knives were subject may have been much less of a problem with scramasaxes which probably needed sharpening much less often.

This construction contrasts markedly with the Early Saxon sword blades examined here (see below) which nearly all were pattern-welded and consisted mostly of iron, and only in their cutting edges can they be compared with one of the four knife construction types. In these sword edges, where steel has been incorporated, a Type 1 or 2 construction has been used (Fig. 1). No example of Type 4 construction was found among any of the Early Saxon sword blades.

The remaining six scramasaxes (Table M) all belonged to the Late Saxon period and differed from the earlier weapons. One example is of Type 1/2 while another is of Type 1, a popular knife construction, and the edge hardnesses are not unlike the best quality knives of the period (see Table A3). The other four blades are pattern-welded, the pattern welded areas being inserted as decoration between the back and the cutting edge.

The combination of the twisted copper alloy wire inlay with the simple sine wave pattern created along the weld between cutting edge and back in the scramasax from the Thames at Hampton may be seen as an example of a simpler and cheaper non-ferrous alternative to a more elaborate form of pattern-welding. Here the back of the blade into which non-ferrous wires have been inlaid was a coarse slaggy iron. The metal of the cutting edge could not be examined so it is uncertain whether or not it is of a similar, poor, quality.

Two of the other pattern-welded examples were similar, both with pattern-welded pieces inserted between a cutting edge and back consisting of low carbon iron. The structure of these two is much closer to the majority of earlier Anglo-Saxon swords examined here than the later pattern-welded examples.

The scramasax from Dorset is more elaborately made and has both back and edge made of medium-high carbon steel which has been fully heat-treated showing that the whole blade was heated to 800°C and quenched in water. The ferritic bands in the pattern-welded composite part would have lent some ductility to the blade, but the average hardness was high. If the way in which steel has been incorporated into these blades, as well as their overall complexity and standard of construction, is used to judge their quality, then we do seem to have some variation in quality here. The Dorset blade can, thus, be seen as a much better quality weapon than the two previous examples which themselves, on the evidence available, appear to be of a better standard than the Hampton blade.

Salin[54] shows similar weapons and one illustration shows an edge with a hardness of 557 HV scarf welded to a piled back. None of the 8 scramasaxes he examined were pattern-welded and Salin knew of only one pattern-welded blade and this came from Zürich.

Anteins [79] shows one from Grobina which has been pattern-welded with two patterned strips like S 34. This is dated to the 8th - 9th century. Unfortunately it is almost totally disintegrated. The carbon content of the points was 0.2-0.3% and it is believed that they were quench hardened.

## Swords

In all, 39 sword blades have been examined dating from the pre-Roman Iron Age to late Medieval times. Thirty-three of these date from the Anglo-Saxon period (5th - 11th centuries). The swords from this large group fall fairly well into three main sub-divisions with twenty swords from the Early Saxon period (5th - 7th centuries), two from the Middle Saxon period (7th - 9th centuries) and eleven from the Late Saxon period (9th - 11th centuries).

Twenty-five of the 33 Anglo-Saxon period swords had been pattern-welded, giving one of the largest corpora of metallographically examined pattern-welded blades from which we can therefore draw details and conclusions of the techniques by which these blades were made.

We must start, however, with the pre-Roman Iron Age in which it would appear that blades were mostly of wrought iron or low carbon steel with no attempt at pattern-welding or heat treatment. However, some examples from this period have shown evidence of more complicated techniques[81]. We can already see from these the beginnings of the techniques developed later to produce the better known pattern-welded types of sword.

Only three blades in this study could be attributed to the Iron Age. The most interesting would appear to be an Early Iron Age copy of a Late Bronze Age leaf-shaped sword. The other two were parts only of blades of long swords and could not be reliably categorised but it was felt they most resembled other examples of Late Iron Age swords, rather than later weapons. All three blades were predominantly of wrought iron or very low carbon steel with no attempt at pattern-welding or heat treatment.

Only one Roman sword was examined in the programme but we can compare it with a few others from Britain and the Continent although the body of comparative material is not large. What is perhaps surprising in the *spatha* from Whittlesey is that it is so uniform in composition and structure, with little evidence visible of piling and with no attempt at any heat treatment. The structure consists of a fairly uniform ferrite and pearlite, a low carbon steel, within the hardness range 159-193 HV. The lack of heat treatment is at least in keeping with Roman blade smithing practice, as we have seen in the first part of this work. The blade would have been tough and not brittle which was clearly the aim.

Eighteen of the 20 Early Saxon sword blades examined were pattern-welded. Pattern-welding is further discussed below but in general it can be said that the blades which exhibit this technique consist predominantly of wrought iron and vary greatly in quality. They are complex in construction (Fig. 103), consisting on each side of between two and five composite rods welded side by side. The rods have been variously twisted or left straight to give rise to the patterns which would have been visible on the surface of the blades after a suitable etching treatment. Each group of rods was normally either (Type VI) welded to a single central 'core piece' or (Type V) welded back-to-back to the similar group of rods forming the surface on the other side of the blade. Another arrangement seen in three of the Kentish blades was (Type IV) where pairs of composite rods had been welded together back-to-back but were separated by an untwisted composite rod which ran through the thickness of the blade (Fig. 103).

One of the least expected results was the discovery that the cutting edges, which had all been butt-welded on separately, had usually been made of more than one strip. Six of these eighteen blades had edges made of a single piece of low carbon wrought iron. In six other cases the edges were of a similar low carbon iron but had been made in two halves with a central weld seam. This seems a curious technique; one would have thought that one strip or rod would have sufficed. However, in one of these cases (S. 10) a narrow, low or medium carbon steel rod had been welded between the two halves at the outside so as to form the actual tip of the cutting edge; this had not been heat treated. In the remaining six examples a sandwich construction had been used, where a low or medium carbon steel strip had been welded between pieces of iron, fairly clear examples of the Type 1 knife construction. The edges of two of these blades (S.17, S.55) had been hardened by quenching in water, hardnesses of between 550 and 600 HV having been achieved. The carbon contents were probably slightly too low to achieve fully heat-treated hardness of 700HV or over, and there was no indication of any subsequent tempering.

The carbon contents of these Early Saxon pattern-welded sword blades were in any case mostly much too low for quenching to have been any use and for the most part this does not even appear to have been attempted. This probably means that the sword-smiths knew that most of their blades would not benefit from this form of heat treatment. Attempts to produce a harder edge for some of these weapons were clearly being made, but appear to have been uncommon.

The two other Early Saxon sword-blades were not pattern-welded and consisted of piled low carbon iron. Slight variations in carbon and phosphorus content appear to be responsible for the banded structure visible in sections with between six and eight bands visible in each case.

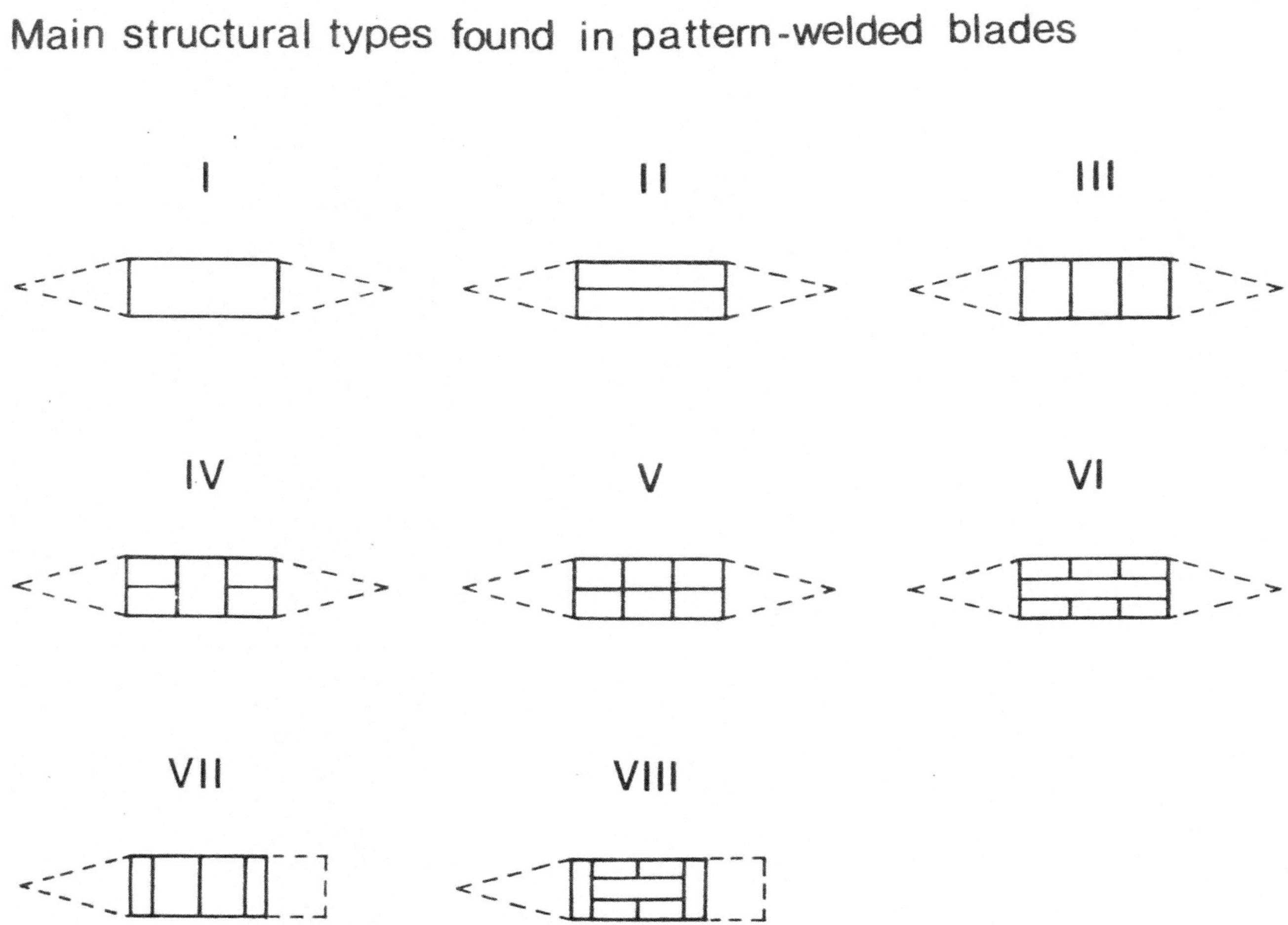


Fig. 103. Types of pattern welded sword structures in section.

Summing up we can say that as weapons these swords were, in the main, little better than those which had proceeded them and they were mostly not as good as the evenly carburised steel of the Roman period spatha from Whittlesey.

The two Middle Saxon sword blades show certain changes in style and techniques of manufacture from the earlier blades. They are the earliest examples seen in this study of blades with a central groove or fuller down each side and the fullers here were rather wide and shallow. Both these blades were pattern-welded, the pattern-welded parts occupying the full width of the fuller. The Reading blade (S.31) shows a Type V form of pattern-welding very similar to that seen on earlier blades whereas the Brentford blade (S.46) shows a variant of Type IV pattern-welded construction where instead of being just a combination of wrought iron pieces the central untwisted 'composite' part was a rod of higher carbon steel which was probably piled and set 'on edge' (see also pattern-welding below). The edges of both blades showed a more advanced technique of construction than seen on nearly all the earlier blades. In both cases a sandwich construcion had been used, the Reading blade showing in a section as a knife Type 1/2 and the Brentford blade a Type 1. The edge of the Reading blade was rather curious in that it had a piece of low carbon steel sandwiched between pieces of iron with an additional narrow rod of high carbon steel also welded in between to form the very tip of the cutting edge. The Brentford blade had been given an edge which was a straightforward sandwich of a medium-carbon steel piece between two iron pieces.

Both sword blades had been quenched, probably in water to judge from the more or less fully martensitic state of the steel parts, and the Brentford blade had also been lightly tempered, the earliest example of this technique found amongst the blades in this study. Although only two sword-blades of this period were examined, already we are seeing a much higher standard of manufacture for (what seem now) outwardly similar blades than we have seen for the sword blades of the Early Saxon period. We now see the economical use of steel to give fully hardened edges, in the range 647-775 HV.

The hardened steel spine and edges of the Brentford blade combined with the ductility of the other, mainly wrought iron components, have produced a very strong blade combining much of the normally mutually exclusive qualities of hardness and flexibility. The ideal aim of the sword-smith, at least where a functional weapon was concerned, must have been to produce a blade which would not bend or break easily but one with hard, sharp edges. The Reading blade is better than previous examples and the Brentford sword a much better weapon again, going a good way toward achieving this aim.

There now seems little doubt that in this country most swords of the Anglo-Saxon period up to the 9th century were pattern-welded. During the Late Saxon period however, this technique of manufacture becomes less common for sword blades and ceases to be used for these possibly early in 11th century, although it continued to be used in scramasaxes or knives for longer. Examples have recently been found (but not yet published) in the City of London from contexts dated to the later 12th century. A new type of sword blade emerged during this period, one with a non pattern-welded blade, into which an inscription or design made of small pattern-welded or plain pieces of iron had been forged into the surface of the blade, either on one or both sides, near the hilt of the weapon. In these plain blades the inlayed design does not form an integral part of the construction.

Although these blades, apart from the inlaid design, appear plain on X-ray, they do not necessarily consist of a homogeneous piece of metal and the results here confirm this.

Eleven sword blades of the 9th - 11th centuries were examined in this study and five of these found to be pattern-welded. Although less common the pattern-welded element showed a variety of form similar to that seen in earlier blades and again this part of these blades was made very largely of low carbon iron. Steel forms an important element of the welded-on cutting edges of these five weapons with three examples of a steel-cored Type 1 knife, sandwich; one example of the opposite, Type 4, iron cored construction with a steel outer casing, not seen earlier; and the final example (S.41 from the Thames at Vauxhall) where the edge was mostly a fairly high carbon steel with a narrow iron strip welded down one side. All five blades had been subjected to some form of heat treatment after final forging. Slack quenching (possibly in oil) may have been used for the Vauxhall blade whereas of the other blades, two (S.24, S.48) appeared to have been quenched in water, and two had been subjected to the almost opposite technique of prolonged heating or sub-critical annealing to toughen rather than harden the steel of the edges. These last two have maximum hardness values in the range 256-272 HV which compare with 490 HV for the Vauxhall blade and 564-647 HV for the water quenched blades. Only the Vauxhall blade gave evidence for auto-tempering.

Three of the remaining six swords of this period appeared on X-ray to have designs formed from short lengths of twisted rods forged into the surface of both sides of the blades near the hilts. These inlaid designs were difficult to make out but in each case appeared to consist of a series of crosses interleaved and flanked by vertical lines (Figs. 91, 95, 99). The parts of the blades with designs were not sampled so the structure of this residual form of pattern-welding cannot be given here. It would however, seem most likely to have consisted of a low carbon iron similar to that found in the twisted elements of the pattern-welded swords examined.

Two of these three inlaid swords had blades of a fairly homogeneous medium-high carbon steel although they had been given quite different final heat treatments. S.23 had been quenched, from about 725-750°C, in water to give an edge hardness of 762 HV whereas the blade S.47 had been given a final prolonged period of sub-critical annealing giving an edge hardness of only 239 HV. The two blades would have possessed very different properties, the first being very hard but inclined to be brittle which may explain how it came to be broken, whereas the other blade would have been much more ductile, although its edges would have been inclined to blunt more easily. The sword from Brentford probably possessed the best overall qualities of these inscribed blades. The fullered part of one side consisted of wrought iron which had been welded to a piece of low carbon steel which formed the fullered part on the other side of the blade. High carbon steel edges had been scarf welded to this duplex core. Hardnesses ranged between 325 HV for the high carbon edge to 157 HV for the ferritic half of the core. The blade appears to have been air cooled without further heat treatment.

The remaining three 'plain' sword blades examined were also quite different to one another in construction. S.7 consisted of a fairly homogenous low to medium carbon steel which showed an incompletely quenched structure, possibly the result of slack quenching, and gave an edge hardness of 477 HV. The sword S8 had a duplex structure, one face of the blade consisting of a high carbon steel and the other of wrought iron. The blade here had been given a prolonged final sub-critical anneal - probably more than was desirable - to toughen the steel. The hardness range here was between 172 HV in the steel half to 81 HV in the ferritic half, a very soft wrought iron. The final sword of this group, S28, is a very good example of a piled or multi-decker sandwich method of construction with its two layers of low carbon steel interleaved between three layers of medium-higher carbon steel. The piled structure may earlier have been one of a higher carbon steel interleaved with pieces of low carbon iron which have become enriched with carbon during forging or by being kept for a time in a non-oxidising part of the smith's hearth. The blade appears to have been finally slack quenched to give hardness values of around 500 HV for the higher carbon areas - similar to the hardness of the other blades which appear to have been given a similar quenching treatment.

The observations made on the eleven swords of the Late Saxon period have been discussed in some detail in this section because of the great variety of their fabrication methods which has come to light in this study.

In summarising the overall changes and developments in sword smithing from the Early to Late Saxon periods, two main points are clear. First, all of the later pattern-welded sword blades, including those ascribed to the 7th-9th centuries, show a much higher standard of overall manufacture with the more extensive and efficient use of steel which would have made these swords more serviceable weapons than those of the 5th-7th centuries, which as we have seen, were mostly of low carbon iron. The actual pattern-welded components, however, did not themselves change much throughout the Anglo-Saxon period. Secondly, the same change in the standard of construction and use of steel is true of the non-pattern-welded swords, the earlier swords being of low-carbon iron, while later on either more or less homogeneous steel or quite elaborate combinations of iron and steel were being used and various heat treatments employed as well.

The major change and improvement in the standard of sword-smithing, as seen in the examples studied from this country, appears to have taken place during the 7th-8th centuries. It is surprising that Early Saxon sword-smiths did not strive harder to produce better blades as they must surely have been capable of doing so. The only Early Saxon scramasax examined was by contrast a good example of a steel cased, iron cored blade - seemingly a much more functional weapon than the contemporary swords. It is possible that the sword was more of a ceremonial weapon at this time and there may simply not have been a demand for the part or wholly steel weapons which would have been both more expensive and difficult to make. The sword may have become a more functional weapon during the 7th-9th centuries creating a demand for the better quality blades. The earlier swords, were, presumably, regarded as adequate for their purpose, although to what extent they were functional or ceremonial is unknown. The hardness values obtained for the earlier blades tend to suggest that these were rather better and more serviceable weapons than their predominantly wrought iron structures would otherwise indicate although still much inferior to the later blades.

Three other sword blades were examined to gain some idea of how the manufacturing techniques might compare in the later medieval period. One of the blades was possibly even later but they do show at least that methods of construction did vary later on. The late medieval sword from Wallingford Bridge was made in a way quite different to the earlier examples, its three pieces of wrought iron interleaved and scarf welded between pieces of medium-high carbon steel. The second late medieval blade (S.28) was probably made from two pieces of wrought iron sandwiched between three pieces of high carbon steel. It had also been quenched. The third blade was a more or less homogeneous medium-high carbon steel. Neither the Wallingford Bridge sword nor this third blade from Sunbury Weir Stream had been quenched but both had been given a sub-critical anneal to toughen the steel of the blades.

## Pattern-welding

Much has been written on the construction of the pattern-welded swords (France-Lanord,[83] Antiens,[79] Maryon, Salin,[54] Ellis-Davidson,[84] Ypey[93] and others) and the technique has been reproduced by John Anstee and discussed by Anstee and Biek,[85] so it is not necessary to go into details of the technique, but merely to summarise the results we have presented here and to discuss the main points which arise from these.

A forerunner of the pattern-welded sword can be seen in a blade from Llyn Cerrig Bach in Anglesey, one of four swords found in a hoard of 2nd century B.C. to first century A.D. date[81]. In this blade several rods of low carbon iron or steel had been welded together side-by-side with the welds running from surface to surface, i.e. at right angles through the thickness of the blade. The technique had developed by the late second century A.D. into the more familiar form with twisted rods welded side-by-side to form the core part of a sword blade. A blade like this from South Shields was found in 1953 beneath contexts dated to c A.D. 200[84] and a similar form of pattern-welding occurs in one of two swords found more recently in a 4th-century grave in Canterbury.[68] These are paralleled by the better known finds from the peat bogs of Jutland. At Vimose, c A.D. 200, a total of 67 swords included fourteen pattern-welded blades and at Nydam in Schleswig, c A.D. 300, 106 swords included 90 pattern-welded blades (Ellis-Davidson,[84] 1962, 32).

Such swords were made by welding and sometimes twisting multiple rods together and forming composite pieces which were laid and welded together in various ways to form the core of the blade to which the edges were added.

A series of pattern-welded types was built up in the course of this study to show the different ways in which these could be constructed (Fig. 103). The Types V and VI were found to be the most common amongst the swords examined although examples of most of the others were encountered and the remainder can be inferred. Further types may come to light in the course of further work.

Nearly all the pattern-welded swords found in this country belong to the Anglo-Saxon period. Several hundred swords of this period exist in museum collections in this country, many found a century or more ago, but it is really only in the last twenty-five years or so with the increasing use of X-radiography that the extent to which these blades were pattern-welded is becoming known. The twisted rod pattern occurs repeatedly on X-ray as shown by Anstee and Biek's survey in their 1961 article[85] which also decribed a series of experiments in which replicas of pattern-welded blades were made with twisted core elements to which cutting edges were welded on separately. A recent radiographic survey by Janet Lang and Barry Ager of 127 swords of the Anglo-Saxon period in the British Museum collections is shortly to be published.

In this, pattern-welding was seen in more than half the swords of the 5th-6th centuries, compared with about two-thirds of the swords of the 6th-7th centuries. Nearly two thirds of the 9th-10th century swords also showed pattern-welding. A similar number of sword and scramasax blades from other collections were examined by X-ray in addition to those metallographically examined and discussed in our work here. Nearly all the Early Saxon blades probably belong to the 6th-7th centuries and a comparison with the earlier blades showed a twisted form of pattern-welding, and about two-thirds of the blades of a broadly similar 9th-10th century period showed pattern welded areas in the centre of the blade.

The characteristic patterns of the blades incorporating twisted components was first noticed on swords from waterlogged sites such as those from Nydam, where slow corrosion had deeply and differentially etched the surface of the weapons to reveal patterns. The patterns seen on X-ray also show up because of the effects of differential corrosion in the ground. The pattern-welded structure of Anstee's modern replica, which was not corroded, did not show on X-ray. The opposite situation may also be true, and buried pattern-welded blades which are totally corroded may not show any of their pattern-welded structure on a radiograph, especially where a dense surface incrustation is present.

We can now see that pattern-welding was very popular in this country for swords during the Anglo-Saxon period, but why was this the case? Was pattern-welding used for structural or decorative reasons? Pattern welded elements of the Nydam blades have been shown to have combined low and high carbon iron[92] suggesting that it might have been intended as a way of combining iron and steel to produce a strong blade. Other parts of the blades also had a varying carbon content so this does not necessarily follow. The pattern-welded elements of the Anglo-Saxon blades examined in the present study were consistent in composition right through the period, and show that the technique served a mainly decorative purpose and that the complexity of the patterns bore little relation to the overall functional qualities of the blade.

The pattern-welded composite parts were found to consist almost invariably of low carbon iron. However, these composite parts were mostly made from alternate pieces of low and high phosphorus iron. The composites appear to have often been lightly carburised giving the low phosphorus parts a carbon content of up to about 0.1%, the high phosphorus parts remaining virtually carbon free as they would not carburise as readily. This may have been done intentionally not so much to strengthen the metal which it would not have done to any appreciable extent, but to reduce the grain size of the low phosphorus parts thereby improving the effect that would have been obtained on etching the finished blade. Phosphorus on the other hand promotes grain growth, and, although the sword-smiths would not have realised this, they were clearly producing composite structures which would have been most impressive when etched.

The phosphorus-containing ferrite stays bright when etched (in nital) whereas the non-phosphorus iron etches darker partly due to the much smaller grain size (where the carbon content is low), so it would appear that the sword-smith by combining high and low phosphorus components and also by the light carburising of the low phosphorus iron was exploiting these grain size and compositional characteristics by empirical means to achieve the best visual qualities he could. The decorative aspect of the patterns must have been the most important factor.

Table 12

Composition of Merovingian pattern-welded sword-blades (After Salin[54]).

| | %C | Si | Mn | S | P | $N_2$ |
|---|---|---|---|---|---|---|
| M 3 Blade | 0.15 | - | 0 | 0.016 | 0.35 | |
| M 7 Blade | 0.12 | - | 0.01 | 0.02 | 0.21 | |
| M 10 Blade | 0.09 | - | 0.05 | 0.02 | 0.30 | 0.006<br>0.0085 |
| M 11 Blade | 0.08 | - | 0 | 0.02 | 0.16 | |
| M 11 Edge | 0.2 | - | 0 | 0.02 | 0.14 | |
| Luneville Blade | 0.01-0.05 | - | - | 0.03 | 0.18 | |
| Luneville Edge | 0.02-0.03 | - | - | - | - | |

This conclusion is further underlined by the fact that pattern-welding continued to be popular in the Late Saxon period at the same time that plain blades with or without inlaid inscriptions (made of pattern-welded pieces) were being made. As we have already seen, the changeover to blades which contained an appreciable steel content and in which further heat treatments were used, occurred in this country between the 7th-9th centuries and bore no relation to the use of pattern-welding.

Pattern-welding for swords appears to have simply gone out of fashion in c A.D. 1000 although for scramasaxes it continued until c A.D. 1200 and later scramasaxes incorporating this technique may yet be found.

The Anglo-Saxon sword was contemporary with the Merovingian blades which were pattern-welded in the same way. Salin has done a considerable amount of work on this subject giving overall analyses of the French blades and some mechanical properties (Table 12).

The results show the usual low Mn and S expected of early charcoal irons and low carbon and higher phosphorus expected from the British metallographic results.

The pattern-welded swords found in Latvia and described by Anteins are believed to have been imported from the Rhineland and are typical of those used by the Rus during the 6th-12th centuries. The structures encountered are in all respects the same as those found in this work. Many have inscriptions such as Ulfbehrt, and there are 115 examples of this type in the Latvian collection. The metal of the inscriptions was itself pattern-welded. Some of the blades had a Type 1 sandwich, the centre piece of which had been heat-treated to give sorbite and troostite with a hardness of 250-330 HV. One blade contained a 'stick' figure of a man and was possibly local. The Viking type swords from Novgorod depicted by Kolchin, of the 9th-13th centuries seem to be of this type, and are not pattern-welded. Carbon contents reached 1.1% in places.

The observations of Al-Biruni (A.D. 973-1050),[84,54] that damascened swords made of wootz were more brittle in cold weather than the pattern-welded swords of the Rus seems to be a reference to brittle fracture. It is well-known today that certain elements and treatments can embrittle ferrite to such an extent that it loses its ductility on impact at temperatures as high as 10°C. Becker and Dick[90] carried out tests on wrought iron from Roman pile-shoes and compared the results with modern steel and wrought iron. The results for the brittle-ductile transition temperature were as follows:

| | Transition temperature °C (20j/cm$^2$). | | |
|---|---|---|---|
| Annealed Roman pile | 40°C, | 20°C | 35°C |
| (Modern) ST 37 steel | 10°C | | |
| Wrought iron | 27°C | | |

The Roman iron contained 0.2% P and 0.12-0.16% Ni as principal impurities. But these are not thought to be responsible for the poor values. This is more likely due to coarse grain size and slag inclusions.

Tavadze _et al_[91] have also examined the brittle fracture properties of early iron with 0.1–0.4% C and came to the conclusion that the notch impact strength was considerably below that of modern material of the same ultimate tensile strength. They also blamed the slag inclusions.

It would seem then that the phosphorus content by itself is not the most important factor. The high carbon content of wootz (1.6% max) would be a factor militating against it in cold climates. It is possible that the banded structure of pattern-welded swords, in spite of the large number of slag inclusions in some blades, may have a beneficial effect.

It would be expected that the high phosphorus iron would have a large grain size while that of the low phosphorus low carbon iron would be finer. Mixing the two structures in alternate bands would produce the high corrosion resistance of the high phosphorus bands and the crack-arresting characteristic of the fine-grained ferrite + carbide structure in the neighbouring bands. It would seem that the pattern-welded swords of the Rus were superior to the oriental swords in the climate in which they had to operate.

## ACKNOWLEDGEMENTS

Much of the work presented here by one of us (RFT) has been done as part of the regular examination of finds from archaeological excavations for the Ancient Monuments Laboratory of the Department of the Environment (now the English Heritage). We are grateful to the excavators and the editors of excavation reports for making the relevant material available.

On the whole most of the material in this monograph has not been published before but some, such as that from Chingley and Goltho, has and we are indebted to D.W. Crossley and Guy Beresford for their permission. The work on the material from Winchester is in course of publication, and more information on the archaeological aspects of the tools briefly mentioned in the tables here will be found in the relevant volume of Winchester Studies. We are grateful to Martin Biddle for allowing us to use this material prior to its final publication by the Oxford University Press. We are indebted to E.W. Holden and the Society for Post-Medieval Archaeology for the drawing shown in Fig. 40 which was published by Butler and Tebbutt[41], and to Dr. E.M. Trent for the hardness measurements shown in Fig. 41.

The second author is responsible for most of the work on edged weapons although we share the conclusions to this part of the work. We are indebted to the keepers and curators who have found us specimens to work on, much of it in bad condition after excvavation in the nineteenth century. Those who especially come to mind are John Clarke of the Museum of London, Leslie Cram, Reading Museum, David Kelly, Maidstone Museum, Rodney Alcock and Roger Peers of the Dorset County Museum, and Brian Oldham of the Institute of Archaeology, London.

Many others have helped with this work in some way and our thanks go out to all those metallographers and archaeologists who have not been mentioned here or in the references. Perhaps one should also thank the inventor of the Vickers Projection Microscope which still seems the best piece of apparatus for taking photographs at a magnification of between 5 and 10 times.

Finally we would like to thank the Historical Metallurgy Society for a grant towards this work when it was in its infancy long ago.

## REFERENCES


1. G. Shaw Scott. Case hardening. JISI, 1907 (3), 120-6.

2. H.R. Schubert. History of the British Iron and Steel Industry (450 B.C. - A.D. 1775). London, 1957

3. Anon. Abbeydale Works - A Museum of Sheffield Industry. (Section of a scythe). E.A. News, 1964, 43, 159-160.

4. P. Ottaway. Personal communication.

5. J.G. McDonnell. Coppergate (York) Ironwork Metallurgy. Rept. Knife 8421, Aston University, Birmingham, 1984.

6. J.P. Chilton and U.R. Evans. The corrosion resistance of wrought iron. JISI, 1955, 183, 113-122.

7. R.F. Tylecote and R. Thomsen. The segregation and surface enrichment of As and P in early iron artifacts. Archaeometry, 1973, 15 (2), 193-198.

8. Cited by H.G. Baron and E. Costello in: Met. Reviews, 1963, 8, 416.

9. E.C. Rollason. Metallurgy for Engineers. London, 1961, p. 199, Fig. 114.

10. B.E. Hopkins and H.R. Tipler. The effect of phosphorus on the tensile and notch impact properties of high purity iron and iron-carbon alloys. JISI, 1958, 188, 218-237.

11. J.E. Stead. Iron, carbon and phosphorus (2). JISI, 1918, 97, 389-409.

12. E. Bramley and C. Lawton. The gaseous cementation of iron and steel. Carnegie Schol. Mem. I.S.I. 1927, 14, 35-100.

13. R.F. Tylecote and J.W.B. Black. The effect of hydrogen reduction on the properties of ferrous materials. Studies in Conservation, 1980, 25, 87-96.

14. R.F. Tylecote. Metallography of early edge tools. In Journées de Paléometallurgie. Université de Technologie de Compiègne. Febr. 1983, pp. 153-179 (Ed. C. Coddet).

16. K. Smith. Excavation at Winklebury Camp, Basingstoke, Hamps. PPS, 1977, 43, 31-129.

17. Ian Stead, R. Goodburn *et al.* Excavations at Winterton Roman Villa and other sites in N. Lincs. 1958-67. D of E. Arch. Rep. No. 9 HMSO, London, 1976.

18. The excavations on Allen and Hanbury's site at Ware, Herts were directed by Clive Partridge. Site details in Roman Britain in 1977

(Ed R. Goodburn), Britannia, 1978, 9, 445.

19. R.F. Tylecote. Metallurgical examination of iron work from Wanborough, Wilts. (Excav. direct. by J.S. Wacher et al). forthcoming.

20. Jo Draper, Excavations at Hill Farm, Gestingthorpe, Essex. East Angl. Arch. No. 25, 1985.

21. C.J.S. Green. Interim Rep. on Excavations at Poundbury, Dorset. Proc. Dorset Nat. Hist. and Arch. Soc. 1975, 97, 53-4. (See also A.M. Lab. Repts. 2685 and 3689).

22. R. F. Tylecote, Brancaster. AM Lab. Rep. No. 1243.

23. R. Leech. Excavations at Catsgore, Somerset Wilts. Arch. Trans. Monograph, 1982 (see also Britannia, 1980, 11, 388) AM Lab. Rep. No. 2873.

24. M. Biddle (Ed.) Winchester Studies, Vol. 7 (2) Arts and Crafts. (Forthcoming).

25. A. Selkirk. West Stow, Suffolk. Current Arch. 1973 (40), 151-158. The Anglo-Saxon Village, East Anglian Arch. No. 24, 1985.

26. J. Haslam, L. Biek and R.F. Tylecote. A middle-Saxon iron smelting site at Ramsbury, Wilts. Med. Arch. 1980, 24, 1-68.

27. D.W. Crossley. The Bewl Valley ironworks, Kent c. 1300-1730 A.D. Roy. Arch. Inst. Monogr. 1975.

28. G. Beresford. The Medieval clay-land village. Soc. Med. Arch. Monog. No. 6, 1975.

29. G. Beresford. Barton Blount. (SK 209346). Med. Arch. 1969, 13. 276-7.

30. The site of Holyoak was excavated by Mrs. Gwen Brown. Report forthcoming.

31. G.C. Boon. Roman Silchester. London, 1957, pp. 184-186.

32. We are indebted to the keepers of the Salisbury and South Wiltshire Mus. at Salisbury for permission to examine these items from their collection.

33. B.G. Scott. Metallographic study of some early iron tools and weapons from Ireland. Proc. Roy. Irish Acad. 77 (C), No. 12, 301-317.

34. V.G. Childe and W. Thorneycroft. The vitrified fort at Rahoy, Morvern, Argyll. P.S.A. Scot. 1937/8, 72, 23-43.

35. W.H. Manning and C. Saunders. A socketed iron axe from Maids Morton, Bucks. Ant.J. 1972, 52 (2), 276-292.

36. R.W. Clarke and S.M. Blackshaw (Eds.). The conservation of iron.

National Maritime Mus. 1982. Monog. and Repts. No. 53.

37. B. Johnson and P. Bearpark (Forthcoming). The site was excavated in 1973 and there is a short report in JHMS 1976, 10 (2), 83, by B. Johnson and P. Bearpark.

38. M.G. Spratling et al. An Iron Age bronze foundry at Gussage All Saints, Dorset. In: E.A. Slater and J. Tate (Eds.). pp. 268-292. (Ref. 94).

39. Brian Hobley. A Neronian - Vespasianic military site at The Lunt, Baginton, Warwicks. Trans. B'ham Arch. Soc. 1969, 83, 65-129.

40. Brit. Standards Inst. B. Standard 876-1957. Hand Hammers.

41. D.S. Butler and C.F. Tebbutt. A Wealden cannon-boring bar. Post-Med. Arch. 1975, 9, 38-41.

42. R. Pleiner. Stare Evropske Kovarstvi, Prague 1962.

43. E. Tholander. Evidence of the use of carburised steel and quench-hardening in LBA Cyprus. J. Swed. Inst. in Athens 1971, Stockholm, 18(3), 15-23.

44. T. Stech-Wheeler, J. Muhly, R. Maxwell-Hyslop, R. Maddin. Iron at Taanach and early iron metallurgy in the East Mediterranian. A.J.A. 1981, 85, 245-268.

45. G.I.H. Lloyd. The Cutlery Trades. 1913. Reprint, Cass, London, 1968.

46. B.G. Scott. Metallographic and chemical studies on a group of iron artifacts from the excavations at Greencastle, Co. Down. Ulster Arch. J. 39, 1976, 42-52.

47. B.A. Kolchin. Chernaya metallurgiya ye metalloobrakta u drevnii Russi. Moscow, 1953.

48. M.W. Thompson. Novgorod the Great (10th-15th cent.). London, 1967.

49. J. Piaskowski. An interesting example of early technology: A socketed axe from Wietrazno-Bobrka in the Carpathians. JISI, 1960, 194, 336-340.

50. J.A. Roper. Early North-West scythe smiths and scythe grinders. 1541-1647. West Med. Studies, 1969, 3, 73-86.

51. Salisbury was a cutlery centre at least up to 1862 when it exhibited in London.

52. H.H. Coghlan. Notes on prehistoric and early iron in the Old World. Oxford, 1977.

53. J.D. Light and H. Unglik. A frontier fur trade blacksmith shop, 1796-1812. Quebec, 1984.

54. E. Salin. La civilisation Mérovingienne d'après les sépultures, les

textes et le laboratoire, (Pt.3 - Les Techniques, 1957, p. 311). A & J Picard, Paris.

55. R. Thomsen. Metallographische Untersuchungen an drei Wikinger-zeitlichen Eisenäxten aus Haithabu. In Ber. Ausgrabung in Haithabu (Ed. K. Schietzel) No. 5, 1971, 30-57.

56. Sian Rees. Ancient agricultural implements. Shire Archaeology, Bucks. 1981.

57. R. Pleiner. Die Technologie des Schmiedes in der Gross-mährischen. Kultur. Slovenska Arch. 1967, 15. (1), 174.

58. C.E. Pearson and J.A. Smythe. Examination of a Roman chisel from Chesterholm. P. Univ. Durham Phil. Soc. 1938, 9 (3), 141-145.

59. Sten Modin and R. Pleiner. The metallographic examination of locks, keys and tools. Excav. at Helgo, Vol. 5 (1), 1978 (Eds. K. Lamm and A. Lunderström) pp. 81-109.

60. R. Pleiner. Smithing techniques in Roman Bavaria. Bayrische Vorgeschichtsblätter, 1970, 35, 113-141.

61. We are indebted to Vanessa Fell of Salisbury for this information.

62. P.N. Jones. A short history of the attack of armour. Metallurgist and Mat. Tech. 1984, May, 16 (5), 247-250.

63. H.H. Coghlan and R.F. Tylecote. Medieval iron artifacts from the Newbury area of Berkshire; metallurgical examinations. JHMS. 1978, 12 (1), 12-17.

64. M.J. Swanton. A corpus of Pagan Anglo-Saxon Spear Types. B.A. Reports, No. 7. 1974.

65. J. Petersen. De Norske Vikingsverd, Oslo, 1919.

66. R.E.M. Wheeler. London and the Vikings. London Mus. Cat. No. 1, 1927.

67. C.W. Brewer. Metallurgical examination of six ancient steel weapons. JHMS. 1976, 10 (1), 1-9.

68. G. Webster, J. Watson, J. Anstee and L. Biek. The swords and pieces of equipment from 'Excavations at Canterbury Castle' by P. Bennett, S.S. Freare and S. Stow. Canterbury, 1982, pp. 185-190.

69. D.B. Kelly, Maidstone. Personal Communication.

70. D.B. Kelly, Maidstone. Personal Communication. (see also R.H. Goodsall, A Kentish Patchwork, 1966, p. 13-16).

71. John Brent. Account of the Society's researches in the Anglo-Saxon cemetery at Sarre. Arch. Cantiana, 1862-63, 5, 303-322.

72. (a) do. " " 1864-65, 6, 157-185.
(b) do. " " 1865, 7, 307-321.

73. George Dowker. A Saxon cemetery at Wickhambreux. Arch. Cantiana, 1887, 17, 6-9.

74. D.M. Wilson and J. Hurst (Eds.). Medieval Britain 1956. (Exc. by K.R. Fennell). Med. Arch. 1957. 1, 148.

75. Vera I. Evison. An Anglo-Saxon cemetery at Holborough, Kent. Arch. Cantiana. 1956, 70, 84-131.

76. L. Thälin-Bergman In: (Ed. Helen Clarke) Blacksmithing in prehistoric Sweden. Iron and Man in prehistoric Sweden. Stockholm, 1979, pp. 99-133.

77. T.G. Godfrey-Fausset. The Saxon cemetery at Bifrons. Arch. Cantiana, 1876, 10, 298-315. (concluded in 1879, 13, 552-556).

78. M. Swanton. The spear in Anglo-Saxon times. Thesis Univ. Durham. 1966.

79. A.K. Anteins. Structure and manufacturing technique of pattern-welded objects found in the Baltic States. JISI, 1968, June, 26, 563-571.

80. C.W. Brewer. Metallurgical examination of medieval and post-medieval armour. JHMS. 1981, 15 (1), 1-8.

81. J.N McGrath. A preliminary report on the metallographic examination of four fragmentary EIA sword blades from Llyn Cerrig Bach, Anglesey. JHMS. 1968, 2 (2), 78-80.

82. A.R. Williams. Roman Arms and Armour: a technical note. J. Arch. Science, 1977, 4, 77-87.

83. A. France-Lanord. La fabrication des épées damascées aux époques mérovingiennes et carolingiennes. Le Pays Gaumois 1949, 1-3, 19-45.

84. H.R. Ellis-Davidson. The Sword in Anglo-Saxon England. Oxford. 1962.

85. J.W. Anstee and L. Biek. A study in pattern-welding. Medieval Arch. 1961. 5, 71-93.

86. H. O'Neill. Metallurgical features in welded steels. Trans. Inst. Welding. 1946, 9, 1-9.

87. M.W. Thompson. Novgorod the Great. London 1967.

88. B.A. Kolchin. Chernaya metallurgiya ye metalloobratka u drevnii Russi. Moscow, 1953.

89. A. Zeki Validi. Die Schwerter der Germanen nach arabischen Berichten des 9-11 Jahrhunderts. Zeitschrift der Deutschen Morgenländischen Gesellschaft. 1936, 90 (NPXV) 19 ff. pages.

90. G. Becker and W. Dick. Iron pile shoe from the Roman bridge at Trier. Arch. Eisenhüttenwesen, 1971, 42 (3), 223-227.

91. F.N. Tavadze, U.I. Sarrak, G.V. Svanishvili. The mechanical properties and the nature of the fracture of archaeological iron. Soobshh. Akad. Nauk. Gruz SSR, 1975, 80 (2), 409–412.

92. E. Schurmann and H. Schroer. Härte und Glühversuche an dem Klingenbruchstuck eines Nydam Schwertes. Arch. Eisenhüttenwesen. 1959, 30, (3), 127-130.

93. J. Ypey. European Pattern-welded weapons. Arch. Korrespondenblatt, 1982, 12, 381-388.

94. E.A. Slater and J.O. Tate (Eds.). Proceedings of the 16th International Symposium on Archaeometry and Archaeological Prospection, Edinburgh. Edinburgh, 1980.

95. M.D. Howe. From the Museum. Durobrivae, 6, 1978.

APPENDIX

## GLOSSARY

### PHASES IN Fe-C ALLOYS

1. Ferrite - $\alpha$-Fe, containing not more than 0.02% C at 723°C and 0.0001% C at 20°C. Body-centred cubic structure.

2. Austenite - $\gamma$-Fe, containing not more than 1.75%C at 1130°C and 0.89% C at the eutectoid temperature of 723°C. Face-centred cubic structure.

3. Delta-ferrite - almost pure body-centred cubic iron, stable between 1403°C and the melting point (1535°C) in the absence of carbon. Maximum carbon concentration 0.09% at 1490°C.

4. Martensite - diffusionless transformation product of quenching austenite. Metastable. Body-centred tetragonal structure the carbon atoms being responsible for the tetragonality. Forms only between $M_s$ and $M_f$ temperatures, as plates (needles) in austenite crystals. Decomposes on heating to temperatures at which carbon can diffuse freely.

5. Cementite - $Fe_3C$ - commonest iron carbide. Frequently encountered as the form of 'combined carbon' in these alloys. Carbon content 6.7%.

6. Pearlite - eutectoid mixture of ferrite and cementite formed from austenite on cooling at slow enough rate to allow diffusion-controlled transformation. Usually lamellar in form, but may be 'spheroidised' if formed by very slow cooling. In this case, the $Fe_3C$ is in the form of spheroids in a ferrite matrix.

7. Sorbite - fine spheroidised structure of $Fe_3C$ in Ferrite formed by tempering martensite between 500°C and 700°C.

8. Troostite - dark-etching constituent, barely or not resolvable under the optical microscope, formed on cooling austenite at rates intermediate between those giving pearlite and martensite. Very fine lamellar mixture of ferrite and iron carbide (normally, if not always, $Fe_3C$). Electron microscopy has now confirmed that this phase consists of very fine pearlite and many authorities prefer to call it by this name using the term 'rosette' to describe the often rounded appearance of the nodules. But it is a common phase in early artifacts and it is convenient and descriptive to retain the term 'nodular troostite' for the dark-etching nodular phase which occurs in conjunction

with martensite under slack (slow) quenching conditions (see Fig. 42).

9. Tempered Martensite  dark etching constituent of martensitic steels tempered below the temperature range in which sorbite forms. Form varies from needles of 'black martensite' to irresolvable mass of ferrite and carbide lamellae, as tempering temperature rises. At low tempering temperatures carbide is 'epsilon carbide', which is commonly stated to have a close-packed hexagonal structure, and whose formula is variously reported as $Fe_4C$, $Fe_3C$, and $Fe_2C$. The former are the more likely.

At higher tempering temperatures, the carbide phase transforms to cementite, which later spheroidises. (see Sorbite).

10. Bainite - phase usually found only (in binary alloys) in isothermally transformed material formed between about 500°C and the $M_s$ temperature of the alloy concerned. In general, two forms are found - 'Upper Bainite', which has a feathery appearance, and 'Lower Bainite', with an acicular structure. Both are very fine mixtures of ferrite and carbide, the carbide in the former definitely being cementite, and probably also in the latter, the phases differing mainly in the kinetics of formation.

11. Epsilon Carbide - metastable phase of unknown composition and somewhat doubtful structure forming as an intermediate between martensite and cementite on tempering the former. Also found on ageing ferrite supersaturated with carbon.

# APPENDIX

## Hardness of Modern Tools and Materials.

| | |
|---|---|
| Mild Steel | 150 HV1 |
| Wrought Iron | 102 |
| " " (+P) | 163-173 |
| Axe | 600 |
| Fork/rake | 700 |
| Kitchen knife | 680 |
| Stainless table knife | 550 |
| Hammer face | 666 |